Measure and Integration

Hari Bercovici • Arlen Brown • Carl Pearcy

Measure and Integration

 Springer

Hari Bercovici
Department of Mathematics
Indiana University
Bloomington, Indiana, USA

Arlen Brown
Department of Mathematics
Indiana University
Bloomington, Indiana, USA

Carl Pearcy
Department of Mathematics
Texas A&M University
College Station, Texas, USA

ISBN 978-3-319-29044-7   ISBN 978-3-319-29046-1   (eBook)
DOI 10.1007/978-3-319-29046-1

Library of Congress Control Number: 2016933772

Mathematics Subject Classification (2010): 28-01, 42-01, 28A05, 54G05

Springer Cham Heidelberg New York Dordrecht London
© Springer International Publishing Switzerland 2016

Printed on acid-free paper

Springer International Publishing AG Switzerland is part of Springer Science+Business Media (www. springer.com)

To our good friend Ciprian Foiaş

Preface

This book was written expressly to serve as a textbook for a one- or two-semester introductory graduate course in the theory of measure and Lebesgue integration, usually designated in graduate programs as *real analysis*. In writing this book we have naturally been concerned with the level of preparation of the prospective reader. Such a reader who has mastered the majority of our earlier book *An Introduction to Analysis* (which is referred to in the text as [**I**]) will be more than adequately prepared to profitably read this book. However, at the other end of the spectrum, any beginning graduate student who has reached a certain level of mathematical maturity, which may be taken to mean the ability to follow and construct ε-δ arguments, should have no difficulty in mastering the material in this book. We have deliberately made this book as self-contained as possible.

In keeping with our pedagogical intent, we have provided in each chapter a copious supply of examples and a lengthy collection of problems. *Some of the problems that appear in these problem sets are stated simply as facts (see, for example the first sentence of Problem 1D). In this case, the reader is supposed to supply a proof of the stated fact.* Hints are provided for the more challenging problems. These problem sets constitute an integral part of the book and the reader should study them along with the text. Working problems is, of course, critically important in the study of mathematics because that is how mathematics is learned. In this textbook it is particularly important because many topics of interest are first introduced in the problem sets. Not infrequently, the solution of a problem depends on material in one or several previous problems, a fact that instructors should keep in mind when assigning problems to a class.

While, as noted, this book is intended to serve as a textbook for a course, it is our hope that the wealth of carefully chosen examples and problems will also make it useful to the reader who wishes to study *real analysis* individually. This certainly has proved to be the case for our earlier books [**I**] and [**II**].

One novel feature of our development of integration theory in this book is that the Lebesgue integral is defined axiomatically (see Chapter 3), and only after its usual properties have been revealed is it proved that there is a one-to-one correspondence between measures and Lebesgue integrals. To our knowledge, this approach is original with the authors and it was sketched in our earlier text [**II**]. Our exposition has the advantage that the Lebesgue integral is developed before, and independently of, any discussion of measure spaces. This order of doing things seemed to be quite successful in classroom use. In particular, preliminary versions of this book have been used successfully in graduate courses at The University of Michigan, Indiana University, and Texas A&M University. We take this opportunity to thank the many students in these courses who pointed out inaccuracies in the text and problem sets. Any remaining mistakes are entirely the authors' responsibility.

In writing this book no systematic effort has been made to attribute results or to assign historical priorities. The association of particular names to theorems serves primarily as a memory aid.

The notation and terminology used throughout the book are in essential agreement with those to be found in contemporary textbooks. In particular, the symbols $\mathbb{N}, \mathbb{N}_0, \mathbb{Z}, \mathbb{Q}, \mathbb{R}$, and $\mathbb{C}$ always represent the systems of positive integers, nonnegative integers, integers, rational numbers, real numbers, and complex numbers, respectively. The closure of a set A in a topological space is denoted $\overline{A}$ or A^-. One basic convention valid throughout the book is that all vector spaces encountered are either real or complex, and the relevant field of scalars should be clear from the context.

The numbering system for examples, propositions, theorems, remarks, etc., is somewhat standard. For instance, Proposition 4.21 is the 21st fact stated in Chapter 4 and it is followed by Example 4.22. Also, Problem 3Z is the 26th problem at the end of Chapter 3 and is followed by Problem 3AA.

The reader who is interested in the historical development of measure and Lebesgue integration theory would do well to consult von Neumann [**vN**] and Halmos [**H**]. Excellent comprehensive reference books on this subject are Fremlin [**F**] and Bogachev [**B**]. Useful treatments of analytic sets are to be found in Lusin [**L**] and Kuratowski [**CK**] (historical) and Kechris [**AK**] (contemporary). For further information on functional analysis, one might consult Brown-Pearcy [**II**], Dunford-Schwartz [**DS**], and Rudin [**R**]. Similarly, for Fourier analysis, Zygmund [**Z**], Stein [**S**], and Muscalu-Schlag [**MS**] are excellent references.

Of course, of necessity, many topics in measure theory and integration are not dealt with in this book; for instance, vector measures and Haar measure.

Bloomington, IN, USA   Hari Bercovici
Bloomington, IN, USA   Arlen Brown
College Station, TX, USA   Carl Pearcy
February, 2016

Bibliography

[I]   A. Brown and C. Pearcy, *An introduction to analysis*. Graduate Texts in Mathematics, 154. Springer Verlag, New York, 1995.

[II]   ———, *Introduction to operator theory. I. Elements of functional analysis*. Graduate Texts in Mathematics, No. 55. Springer Verlag, New York-Heidelberg, 1977.

[vN]   J. von Neumann, *Functional Operators. I. Measures and Integrals*. Annals of Mathematics Studies, no. 21. Princeton University Press, Princeton, N. J., 1950.

[H]   P. R. Halmos, *Measure Theory*. D. Van Nostrand Company, Inc., New York, N. Y., 1950.

[F]   D. H. Fremlin, *Measure theory. Volumes 1–5*. Torres Fremlin, Colchester, 2004–2006.

[B]   V. I. Bogachev, *Measure theory. Volumes I, II*. Springer-Verlag, Berlin, 2007.

[L]   N. Lusin, *Leçons sur les ensembles analytiques et leurs applications. Avec une note de W. Sierpiński. Preface de Henri Lebesgue*. Reprint of the 1930 edition. Chelsea Publishing Co., New York, 1972.

[CK]   C. Kuratowski, *Topologie. I et II*. Part I with an appendix by A. Mostowski and R. Sikorski. Reprint of the fourth (Part I) and third (Part II) editions. éditions Jacques Gabay, Sceaux, 1992.

[AK]   A. Kechris, *Classical descriptive set theory*. Graduate Texts in Mathematics, 156. Springer Verlag, New York, 1995.

[DS]   N. Dunford and J. T. Schwartz, *Linear operators. Parts I–III*. Wiley Classics Library. John Wiley & Sons, Inc., New York, 1988.

[R]   W. Rudin, *Functional analysis. Second edition*. International Series in Pure and Applied Mathematics. McGraw-Hill, Inc., New York, 1991.

[Z]   A. Zygmund, *Trigonometric series. Volumes I, II. Third edition*. With a foreword by Robert A. Fefferman. Cambridge Mathematical Library. Cambridge University Press, Cambridge, 2002.

[S]   E.M. Stein, *Harmonic analysis: real-variable methods, orthogonality, and oscillatory integrals*. With the assistance of Timothy S. Murphy. Princeton Mathematical Series, 43. Monographs in Harmonic Analysis, III. Princeton University Press, Princeton, NJ, 1993

[MS]   C. Muscalu and W. Schlag, *Classical and multilinear harmonic analysis. Volumes I, II*. Cambridge Studies in Advanced Mathematics, 137. Cambridge University Press, Cambridge, 2013.

Contents

Chapter 1
Rings of sets

It is a familiar fact of elementary calculus that the integral of a function exists only if the function is continuous, or nearly so. In the theory of the Lebesgue integral, with which we are concerned in this book, continuity is replaced by a significantly less stringent requirement known as *measurability*. This concept, in turn, is defined in terms of a certain type of collection of sets, called a *σ-algebra*, and so we begin with a brief look at this and some related concepts.

One says of a collection $\mathcal{C}$ of sets that it is *closed* with respect to one or another operation on sets if, whenever the operation is performed on sets belonging to $\mathcal{C}$, the resulting set also belongs to $\mathcal{C}$. For example, if $A \cup B$ belongs to $\mathcal{C}$ whenever A and B do, then $\mathcal{C}$ is *closed with respect to* (the formation of) *unions*. Likewise, if $\mathcal{C}$ is a collection of subsets of a fixed set X, and $X \setminus A$ belongs to $\mathcal{C}$ whenever A does, then $\mathcal{C}$ is said to be *closed with respect to* (the formation of) *complements*, or, more simply, to be *complemented*. More generally, if $A \setminus B$ belongs to $\mathcal{C}$ whenever A and B do, then $\mathcal{C}$ is *closed with respect to* (the formation of) *differences*. These examples bring us at once to one of the most important concepts with which we are concerned.

Definition 1.1. A nonempty collection **R** of subsets of a set X is a *ring* (of sets) in X (or a *ring of subsets* of X) if **R** is closed with respect to the formation of both unions and differences. A ring **S** of sets in X is a *σ-ring* in X if it is closed with respect to the formation of *countable unions*, that is, if the union $\bigcup \mathcal{C}$ belongs to **S** whenever $\mathcal{C}$ is a countable subcollection of **S**. A ring **S** of sets in X is an *algebra* if it is complemented. A *σ-ring* **S** of sets in X is a *σ-algebra* if it is complemented.

If **R** is a ring of sets in X, and if A, B, and C belong to **R**, then it is clear that $A \cup B \cup C$ does so too. More generally, by mathematical induction, if $A_1, \ldots, A_n$ all belong to **R**, then so does $A_1 \cup \cdots \cup A_n$. This modest observation provides the proof of a key part of the following elementary proposition.

© Springer International Publishing Switzerland 2016
H. Bercovici et al., *Measure and Integration*,
DOI 10.1007/978-3-319-29046-1_1

Proposition 1.2. *If* **R** *is a ring* [σ-*ring*] *of subsets of a set* X, *then* **R** *is closed with respect to the formation of finite* [*countable*] *unions and intersections. Furthermore, the empty set* $\varnothing$ *belongs to* **R**, *and* **R** *is complemented if and only if* X *belongs to* **R**. *Every* σ-*ring is a ring, and a ring* **R** *is a* σ-*ring if and only if* $\bigcup_n A_n \in$ **R** *for every* (pairwise disjoint) *sequence* $\{A_n\}_{n=1}^{\infty}$ *of sets in* **R**.

Proof. A ring **R** of sets contains at least one set E, and therefore also contains $\varnothing = E \setminus E$. Thus **R** contains X if it is complemented.

Suppose now that $\mathcal{F}$ is a finite collection of sets in a ring **R**. Then either $\mathcal{F}$ is empty, whereupon $\bigcup \mathcal{F} = \varnothing$, or $\mathcal{F}$ can be enumerated as a finite sequence $\{A_1, \ldots, A_n\}$, in which case $A_1 \cup \ldots \cup A_n \in$ **R** as noted above. Thus $\bigcup \mathcal{F} \in$ **R** in any case, so **R** is closed with respect to finite unions.

Suppose next that **R** is a ring of sets in X, and that **R** has the property that $\bigcup_n A_n \in$ **R** whenever $\{A_n\}_{n=1}^{\infty}$ is a (pairwise disjoint) sequence in **R**. A countable subcollection $\mathcal{C}$ of **R** is either finite (in which case $\bigcup \mathcal{C}$ is already known to belong to **R**) or can be enumerated as an infinite sequence $\{A_n\}_{n=1}^{\infty}$. Along with the sets A_n the ring **R** contains the pairwise disjoint sets B_n defined as:

$$B_n = A_n \setminus \bigcup_{j<n} A_j, \qquad n \in \mathbb{N},$$

(see [**I**, Problem 1U]). But then

$$\bigcup_{n=1}^{\infty} B_n = \bigcup_{n=1}^{\infty} A_n \in \mathbf{R}$$

as well. Thus $\bigcup \mathcal{C} \in$ **R** in any case, and **R** is a σ-ring.

The remaining assertions of the proposition that are not completely evident have to do with intersections, and follow from the identity

$$\bigcap_n A_n = A \setminus \bigcup_n (A \setminus A_n),$$

where $A = \bigcup_n A_n$. $\qquad\qquad\qquad\qquad\qquad\qquad\qquad\qquad\qquad\qquad\qquad\qquad$ $\square$

Remark 1.3. The foregoing argument turns in part on a small logical point that should not go unmentioned. Since the empty set is finite, and since the empty union is empty, a collection of sets that is closed with respect to finite (or countable) unions must necessarily contain the empty set. Thus, while the distinction is of little practical significance, we need to distinguish between being closed with respect to *unions* (of two or more sets) and being closed with respect to *finite unions*. (No such distinction arises for intersections, since the empty intersection is undefined.)

Example 1.4. The collection of all finite unions of intervals of all kinds, closed, open, half-open, and degenerate, is a ring of sets in $\mathbb{R}$. A somewhat more subtle example is the collection of all finite (disjoint) unions of half-open intervals of the single form $(a, b]$, $a < b$; see also Problem 1B.

Example 1.5. For any set X the singleton $\{\varnothing\}$ is a σ-ring in X—the smallest σ-ring of all, known henceforth as the *trivial* σ-ring in X. Likewise, if $X \neq \varnothing$, the doubleton $\{X, \varnothing\}$ is the smallest σ-algebra of subsets of X—the *trivial* σ-algebra in X. The collection of *all* subsets of X, that is, the power class 2^X, is a σ-ring in X, as is the collection of all subsets of X not containing some one particular element x_0 of X. More generally, if E_0 is a fixed subset of X, then the collection of all subsets of X that are disjoint from E_0 is a σ-ring in X.

Example 1.6. The collection of all finite subsets of a set X is a ring in X. The collection of all countable subsets of X is a σ-ring in X. More generally, if a is an arbitrary infinite cardinal number, then the collection $\mathcal{S}_a$ of all subsets E of X such that $\operatorname{card}(E) \leq a$ is a σ-ring in X. (See [**I**, Proposition 4.6]. This σ-ring is just 2^X, of course, unless $a < \operatorname{card}(X)$.)

Example 1.7. Let $\mathcal{P}$ be an arbitrary partition of a nonempty set X (see [**I**, p. 7]). The collection of all finite unions of sets belonging to $\mathcal{P}$ is a ring in X, and the collection of all countable unions of sets in $\mathcal{P}$ is a σ-ring in X.

Example 1.8. Let X and Y be sets, and let $\varphi : X \to Y$ be a mapping of X into Y. Then the inversely induced mapping φ^{-1} of 2^Y into 2^X preserves arbitrary unions and intersections as well as complements, differences, and symmetric differences. (See [**I**, Problem 1F]). This is sometimes expressed by saying that φ^{-1} is a *complete Boolean homomorphism*.) Thus if $\mathbf{R}$ is a ring [σ-ring] in Y, then $\varphi^{-1}(\mathbf{R}) = \{\varphi^{-1}(E) : E \in \mathbf{R}\}$ is a ring [σ-ring] in X. Dually, if $\mathbf{R}$ is a ring [σ-ring] in X, then $\{B \subset Y : \varphi^{-1}(B) \in \mathbf{R}\}$ is a ring [σ-ring] in Y.

The σ-rings presented in Examples 1.5, 1.6, and 1.7 have an air of superficiality about them. As a matter of fact, most of the interesting σ-rings with which we are concerned are obtained by enlarging simpler collections of sets. In this connection the following idea is important.

Definition 1.9. If $\mathbf{R}_1$ and $\mathbf{R}_2$ are two rings of subsets of the same set X such that $\mathbf{R}_1 \subset \mathbf{R}_2$, then we say that $\mathbf{R}_2$ *refines*, or *is finer than*, $\mathbf{R}_1$, and also that $\mathbf{R}_1$ *is coarser than* $\mathbf{R}_2$.

Theorem 1.10. *Given a set X and an arbitrary collection $\mathcal{C}$ of subsets of X, there exists a coarsest ring $\mathbf{R} = \mathbf{R}(\mathcal{C})$ [σ-ring $\mathbf{S} = \mathbf{S}(\mathcal{C})$] in X containing the collection $\mathcal{C}$. The ring $\mathbf{R}(\mathcal{C})$ [σ-ring $\mathbf{S}(\mathcal{C})$] is said to be* generated by $\mathcal{C}$.

Proof. Clearly, the intersection of an arbitrary nonempty family of rings [σ-rings] is itself a ring [σ-ring]. In particular, the intersection of the family of all the rings [σ-rings] in X containing the collection $\mathcal{C}$ (not empty by virtue of Example 1.5) is a ring [σ-ring] containing $\mathcal{C}$. This is clearly the coarsest ring [σ-ring] in X that contains $\mathcal{C}$.   □

Example 1.11. If $\mathcal{J}$ is the collection of all subintervals of the real line $\mathbb{R}$—open, closed, half-open, or degenerate—then $\mathbf{R}(\mathcal{J})$ is the ring of Example 1.4. The most general element of $\mathbf{R}(\mathcal{J})$ is a finite (disjoint) union of intervals in $\mathcal{J}$. See Problem 1A.

Example 1.12. Let $\mathcal{A}$ denote the collection of all singletons in a set X. Then $\mathbf{S}(\mathcal{A})$ is the collection of countable subsets of X (see Example 1.6).

Example 1.13. Let $\mathbf{R}_0$ be a ring of sets in a set X. If $X \notin \mathbf{R}_0$, so that $\mathbf{R}_0$ is not complemented, then $\mathbf{R}_1 = \mathbf{R}(\mathbf{R}_0 \cup \{X\})$ is the coarsest algebra of sets in X containing the given ring $\mathbf{R}_0$. (If $\mathbf{R}_0$ itself is complemented, then $\mathbf{R}_1 = \mathbf{R}_0$, of course.) As it turns out,

$$\mathbf{R}_1 = \mathbf{R}_0 \cup \{X \setminus A : A \in \mathbf{R}_0\}; \tag{1.1}$$

that is, $\mathbf{R}_1$ is just the collection of those subsets A of X such that either A or $X \setminus A$ belongs to $\mathbf{R}_0$. Indeed, the collection of sets in (1.1) contains $\mathbf{R}_0$, is complemented, and is clearly contained in any complemented collection of subsets of X that contains $\mathbf{R}_0$. Thus it suffices to verify that it is, in fact, a ring of sets. To see this, suppose A and B are elements of $\mathbf{R}_0$. Then

$$A \cup (X \setminus B) = X \setminus (B \setminus A) \quad \text{and} \quad (X \setminus A) \cup (X \setminus B) = X \setminus (A \cap B)$$

are both complements of elements of $\mathbf{R}_0$, while $A \cup B$ is itself an element of $\mathbf{R}_0$. Thus the collection (1.1) is closed with respect to the formation of unions, and a similar consideration of cases shows that it is also closed with respect to the formation of differences. Thus it is indeed a ring. Similar reasoning reveals that if $\mathbf{R}_0$ is a σ-ring, then (1.1) yields $\mathbf{S}(\mathbf{R}_0 \cup \{X\})$—the coarsest σ-algebra containing $\mathbf{R}_0$.

Example 1.14. In any metric space (X, ρ) ([**I**, Chapter 6]) the topology $\mathcal{G}$, that is, the collection of all open sets in X, generates a σ-ring $\mathbf{B}_X$, known as the ring of *Borel sets* in X. These σ-rings are by far the most important, and will occupy our attention repeatedly in the sequel. Since the whole space X is an open set, the ring $\mathbf{B}_X$ is a σ-algebra. In fact, $\mathbf{B}_X$ coincides with the σ-algebra generated by the collection of all *closed* sets in X.

Example 1.15. Let $\{X_1, X_2\}$ be a partition of a set X, and let $\mathbf{R}_1$ and $\mathbf{R}_2$ be rings of subsets of X_1 and X_2, respectively. Then the collection $\mathbf{R}$ of all unions of the form $A_1 \cup A_2$, where $A_i \in \mathbf{R}_i, i = 1, 2$, is clearly a ring in X. This ring is called the *direct sum* of the rings $\mathbf{R}_1$ and $\mathbf{R}_2$, and is denoted by $\mathbf{R}_1 \oplus \mathbf{R}_2$. Note that $\mathbf{R}_1 \cup \mathbf{R}_2 \subset \mathbf{R}_1 \oplus \mathbf{R}_2$, and also that if the summands $\mathbf{R}_1$ and $\mathbf{R}_2$ are both σ-rings, then their direct sum is also a σ-ring.

Suppose now that $\mathcal{C}_1$ and $\mathcal{C}_2$ are collections of subsets of X_1 and X_2, respectively, and set $\mathcal{C} = \mathcal{C}_1 \cup \mathcal{C}_2$. Then the ring $\mathbf{R}(\mathcal{C})$ contains both $\mathbf{R}(\mathcal{C}_1)$ and

$\mathbf{R}(\mathcal{C}_2)$, and therefore contains $\mathbf{R}(\mathcal{C}_1) \oplus \mathbf{R}(\mathcal{C}_2)$. On the other hand, this latter ring contains $\mathcal{C}$ and therefore contains $\mathbf{R}(\mathcal{C})$. Thus $\mathbf{R}(\mathcal{C}) = \mathbf{R}(\mathcal{C}_1) \oplus \mathbf{R}(\mathcal{C}_2)$; similarly $\mathbf{S}(\mathcal{C}) = \mathbf{S}(\mathcal{C}_1) \oplus \mathbf{S}(\mathcal{C}_2)$.

Example 1.16. Let $\mathcal{P} = \{X_\gamma\}_{\gamma \in \Gamma}$ be an indexed partition of a nonempty set X, and for each index γ suppose given a ring $\mathbf{R}_\gamma$ of subsets of X_γ. Then the collection of all unions $\bigcup_{\gamma \in \Gamma} A_\gamma$ of indexed families $\{A_\gamma\}$, where $A_\gamma \in \mathbf{R}_\gamma, \gamma \in \Gamma$, is a ring $\mathbf{R}$ in X, called the *full direct sum* of the family $\{\mathbf{R}_\gamma\}$ and denoted by $\bigoplus_{\gamma \in \Gamma} \mathbf{R}_\gamma$. Likewise, the collection of all unions of families $\{A_\gamma\}$, where $A_\gamma \in \mathbf{R}_\gamma$ for each γ and $A_\gamma = \varnothing$ except for those indices in some finite subset Δ of Γ (which may vary with the family $\{A_\gamma\}$) is a ring $\mathbf{R}_f$ in X. This ring is called the *finite direct sum* of the family $\{\mathbf{R}_\gamma\}$ and is denoted by $\bigoplus_{\gamma \in \Gamma}^{(f)} \mathbf{R}_\gamma$. (If Γ is finite, as in the preceding example, these two concepts coincide, of course.)

If each of the given rings $\mathbf{R}_\gamma$ is a σ-ring, the full direct sum $\mathbf{R}$ is also a σ-ring, but (except in trivial cases) the finite direct sum $\mathbf{R}_f$ is not. Between these two notions lies a third, the *countable direct sum* $\mathbf{R}_\omega = \bigoplus_{\gamma \in \Gamma}^{(\omega)} \mathbf{R}_\gamma$ of the family $\{\mathbf{R}_\gamma\}$, consisting of all unions of families $\{A_\gamma\}_{\gamma \in \Gamma}$, where $A_\gamma \in \mathbf{R}_\gamma$ for each $\gamma \in \Gamma$ and $A_\gamma = \varnothing$ except for indices γ in some *countable* subset Δ of Γ (which again may vary with $\{A_\gamma\}$). If all of the rings $\mathbf{R}_\gamma$ are σ-rings, then $\mathbf{R}_\omega$ is also a σ-ring. (When Γ is countable, then $\mathbf{R}_\omega = \mathbf{R}$. If each of the rings $\mathbf{R}_\gamma$ is taken to be the trivial complemented σ-ring in its set X_γ, this example reduces to Example 1.6.)

Suppose now that X and $\{X_\gamma\}$ are as above, let $\mathcal{C}_\gamma$ be a collection of subsets of $X_\gamma, \gamma \in \Gamma$, and set $\mathcal{C} = \bigcup_\gamma \mathcal{C}_\gamma$. Then it is readily verified that the ring $\mathbf{R}(\mathcal{C})$ is simply the finite direct sum $\bigoplus_{\gamma \in \Gamma}^{(f)} \mathbf{R}(\mathcal{C}_\gamma)$. Likewise $\mathbf{S}(\mathcal{C})$ coincides with the countable direct sum $\bigoplus_{\gamma \in \Gamma}^{(\omega)} \mathbf{S}(\mathcal{C}_\gamma)$, while the full direct sum of $\{\mathbf{R}(\mathcal{C}_\gamma)\}[\{\mathbf{S}(\mathcal{C}_\gamma)\}]$ is, in general, a strictly finer ring [σ-ring] in X, unless Γ is finite [countable].

Theorem 1.10 answers the questions of existence concerning rings and σ-rings we shall encounter. It has the drawback, however, of telling us nothing at all concerning two important questions: Given a collection of sets $\mathcal{C}$, which sets belong to $\mathbf{R}(\mathcal{C})$ and which to $\mathbf{S}(\mathcal{C})$? As the preceding examples clearly show, these questions vary greatly in depth from one case to another. Thus in Example 1.12 the description of the general element of the generated σ-ring is almost trivial, while in Example 1.11 the description of the general element of the generated ring is longer and more complex, but still complete and informative. On the other hand, Example 1.14 provides no hint at all as to what the general Borel set in the real line looks like.

The question of the nature of the ring $\mathbf{R}(\mathcal{C})$ can be answered quite satisfactorily in various elementary ways (see Problems 1E, 1F, and 1G). The more interesting case of the σ-ring $\mathbf{S}(\mathcal{C})$ is much deeper and requires careful investigation. (The special cases set forth in Examples 1.13, 1.15, and 1.16 are unusually simple.) In this connection the following construction is relevant.

To begin with, for arbitrary collections $\mathcal{C}_1$ and $\mathcal{C}_2$ of sets we write $\mathcal{C}_1 \# \mathcal{C}_2$ for the collection of all differences $C \setminus D$ where C and D belong to $\mathcal{C}_1$ and $\mathcal{C}_2$, respectively. Secondly, for a collection $\mathcal{C}$ we define $\mathcal{C}_\sigma$ to consist of the collection of all unions of countable subcollections of $\mathcal{C}$. Finally, we define

$$\mathcal{C}^* = (\mathcal{C} \# \mathcal{C})_\sigma.$$

In terms of these operations it is possible to give a description of the generated σ-ring $\mathbf{S}(\mathcal{C})$. (Another slightly different version of this construction is found in Problem 1H.) To begin with, we may assume that $\mathcal{C}$ contains the empty set $\varnothing$ (since $\mathcal{C}$ and $\mathcal{C} \cup \{\varnothing\}$ clearly generate the same σ-ring). We set $\mathcal{C}^0 = \mathcal{C}$ and employ the principle of transfinite definition ([**I**, Theorem 5.12]) to obtain an indexed family of collections $\{\mathcal{C}^\alpha\}_{\alpha \in \Omega}$ of sets such that

$$\mathcal{C}^\alpha = \left(\bigcup_{\xi < \alpha} \mathcal{C}^\xi \right)^*$$

for every ordinal number α such that $0 < \alpha < \Omega$, where Ω denotes, as always, the smallest uncountable ordinal number ([**I**, Example 5I]). (Note that the transfinite sequence $\{\mathcal{C}^\alpha\}_{\alpha < \Omega}$ is monotone increasing, and thus $\mathcal{C}^{\alpha+1} = (\mathcal{C}^\alpha)^*$ for each countable ordinal number α.)

Theorem 1.17. *Let $\mathcal{C}$ be a collection of subsets of a set X such that the empty set $\varnothing$ belongs to $\mathcal{C}$, and let $\mathcal{C}^\alpha, \alpha < \Omega$, be defined as above. Then*

$$\mathbf{S}(\mathcal{C}) = \bigcup_{\alpha < \Omega} \mathcal{C}^\alpha.$$

Proof. It is clear that $\mathcal{C}^* \subset \mathbf{S}(\mathcal{C})$ and hence by transfinite induction ([**I**, Theorem 5.11]) $\mathcal{T} = \bigcup_{\alpha < \Omega} \mathcal{C}^\alpha$ is contained in $\mathbf{S}(\mathcal{C})$. Thus it suffices to verify that $\mathcal{T}$ is a σ-ring. Suppose that C and D belong to $\mathcal{T}$. Then there exists an ordinal number $\alpha < \Omega$ such that C and D both belong to $\mathcal{C}^\alpha$. But then $C \setminus D \in \mathcal{C}^{\alpha+1}$. Similarly, if $\{C_n\}$ is an arbitrary sequence of sets belonging to $\mathcal{T}$, then there exists an ordinal number $\alpha < \Omega$ such that all of the sets in the sequence $\{C_n\}$ belong to $\mathcal{C}^\alpha$ ([**I**, Exercise 5L]), whence it follows that $\bigcup_n C_n$ belongs to $\mathcal{C}^{\alpha+1}$. Thus $\mathcal{T}^* = \mathcal{T}$, and the theorem follows. $\square$

Example 1.18. Let us consider the application of Theorem 1.17 to the generation of the Borel sets in a metric space (X, ρ). Since both X and $\varnothing$ belong to the topology $\mathcal{G}$ on X, the collection $\mathcal{G} \# \mathcal{G}$ contains not only the open subsets of X but all the closed subsets as well. Hence $\mathcal{G}^* = \mathcal{G}^1$ contains all of the sets known as F_σ sets (see [**I**, Problem 6P]), whereupon it follows that $\mathcal{G}^2$ contains all the G_δ sets in X. Continuing in this way we see that the classes of sets $G_\delta, G_{\delta\sigma}, \ldots$ as well as the classes $F_\sigma, F_{\sigma\delta}, \ldots$ consist entirely of Borel sets in every metric space X.

Remark 1.19. In certain cases the construction of Theorem 1.17 is over-complicated. Thus in a discrete metric space ([**I**, Problem 6O]) the ring of Borel sets is simply the power class $\mathcal{G} = 2^X$ itself. Likewise, if the metric space is countable, then every subset of X is an F_σ, and we have $\mathbf{B} = \mathcal{G}^2 (= 2^X)$. On the other hand, as we shall see in due course, in any sufficiently rich metric space (for instance, one containing a homeomorphic image of the Cantor set), each of the classes $G_\delta, G_{\delta\sigma}, \ldots [F_\sigma, F_{\sigma\delta}, \ldots]$ contains sets not to be found in its predecessors. The class $\mathbf{B}_X$ itself consists of the union of the classes $G_\delta, G_{\delta\sigma}, \ldots [F_\sigma, F_{\sigma\delta}, \ldots]$.

The notion of a Borel set makes sense in a general topological space, and it is particularly important when X is a locally compact space. We return to this case in a later chapter.

The description of $\mathbf{S}(\mathcal{C})$ given in Theorem 1.17 is complicated, while that given in Theorem 1.10 is uninformative. There is ample room between these two extremes for other theorems concerning the generated σ-ring $\mathbf{S}(\mathcal{C})$, and in Theorem 1.23 below we present one such result. First we require a new concept.

Definition 1.20. A nonempty collection $\mathbf{Q}$ of subsets of a set X is a *quasiring* in X if

(1) $P \setminus Q$ belongs to $\mathbf{Q}$ whenever $P, Q \in \mathbf{Q}$ and $Q \subset P$, and
(2) $\bigcup \mathcal{F}$ belongs to $\mathbf{Q}$ whenever $\mathcal{F}$ is a finite disjoint subcollection of $\mathbf{Q}$.

A quasiring $\mathbf{Q}$ in X is a *σ-quasiring* in X if $\bigcup_n Q_n$ belongs to $\mathbf{Q}$ whenever $\{Q_n\}_{n=1}^\infty$ is a pairwise disjoint sequence of sets belonging to $\mathbf{Q}$.

In the standard terminology introduced above, a quasiring [σ-quasiring] is closed with respect to the formation of *proper* differences and finite [countable] *pairwise disjoint* unions. In the probability literature, a complemented σ-quasiring is called a *λ-system*, while a collection of sets which is closed under finite intersections is called a *π-system*.

The following result is the counterpart of Proposition 1.2 for quasirings. The proof is trivial and is therefore omitted.

Proposition 1.21. *Every ring [σ-ring] of subsets of a set X is a quasiring [σ-quasiring] in X. Every σ-quasiring is a quasiring, and every quasiring contains the empty set.*

The basic existence theorem for quasirings is as simple as that for rings. The proof of the following result parallels that of Theorem 1.10 and will also be omitted.

Proposition 1.22. *Given a set X and an arbitrary collection $\mathcal{C}$ of subsets of X, there exists a smallest quasiring $\mathbf{Q}(\mathcal{C})$ [σ-quasiring $\mathbf{Q}_\sigma(\mathcal{C})$] in X containing $\mathcal{C}$. This quasiring $\mathbf{Q}(\mathcal{C})$ [σ-quasiring $\mathbf{Q}_\sigma(\mathcal{C})$] is said to be generated by $\mathcal{C}$.*

In many situations the following theorem provides the most effective characterization of a generated σ-ring. This result, referred to as the *monotone class theorem* (or the π-λ *theorem* in the probability literature), is due to Sierpiński.

Theorem 1.23. *Let $\mathcal{D}$ be a collection of subsets of a set X that is closed with respect to the formation of (finite) intersections. Then any σ-quasiring in X that contains $\mathcal{D}$ also contains the σ-ring $\mathbf{S}(\mathcal{D})$. Equivalently, the σ-quasiring $\mathbf{Q}_\sigma(\mathcal{D})$ generated by $\mathcal{D}$ coincides with $\mathbf{S}(\mathcal{D})$.*

Proof. Since every σ-ring is automatically a σ-quasiring, it is clear that $\mathbf{Q} = \mathbf{Q}_\sigma(\mathcal{D})$ is contained in $\mathbf{S}(\mathcal{D})$. Hence the proof will be complete if we show that $\mathbf{Q}$ is a σ-ring, and to this end it suffices to verify that $\mathbf{Q}$ is closed with respect to the formation of intersections (Problem 1Y). To see that this is indeed so, for each subset E of X let us write $\mathcal{M}_E$ for the collection of all those subsets M of X such that $M \cap E \in \mathbf{Q}$. If M and N are two sets belonging to $\mathcal{M}_E$, and if $N \subset M$, then, since

$$(M \setminus N) \cap E = (M \cap E) \setminus (N \cap E),$$

the difference $M \setminus N$ also belongs to $\mathcal{M}_E$. Similarly, the union of any pairwise disjoint sequence of sets in $\mathcal{M}_E$ belongs to $\mathcal{M}_E$. In other words, $\mathcal{M}_E$ is a σ-quasiring for every E in 2^X. But now if $E = D$ is an element of $\mathcal{D}$, then $\mathcal{D} \subset \mathcal{M}_D$, and it follows that $\mathbf{Q} \subset \mathcal{M}_D$. Thus for each set Q in $\mathbf{Q}$ and each D in $\mathcal{D}$ we have $Q \cap D \in \mathbf{Q}$. In other words, for each Q in $\mathbf{Q}$ we have $\mathcal{D} \subset \mathcal{M}_Q$, and therefore $\mathbf{Q} \subset \mathcal{M}_Q$. $\square$

According to Problem 1J(i), the ring $\mathbf{R}(\mathcal{C})$, generated by a complemented collection $\mathcal{C}$, consists precisely of the finite unions of finite intersections of elements in $\mathcal{C}$. It is desirable to have a similar result for the sets in the Borel σ-ring $\mathbf{B}_X$ of a metric space X.

Definition 1.24. Let (X, ρ) be a metric space, and assume given for every sequence $(n_1, \ldots, n_k)$ of natural numbers a set $F_{n_1, \ldots, n_k} \subset X$. Given an infinite sequence $\mathbf{n} \in \mathbb{N}^{\mathbb{N}}$, we set

$$F_{\mathbf{n}} = F_{n_1} \cap F_{n_1, n_2} \cap \cdots \cap F_{n_1, \ldots, n_k} \cap \cdots = \bigcap_{k=1}^{\infty} F_{\mathbf{n}|k},$$

where $\mathbf{n}|k$ stands for the sequence consisting of the first k elements of $\mathbf{n}$. The *Suslin operation* applied to the system of sets $\{F_{n_1, \ldots, n_k}\}$ is the set

$$A = \bigcup_{(n_1, n_2, \ldots) \in \mathbb{N}^{\mathbb{N}}} (F_{n_1} \cap F_{n_1, n_2} \cap \cdots \cap F_{n_1, \ldots, n_k} \cap \cdots) = \bigcup_{\mathbf{n} \in \mathbb{N}^{\mathbb{N}}} F_{\mathbf{n}}.$$

A subset $A \subset X$ is said to be *analytic* if it is the result of the Suslin operation applied to a system $\{F_{n_1, \ldots, n_k}\}$ of *closed* sets. We denote by $\mathcal{A}_X$ the collection of analytic sets in X.

Observe that an analytic set A is constructed from a countable collection of closed sets, but it is not necessarily an F_σ set because $\mathbb{N}^\mathbb{N}$ is not countable. When X contains, for instance, a homeomorphic copy of the Cantor set C, it is known that there exist analytic sets in X which are not Borel sets (see Problems 11P and 11R). The Suslin operation does however provide a representation for all Borel sets.

Theorem 1.25. *The collection $\mathcal{A}_X$ of analytic sets in a metric space X is closed under countable unions and countable (nonempty) intersections. Therefore $\mathcal{A}_X$ contains the Borel σ-algebra $\mathbf{B}_X$.*

Proof. Consider a sequence $\{A_m\} \subset \mathcal{A}_X$, and assume that

$$A_m = \bigcup_{\mathbf{n} \in \mathbb{N}^\mathbb{N}} F_\mathbf{n}^{(m)}, \quad F_\mathbf{n} = \bigcap_{k=1}^{\infty} F_{\mathbf{n}|k}^{(m)}, \quad m \in \mathbf{N},$$

where all the sets $F_{\mathbf{n}|k}^{(m)}$ are closed. It is immediate that $\bigcup_{m=1}^{\infty} A_m$ is obtained by applying the Suslin operation to the closed sets

$$K_{n_1, n_2, \ldots, n_k} = \begin{cases} X & \text{if } k = 1, \\ F_{n_2, \ldots, n_k}^{(n_1)} & \text{if } k \geq 2. \end{cases}$$

On the other hand, distributivity of the set operations implies that

$$\bigcap_{m=1}^{\infty} A_m = \bigcup_{\mathbf{n}(1), \mathbf{n}(2), \cdots \in \mathbb{N}^\mathbb{N}} \bigcap_{m=1}^{\infty} F_{\mathbf{n}(m)}^{(m)}.$$

Fix a bijection $\varphi : \mathbb{N} \times \mathbb{N} \to \mathbb{N}$, for instance $\varphi(p, q) = 2^{p-1}(2q - 1)$, $p, q \in \mathbb{N}$. From each sequence $\{\mathbf{n}(m)\}_{m=1}^{\infty} \subset \mathbb{N}^\mathbb{N}$ one can then construct a single sequence $\mathbf{n} \in \mathbb{N}^\mathbb{N}$ by setting $\mathbf{n}_{\varphi(p,q)} = \mathbf{n}(p)_q$. The double intersection

$$\bigcap_{m=1}^{\infty} F_{\mathbf{n}(m)}^{(m)} = \bigcap_{p,q=1}^{\infty} F_{\mathbf{n}(p)|q}^{(p)}$$

equals then $\bigcap_{k=1}^{\infty} F_{\mathbf{n}|k}$, where $F_{\mathbf{n}|\varphi(p,q)} = F_{\mathbf{n}(p)|q}^{(p)}$ only depends on $\mathbf{n}|\varphi(p, q)$.

The collection $\mathcal{A}_X$ clearly contains all closed sets, and by the preceding argument it contains the F_σ sets as well, in particular the open sets in X. The last assertion follows from these facts.   □

Problems

1A. Let $\mathcal{D}$ be a nonempty collection of subsets of a set X that is closed with respect to the formation of (finite) intersections and that also possesses the property that if C and D are sets in $\mathcal{D}$ such that $D \subset C$, then there exists a finite disjoint subcollection

$\mathcal{F}$ of $\mathcal{D}$ such that $C \setminus D = \bigcup \mathcal{F}$. Prove that the collection of all finite disjoint unions of sets in $\mathcal{D}$ is the ring $\mathbf{R}(\mathcal{D})$ generated by $\mathcal{D}$. (Hint: If $\mathcal{C}$ denotes the collection of all finite disjoint unions of sets in $\mathcal{D}$, then $\mathcal{C}$ is also closed with respect to intersections and $\mathcal{C} \# \mathcal{D} \subset \mathcal{C}$.)

1B. Consider in Euclidean space $\mathbb{R}^d$ the collection of all half-open cells of the form

$$H = (a_1, b_1] \times \cdots \times (a_d, b_d].$$

Verify that the collection of all finite disjoint unions of such cells is a ring of sets in $\mathbb{R}^d$. We denote this ring by $\mathbf{H}_r$. Similarly, the collection of sets in $\mathbb{R}^d$ that can be written as finite disjoint unions of half-open cells of the form

$$[a_1, b_1) \times \cdots \times [a_d, b_d)$$

is a ring of sets in $\mathbb{R}^d$, which we denote by $\mathbf{H}_\ell$.

1C. Let $\{E_1, \ldots, E_n\}$ be a finite collection of subsets of a set X, and let $\mathcal{P}_0$ denote the collection of sets obtained by deleting from the partition of X determined by $\{E_1, \ldots, E_n\}$ the one set $X \setminus (E_1 \cup \cdots \cup E_n)$. (See [**I**, Problem 1S] for an explanation of this terminology.)

   (i) Show that the ring of sets generated by $\{E_1, \ldots, E_n\}$ consists of all (finite) unions of sets belonging to $\mathcal{P}_0$. Show also that the ring $\mathbf{R}(\mathcal{C})$ generated by a collection $\mathcal{C}$ of subsets of X is the union of the rings generated by the various finite subcollections of $\mathcal{C}$, and thus that every set in $\mathbf{R}(\mathcal{C})$ is covered by some finite number of sets in $\mathcal{C}$. (Hint: This union forms a ring of sets.)
   (ii) When $X = \mathbb{R}$ and $\mathcal{C} = \{[0,1], (\frac{1}{2}, \frac{3}{2}), \{1\}, \{4\}\}$, what is the cardinal number of $\mathbf{R}(\mathcal{C})$?
   (iii) Show, in a like vein, that the σ-ring $\mathbf{S}(\mathcal{C})$ is the union of the σ-rings generated by the various countable subcollections of $\mathcal{C}$, and that every set in $\mathbf{S}(\mathcal{C})$ is covered by some countable collection of sets in $\mathcal{C}$.

Remark 1.26. Recall from the preface the convention, in force throughout the book, that sometimes in the problem sets, to avoid repetitiveness, what is to be shown is simply stated as a fact. For instance, in the first part of Problem 1D below, to work the problem, the reader must show that if $\mathbf{R}$ is a ring of sets with the property that $\bigcup_{n=1}^{\infty} A_n$ belongs to $\mathbf{R}$ whenever $\{A_n\}_{n=1}^{\infty}$ is an increasing sequence of sets in $\mathbf{R}$, then $\mathbf{R}$ is a σ-ring.

1D. A ring of sets $\mathbf{R}$ is a σ-ring if and only if $\bigcup_n A_n \in \mathbf{R}$ whenever $\{A_n\}_{n=1}^{\infty}$ is an increasing sequence of sets belonging to $\mathbf{R}$. Give an example of a ring of sets $\mathbf{R}$ with the property that $\bigcap_n A_n \in \mathbf{R}$ for every *decreasing* sequence of sets in $\mathbf{R}$ but that is, nevertheless, not a σ-ring.

1E. We recall the notation $A \triangle B = (A \setminus B) \cup (B \setminus A)$ for the *symmetric difference* of the sets A and B. Show that the binary set operation $\triangle$ (symmetric difference) is *associative*, that is, that $A \triangle (B \triangle C) = (A \triangle B) \triangle C$ for any three sets A, B, C, and conclude that the symmetric difference $E_1 \triangle \cdots \triangle E_n$ of an arbitrary finite sequence of sets $\{E_1, \ldots, E_n\}$ can be defined without regard to either order or bracketing. (Hint: If $E_1, \ldots, E_n$ are subsets of a set X, then an element x of X belongs to $E_1 \triangle \cdots \triangle E_n$ if and only if $x \in E_i$ for an odd number of indices i.)

(i) A nonempty collection $\mathbf{R}$ of subsets of a set X is a ring [σ-ring] in X if and only if it is closed with respect to the formation of finite [countable] unions and the formation of (finite) symmetric differences

$$E_1 \triangle \cdots \triangle E_n.$$

What is the status of these assertions if the word "unions" is replaced by "intersections"?

(ii) For any collection $\mathcal{C}$ of subsets of a set X we write $\mathcal{C}_d$ for the collection of all finite intersections of sets belonging to $\mathcal{C}$. Show that the ring $\mathbf{R}(\mathcal{C})$ generated by $\mathcal{C}$ coincides with the collection of all finite symmetric differences

$$D_1 \triangle \cdots \triangle D_n$$

of sets in $\mathcal{C}_d$.

Remark 1.27. If X is a set and the binary set operations $\triangle$ and $\cap$ are interpreted as "addition" and "multiplication," respectively, on 2^X, then 2^X becomes a ring in the sense of abstract algebra. The sets $\varnothing$ and X play the roles of 0 and 1, respectively. Moreover, the *rings of sets* discussed in this chapter and below are precisely the *subrings* of 2^X in the algebraic sense (hence the terminology). Viewed in this light, Problem 1E(ii) can be seen to be simply the translation into set-theoretic terms of the standard description in ring theory of the subring generated by a subset of a ring.

Remark 1.28. Viewed algebraically, rings of sets have the curious property that every nonzero element has additive order two ($E \triangle E = \varnothing$) and every element is idempotent ($E \cap E = E$). Such rings have been called *Boolean rings* by M. H. Stone.

1F. (The counterpart of Theorem 1.17 for rings of sets.) For an arbitrary collection $\mathcal{C}$ of subsets of a set X we write $\mathcal{C}_s$ for the collection of all finite unions of sets in $\mathcal{C}$, and we set $\mathcal{C}^\dagger = (\mathcal{C}\#\mathcal{C})_s$. Suppose that $\mathcal{C}$ contains the empty set and define inductively

$$\mathcal{C}_0 = \mathcal{C},$$

and

$$\mathcal{C}_{n+1} = (\mathcal{C}_n)^\dagger, \quad n \in \mathbf{N}_0.$$

Then $\mathbf{R}(\mathcal{C})$ coincides with the union $\bigcup_{n=1}^\infty \mathcal{C}_n$.

1G. The inductive construction in the preceding problem is vastly over-complicated. In justification of this assertion, show that for an arbitrary nonempty collection of sets $\mathcal{C}$ the collection

$$\mathcal{C}^\ddagger = \mathcal{C}_d \# \mathcal{C}_s$$

(in the notation of Problems 1E and 1F) is closed with respect to the formation of intersections, and also that the difference of any two sets in $\mathcal{C}^\ddagger$ is a finite union of sets in $\mathcal{C}^\ddagger$. Use these facts to prove that the difference of any two sets in $(\mathcal{C}^\ddagger)_s$ is in $(\mathcal{C}^\ddagger)_s$, and hence that $(\mathcal{C}^\ddagger)_s$ is a ring of sets. Conclude that $(\mathcal{C}^\ddagger)_s = \mathbf{R}(\mathcal{C})$ whenever $\varnothing \in \mathcal{C}$. Show also that

$$\mathcal{C}^\ddagger \subset \mathcal{C}\#\mathcal{C}^\dagger,$$

and hence $\mathbf{R}(\mathcal{C})$ is, in fact, equal to $\mathcal{C}^{\dagger\dagger} = \mathcal{C}_2$ (assuming always that $\varnothing \in \mathcal{C}$). (Hint: The most general element of $\mathcal{C}^\ddagger$ can be written as

$$A_1 \cap \ldots \cap A_p \cap (X \setminus B_1) \cap \ldots \cap (X \setminus B_q),$$

where the A_i and B_j belong to $\mathcal{C}$, $p, q \in \mathbb{N}_0$.)

1H. There are a number of versions of the transfinite construction in Theorem 1.17. Here is another, slightly different, one. Let X be a set and let $\mathcal{C}$ be a collection of subsets of X such that $\varnothing \in \mathcal{C}$. We set

$$\mathcal{C}_{(0)} = \mathcal{C}$$

and, supposing $\mathcal{C}_{(\xi)}$ has been defined for all ordinal numbers ξ in $W(\alpha) = \{\beta : \beta < \alpha\}$ (where α denotes a countable ordinal number), we define $\mathcal{C}_{(\alpha)}$ to be either

$$\left(\bigcup_{\xi < \alpha} \mathcal{C}_{(\xi)} \right) \# \left(\bigcup_{\xi < \alpha} \mathcal{C}_{(\xi)} \right)$$

or

$$\left(\bigcup_{\xi < \alpha} \mathcal{C}_{(\xi)} \right)_{\sigma},$$

according as α is odd or even, respectively. (Every infinite ordinal number α can be written in one and only one way as $\alpha = \lambda + n$ where λ is a limit number and n is a nonnegative integer ([**I**, Prob. 5G]); we say that α is *odd* or *even* according as n is odd or even. In particular, every limit ordinal is even.) Verify that $\mathcal{C}_{(\alpha+1)}$ coincides with $\mathcal{C}_{(\alpha)} \# \mathcal{C}_{(\alpha)}$ or $(\mathcal{C}_{(\alpha)})_{\sigma}$ for every countable ordinal number α according as α is even or odd, and also that

$$\mathbf{S}(\mathcal{C}) = \bigcup_{\alpha < \Omega} \mathcal{C}_{(\alpha)}.$$

For the following problem we use the notation $\mathfrak{c} = 2^{\aleph_0}$ for the power of the continuum. (This was denoted $\aleph$ in [**I**, Example 4D].)

1I. Let $\mathcal{C}$ be a collection of subsets of a set X, and let $\mathbf{R}$ and $\mathbf{S}$ denote, respectively, the ring and σ-ring of sets generated by $\mathcal{C}$. If $\mathcal{C}$ is finite, then so is $\mathbf{R}$. What about $\mathbf{S}$? If $\mathcal{C}$ is countable, then so is $\mathbf{R}$, while for $\mathbf{S}$ we have the estimate card $\mathbf{S} \leq \mathfrak{c}$. Indeed, card $\mathcal{C} \leq \mathfrak{c}$ implies card $\mathbf{S} \leq \mathfrak{c}$. Conclude that in the real line $\mathbf{R}$ there are many subsets that are not Borel sets. (Hint: If $\mathcal{C}$ consists of k sets, then $\mathbf{R}$ cannot contain more than 2^{2^k} sets, and if card $\mathcal{C} \leq \mathfrak{c}$, then card $\mathcal{C}_{\sigma} \leq \mathfrak{c}$.)

1J. Let X be a set and let $\mathcal{C}$ be a collection of subsets of X. The symbol $\mathcal{C}_{\sigma}$ was introduced above, in connection with Theorem 1.17, to denote the collection of all unions of countable subcollections of $\mathcal{C}$, and similar symbols $\mathcal{C}_s$ and $\mathcal{C}_d$ were introduced in Problems 1E and 1F. Here we complete this notational cycle by agreeing to write $\mathcal{C}_{\delta}$ for the collection of all intersections of (nonempty) countable subcollections of $\mathcal{C}$. Thus $\mathcal{C} \subset \mathcal{C}_s \subset \mathcal{C}_{\sigma}$ and $\mathcal{C} \subset \mathcal{C}_d \subset \mathcal{C}_{\delta}$ for arbitrary $\mathcal{C}$, while $\mathcal{C}$ is closed with respect to finite unions [intersections] if and only if $\mathcal{C} = \mathcal{C}_s [\mathcal{C} = \mathcal{C}_d]$. Similarly, $\mathcal{C}$ is closed with respect to countable unions [intersections] if and only if $\mathcal{C} = \mathcal{C}_{\sigma} [\mathcal{C} = \mathcal{C}_{\delta}]$. In terms of this notation some of the familiar facts of elementary set theory assume a novel appearance.

(i) (The distributive laws; see [**I**, Problem 1D]) Show that for any collection $\mathcal{C}$ we have

$$\mathcal{C}_{sd} \subset \mathcal{C}_{ds} \quad \text{and} \quad \mathcal{C}_{\sigma d} \subset \mathcal{C}_{d\sigma}.$$

Likewise,

$$\mathcal{C}_{ds} \subset \mathcal{C}_{sd} \quad \text{and} \quad \mathcal{C}_{\delta s} \subset \mathcal{C}_{s\delta}.$$

In particular $\mathcal{C}_{sd} = \mathcal{C}_{ds}$ for arbitrary $\mathcal{C}$, and hence $\mathbf{R}(\mathcal{C}) = \mathcal{C}_{ds}$ if $\mathcal{C}$ is complemented. Does the equality $\mathcal{C}_{\sigma d} = \mathcal{C}_{d\sigma}$ hold in general? How about $\mathcal{C}_{\delta s} = \mathcal{C}_{s\delta}$? (Hint: Consider in $X = \mathbb{R}$ the collection $\mathcal{J}$ of all closed intervals $[a, b], a \leq b$.)

(ii) (De Morgan's laws) For any collection $\mathcal{C}$ of subsets of X let $\mathcal{C}^c$ denote the collection $\mathcal{C}^c = \{X \setminus C : C \in \mathcal{C}\}$ of complements of the sets in $\mathcal{C}$. (Thus $\mathcal{C}$ is complemented if and only if $\mathcal{C} = \mathcal{C}^c$.) Show that if $\varnothing$ and X belong to $\mathcal{C}$, then

$$(\mathcal{C}^c)_d = (\mathcal{C}_s)^c \quad \text{and} \quad (\mathcal{C}^c)_s = (\mathcal{C}_d)^c.$$

Likewise,

$$(\mathcal{C}^c)_\delta = (\mathcal{C}_\sigma)^c \quad \text{and} \quad (\mathcal{C}^c)_\sigma = (\mathcal{C}_\delta)^c.$$

(iii) Show also that if $\varnothing$ and X both belong to $\mathcal{C}$, then

$$\mathcal{C}_{\delta\sigma} \subset \mathcal{C}^{**}$$

in the notation of Theorem 1.17. Does the finite counterpart of this assertion hold? That is to say, is it the case that

$$\mathcal{C}_{ds} \subset \mathcal{C}^{\dagger\dagger}$$

in the notation of Problem 1F (provided, of course, that $\mathcal{C}$ contains $\varnothing$ and X)?

1K. A collection $\mathbf{L}$ of subsets of a set X is a *lattice* (of sets) in X (or a *lattice of subsets of X*) if $\mathbf{L}$ is closed with respect to the formation of finite unions and intersections. (The terminology is self-explanatory; such a collection of sets is, in fact, a sublattice of the lattice 2^X. Observe that every lattice of sets contains the empty set $\varnothing$; recall Remark 1.3.)

(i) If $\mathbf{L}$ is a lattice in X, then $\mathbf{L}^c$ is also a lattice in X if (and only if) $X \in \mathbf{L}$. Every ring of sets in X is a lattice in X. Every complemented lattice in X is a ring.
(ii) A nonempty complemented collection $\mathcal{C}$ of subsets of X is a lattice (and therefore a ring) in X if it is closed with respect to the formation of *either* finite unions or intersections.
(iii) For an arbitrary collection $\mathcal{C}$ of subsets of X there is a smallest lattice $\mathbf{L}(\mathcal{C})$ in X that contains $\mathcal{C}$, the lattice *generated by* $\mathcal{C}$. Use Problem 1J to show that, in fact, $\mathbf{L}(\mathcal{C}) = \mathcal{C}_{sd} = \mathcal{C}_{ds}$. Hence if $\mathcal{D}$ is a collection of subsets of X that is closed with respect to finite intersections, then $\mathbf{L}(\mathcal{D}) = \mathcal{D}_s$. (Dually, if $\mathcal{S}$ is a collection of subsets of X that is closed with respect to finite unions, then $\mathbf{L}(\mathcal{S}) = \mathcal{S}_d$.)
(iv) If $\mathcal{C}$ is a collection of subsets of X, then

$$\mathbf{L}(\mathcal{C})^c = \mathbf{L}(\mathcal{C}^c)$$

if (and only if) X belongs to both $\mathbf{L}(\mathcal{C})$ and $\mathbf{L}(\mathcal{C}^c)$. (In particular this is the case if $\mathcal{C}$ contains both $\varnothing$ and X.) If $\mathcal{C}$ is nonempty and complemented, then $\mathbf{L}(\mathcal{C})$ is also complemented, and therefore coincides with $\mathbf{R}(\mathcal{C})$.

1L. In a metric space X the topology $\mathcal{G}$ is a lattice, and so is the collection $\mathcal{F}$ of all closed sets. The collection of all G_δ sets is a lattice in X containing both $\mathcal{G}$ and $\mathcal{F}$, and the same is true of the collection of all F_σ sets. Show that, in the real line, for example, no one of these four lattices is a ring. (Hint: See [**I**, Problem 8J] and recall the Baire category theorem.)

1M. A finite union of closed cells

$$Z = [a_1, b_1] \times \cdots \times [a_d, b_d]$$

in Euclidean space $\mathbb{R}^d$ is called an *elementary figure*. Verify that the collection $\mathbf{E}$ of all elementary figures is a lattice in $\mathbb{R}^d$. Prove that in $\mathbb{R}^d$ the lattice $\mathbf{E}$, the rings $\mathbf{H}_r$ and $\mathbf{H}_\ell$ (Problem 1B), and the lattice $\mathcal{G}$ of open sets, all generate the same σ-ring, namely, the σ-algebra $\mathbf{B}$ of Borel sets. (Hint: Every open set in $\mathbb{R}^d$ is the union of a countable collection of open cells; see [**I**, Example 6U].) Generalize these results to $\mathbb{C}^d$.

1N. A collection $\mathbf{L}$ of subsets of a set X is a σ-*lattice* (*of sets*) in X (or a σ-*lattice of subsets* of X) if it is closed with respect to the formation of countable unions and countable intersections.

   (i) Clearly every σ-ring is a σ-lattice, and every σ-lattice is a lattice. Show that a ring of sets is a σ-ring if and only if it is a σ-lattice, and give an example of a σ-lattice that is not a ring. Show also that a complemented σ-lattice is a σ-ring, and verify that a nonempty complemented collection of subsets of X is a σ-lattice (and therefore a σ-ring) in X if it is closed with respect to *either* countable unions or countable intersections. If $\mathbf{L}$ is a σ-lattice in X, then $\mathbf{L}^c$ is also a σ-lattice in X if (and only if) $X \in \mathbf{L}$.

   (ii) For an arbitrary collection $\mathcal{C}$ of subsets of a set X there is a smallest σ-lattice in X that contains $\mathcal{C}$. (This σ-lattice is said to be *generated by* $\mathcal{C}$, and is denoted by $\mathbf{L}_\sigma(\mathcal{C})$.)

   (iii) Verify that $\mathbf{L}_\sigma(\mathcal{C})^c = \mathbf{L}_\sigma(\mathcal{C}^c)$ if (and only if) X belongs to both $\mathbf{L}_\sigma(\mathcal{C})$ and $\mathbf{L}_\sigma(\mathcal{C}^c)$. Conclude that if $\mathcal{C}$ is nonempty and complemented, then $\mathbf{L}_\sigma(\mathcal{C})$ is also complemented, and therefore coincides with $\mathbf{S}(\mathcal{C})$.

   (iv) Show that if for some nonempty collection $\mathcal{C}$ of subsets of X we have $\mathcal{C}^c \subset \mathbf{L}_\sigma(\mathcal{C})$, then, in fact, $\mathbf{L}_\sigma(\mathcal{C}) = \mathbf{S}(\mathcal{C})$ is a σ-algebra. (Similarly, if $\mathcal{C} \neq \varnothing$ and $\mathcal{C}^c \subset \mathbf{L}(\mathcal{C})$, then $\mathbf{L}(\mathcal{C}) = \mathbf{R}(\mathcal{C})$ is an algebra.) In particular, the σ-lattice generated by the topology on a metric space X coincides with the σ-algebra of Borel sets in X.

1O. Let X be a set, let $\mathcal{C}$ be a collection of subsets of X, and let α be a countable ordinal number.

   (i) We define $\mathcal{C}^{[0]} = \mathcal{C}$, and, assuming $\mathcal{C}^{[\xi]}$ already defined for each ordinal number ξ such that $\xi < \alpha$, we define $\mathcal{C}^{[\alpha]}$ to be either

$$\left(\bigcup_{\xi < \alpha} \mathcal{C}^{[\xi]} \right)_\delta \quad \text{or} \quad \left(\bigcup_{\xi < \alpha} \mathcal{C}^{[\xi]} \right)_\sigma ,$$

   according as α is odd or even, respectively (Problem 1H). Show that $\mathcal{C}^{[\alpha+1]}$ coincides with either $(\mathcal{C}^{[\alpha]})_\delta$ or $(\mathcal{C}^{[\alpha]})_\sigma$ according as α is even or odd, respectively, and also that

$$\bigcup_{\alpha < \Omega} \mathcal{C}^{[\alpha]} = \mathbf{L}_\sigma(\mathcal{C}).$$

   (ii) Interchanging δ and σ leads to another, dual, transfinite construction of $\mathbf{L}_\sigma(\mathcal{C})$. Set $\mathcal{C}_{[0]} = \mathcal{C}$ and define $\mathcal{C}_{[\alpha]}$ to be either

$$\left(\bigcup_{\xi < \alpha} \mathcal{C}_{[\xi]} \right)_\delta \quad \text{or} \quad \left(\bigcup_{\xi < \alpha} \mathcal{C}_{[\xi]} \right)_\sigma ,$$

   according as α is even or odd, respectively. Verify that it is also true that

$$\bigcup_{\alpha < \Omega} \mathcal{C}_{[\alpha]} = \mathbf{L}_\sigma(\mathcal{C}).$$

1P. If $\mathcal{C}$ is any nonempty collection of subsets of a metric space X, then by the *mesh* of $\mathcal{C}$ (notation: mesh $\mathcal{C}$) we mean the supremum

$$\sup\{\operatorname{diam} C : C \in \mathcal{C}\}.$$

(The diameter of a set is, by definition, an extended real number, and so therefore is the mesh of a collection of sets.)

  (i) Show that any metric space X admits open coverings $\mathcal{U}$ with arbitrarily small positive mesh, and, if X is separable, then $\mathcal{U}$ can also be taken to be countable. Conclude that if X is a nonempty separable metric space and ε is a positive number, then there exist countable partitions $\mathcal{P}$ of X with mesh $\mathcal{P} < \varepsilon$ and such that the sets in $\mathcal{P}$ are at once F_σ and G_δ sets. (A partition of a metric space X into Borel sets is called a *Borel partition* of X.)
  (ii) If the space X is compact ([**I**,Chapter 8]), then the open covering $\mathcal{U}$ and the Borel partition $\mathcal{P}$ of part (i) can be taken to be finite.

1Q. Let X be a set and let M be a fixed subset of X. Show that the mapping $A \to A \cap M$ of 2^X onto 2^M is a complete Boolean homomorphism in the sense of Example 1.8. (Indeed, this is a special case of the general situation considered there.) Conclude that if for an arbitrary collection $\mathcal{C}$ of subsets of X we write $\mathcal{C}_M$ for the collection $\{C \cap M : C \in \mathcal{C}\}$ of subsets of M (called the *trace* of $\mathcal{C}$ on M), then the σ-ring generated by $\mathcal{C}_M$ coincides with the trace on M of the σ-ring $\mathbf{S}(\mathcal{C})$: $(\mathbf{S}(\mathcal{C}))_M = \mathbf{S}(\mathcal{C}_M)$.

1R. Let $\mathcal{D}$ be a nonempty collection of subsets of a set X that is closed with respect to the formation of (finite) intersections, and let $\mathbf{E}$ denote the collection of all those subsets E of X such that either E or $X \setminus E$ contains some set D belonging to $\mathcal{D}$. Show that $\mathbf{E}$ is an algebra. Show also that if $\mathcal{D}$ is closed with respect to the formation of *countable* intersections, then $\mathbf{E}$ is a σ-algebra.

1S. Let $\mathbf{R}$ be a ring of subsets of a set X. We say that a subset S of X *splits* $\mathbf{R}$, or is *splitting for* $\mathbf{R}$, if $A \cap S \in \mathbf{R}$ whenever $A \in \mathbf{R}$, that is, if $\mathbf{R}_S \subset \mathbf{R}$.

  (i) Show that if S splits $\mathbf{R}$, then so does $X \setminus S$, and that $\mathbf{R}$ is the direct sum of $\mathbf{R}_S$ and $\mathbf{R}_{X \setminus S}$ (Example 1.15). Show conversely that if $\{X_1, X_2\}$ is a partition of X and $\mathbf{R}_1$ and $\mathbf{R}_2$ are rings in X_1 and X_2, respectively, then X_1 and X_2 split $\mathbf{R}_1 \oplus \mathbf{R}_2$ and $\mathbf{R}_i = (\mathbf{R}_1 \oplus \mathbf{R}_2)_{X_i}, i = 1, 2$.
  (ii) Let S be a splitting set for $\mathbf{R}$, and let $\mathbf{T}$ denote the collection of all those subsets E of X such that $E \cap S \in \mathbf{R}$. Show that $\mathbf{T}$ is a ring in X, and also that $\mathbf{T}$ is a σ-ring whenever $\mathbf{R}$ is. Give a description of the ring $\mathbf{T}$ as a direct sum.
  (iii) Let S be a splitting set for $\mathbf{R}$, and let $\mathbf{U}$ denote the collection of all those subsets E of X such that $E \triangle S \in \mathbf{R}$. Show that $\mathbf{R} \cup \mathbf{U}$ is a ring in X, and give a description of $\mathbf{R} \cup \mathbf{U}$ as a direct sum.

1T. For any ring [σ-ring] $\mathbf{R}$ in a set X, the collection $\widehat{\mathbf{R}}$ of all those subsets of X that split $\mathbf{R}$ is an algebra [σ-algebra] in X that contains $\mathbf{R}$. Give an example of a ring $\mathbf{R}$ for which the ring $\widehat{\mathbf{R}}$ is strictly finer than the complemented refinement of $\mathbf{R}$ constructed in Example 1.13.

1U. Let X be a set and let $\mathcal{C}$ be a nonempty collection of subsets of X that is closed with respect to the formation of countable intersections and countable pairwise disjoint unions. Prove that the collection $\mathcal{E}$ of those subsets E of X such that both E and $X \setminus E$ belong to $\mathcal{C}$ is (if nonempty) a σ-algebra in X. (Hint: If E and F belong to $\mathcal{E}$, then the four sets in the partition of X determined by E and F all belong to $\mathcal{C}$ and therefore to $\mathcal{E}$ as well.)

1V. A collection $\mathcal{M}$ of subsets of a set X is said to be closed with respect to the *formation of limits* if, whenever $\{E_n\}_{n=1}^{\infty}$ is a sequence of sets in $\mathcal{M}$ such that the limit

$$E = \lim_n E_n$$

exists, then $E \in \mathcal{M}$ (see [**I**, Example 1U]).

(i) Show that a lattice of subsets of X is closed with respect to the formation of limits if and only if it is a σ-lattice.

(ii) Prove that if $\mathcal{C}$ is an arbitrary collection of subsets of X, then there exists a smallest collection $\mathcal{M}$ of subsets of X that contains $\mathcal{C}$ and is closed with respect to the formation of limits. Devise a transfinite construction of $\mathcal{M}$. Show too that if $\mathcal{C}$ is a lattice of subsets of X, then $\mathcal{M}$ coincides with the generated σ-lattice $\mathbf{L}_\sigma(\mathcal{C})$. (Hint: For an arbitrary subset A of X, the collection of sets E such that $E \cup A \in \mathcal{M}$ is closed with respect to limits, and likewise with $E \cup A$ replaced by $E \cap A$.)

1W. If $\mathbf{R}$ is a ring of sets, then a nonempty subset $\mathbf{J}$ of $\mathbf{R}$ is an *ideal* in $\mathbf{R}$ if $\mathbf{J}$ is closed with respect to the formation of (finite) unions and if $A \in \mathbf{J}$ whenever $A \subset B$ with A in $\mathbf{R}$ and B in $\mathbf{J}$. If $\mathbf{S}$ is a σ-ring of sets, then an ideal in $\mathbf{S}$ is a σ-*ideal* if it is closed with respect to the formation of countable unions. Verify that an ideal [σ-ideal] in a ring $\mathbf{R}$ [σ-ring $\mathbf{S}$] is itself a ring [σ-ring]. If $\mathbf{R}$ is a ring [σ-ring] of sets, and if $\mathcal{C}$ is an arbitrary subcollection of $\mathbf{R}$, then $\mathcal{C}$ is contained in a smallest ideal $\mathbf{J}(\mathcal{C})$ [σ-ideal $\mathbf{J}_\sigma(\mathcal{C})$] in $\mathbf{R}$ called the ideal [σ-ideal] *generated* by $\mathcal{C}$. Show that $\mathbf{J}(\mathcal{C})[\mathbf{J}_\sigma(\mathcal{C})]$ consists of all the sets A belonging to $\mathbf{R}$ such that there exists a finite [countable] collection of sets in $\mathcal{C}$ that covers A.

Remark 1.29. This terminology is readily understood in algebraic terms. If the rings of subsets of a set X are viewed as the subrings of the Boolean ring consisting of the power class 2^X equipped with the operations $\triangle$ and $\cap$ (see the Remarks 1.27 and 1.28), then a subset of a ring $\mathbf{R}$ is an ideal in $\mathbf{R}$ if and only if is an ideal in $\mathbf{R}$ in the sense of ring theory.

1X. If $\mathcal{C}$ is a nonempty collection of subsets of a set X, and if A and B are subsets of X, then it is customary to say that A and B *differ by a set* in $\mathcal{C}$, or are *congruent modulo $\mathcal{C}$*, if $A \triangle B$ belongs to $\mathcal{C}$. Verify that this holds if and only if B is of the form $A \triangle C$ for some set C in $\mathcal{C}$.

(i) Suppose now that $\mathbf{J}$ is a collection of subsets of X that is closed with respect to unions and possesses the property that any subset of a set in $\mathbf{J}$ is also in $\mathbf{J}$. (In other words, $\mathbf{J}$ is an ideal in the ring 2^X, or an ideal of sets in X. Such an ideal is automatically a ring in X, of course.) Show that if A_1, A_2 differ from B_1, B_2, respectively, by sets in $\mathbf{J}$, then $A_1 \cup A_2, A_1 \cap A_2$ and $A_1 \triangle A_2$ differ from $B_1 \cup B_2, B_1 \cap B_2$ and $B_1 \triangle B_2$, respectively, by sets in $\mathbf{J}$. (Hint: See [**I**, Problem 1E].)

(ii) Let $\mathbf{J}$ be as in (i), let $\mathcal{E}$ be a collection of subsets of X, and write $\widehat{\mathcal{E}}$ for the collection of all subsets of X that are congruent to some set in $\mathcal{E}$ modulo $\mathbf{J}$. Verify that $\widehat{\mathcal{E}}$ is a lattice in X whenever $\mathcal{E}$ is a lattice, and also that $\widehat{\mathcal{E}}$ is a ring when $\mathcal{E}$ is a ring. Show finally that if $\mathbf{J}$ is a σ-ideal in 2^X, and if $\mathbf{S}$ is a σ-ring in X, then so is $\widehat{\mathbf{S}}$.

(iii) The sets in an ideal $\mathbf{J}$ may be thought of as negligible in one context or another. Examples of such ideals already familiar to the reader are the σ-ideal $\mathbf{C}$ of all countable subsets of an arbitrary set X and the σ-ideal $\mathbf{F}$ of all sets of the first category in an arbitrary metric space X. Still another example is constructed as follows. A subset Z of $\mathbb{R}$ is said to have *measure zero* if for each positive number ε there exists a countable covering $\{(a_n, b_n)\}$ of Z such that $\sum_n (b_n - a_n) < \varepsilon$. Show that the collection $\mathbf{Z}$ of all sets of measure zero is a σ-ideal in $2^{\mathbb{R}}$. It is a triviality of cardinal number theory that the ideal $\mathbf{C}$ is proper, that is, does not contain X, in any noncountable set X. On the other hand, it is a result of the Baire category theorem for metric spaces ([$\mathbf{I}$, Theorem 8.20]) that the ideal $\mathbf{F}$ is proper in any nonempty *complete* metric space X. Show, in the same vein, that the ideal $\mathbf{Z}$ is proper in $\mathbb{R}$. (Hint: Show first, by induction on p, that if a closed interval $[c, d]$ is covered by a collection $\{(a_i, b_i)\}_{i=1}^p$ of p open intervals, then $d - c < \sum_{i=1}^p (b_i - a_i)$. Next invoke the Heine-Borel theorem ([$\mathbf{I}$, Corollary 8.33]) to prove that if $[c, d]$ has measure zero, then $c = d$.)

1Y. Let $\mathbf{Q}$ be a quasiring of subsets of a set X.

(i) Verify that $\mathbf{Q}$ is a ring if and only if it is closed with respect to intersections. Show too that if $\mathbf{Q}$ is a σ-quasiring that is closed with respect to intersections, then $\mathbf{Q}$ is a σ-ring. Give an example of a σ-quasiring that is not a σ-ring.

(ii) Let $\widehat{\mathbf{Q}}$ denote the collection of all those subsets E of X such that $\mathbf{Q}_E \subset \mathbf{Q}$. Show that $\widehat{\mathbf{Q}}$ is an algebra in X and that $\widehat{\mathbf{Q}}$ is a σ-algebra if $\mathbf{Q}$ is a σ-quasiring. Show also that $\mathbf{Q} \cap \widehat{\mathbf{Q}}$ is a ring in X and that $\mathbf{Q} \subset \widehat{\mathbf{Q}}$ when and only when $\mathbf{Q}$ is itself a ring.

1Z. Consider a separable, complete metric space (X, ρ).

(i) Show that any nonempty analytic set can be obtained as a result of the Suslin operation applied to a system $\{F_{n_1,\ldots,n_k}\}$ of closed sets satisfying the additional conditions $F_{n_1,\ldots,n_k,n_{k+1}} \subset F_{n_1,\ldots,n_k} \neq \varnothing$ and $\operatorname{diam}(F_{n_1,\ldots,n_k}) \leq 1/k$ for all $k \in \mathbb{N}$ and $n_1, \ldots, n_{k+1} \in \mathbb{N}$. In particular, each set $F_{\mathbf{n}}$, $\mathbf{n} \in \mathbb{N}^{\mathbb{N}}$, contains exactly one point. (Hint: Separability implies that X can be covered with a countable number of balls of diameter $< 1/k$. Use induction on k to break each $F_{n_1,\ldots,n_k}$ into smaller pieces.)

(ii) Endow the space $\mathcal{S} = \mathbb{N}^{\mathbb{N}}$ with the topology of pointwise convergence. With an appropriate metric, $\mathbb{N}^{\mathbb{N}}$ is a complete, separable metric space. Let $A \in \mathcal{A}_X$ be the result of the Suslin operation applied to a system $\{F_{n_1,\ldots,n_k}\}$ of closed sets satisfying the additional conditions in (i). Denote by $f(\mathbf{n})$ the unique element of $F_{\mathbf{n}}$ if $\mathbf{n} \in \mathcal{S}$. Show that $f : \mathcal{S} \to X$ is a continuous map. Thus every set $A \in \mathcal{A}_X$ is the continuous image of a continuous map defined on a complete metric space.

(iii) Conversely, assume that Y is a complete, separable metric space, and $f : Y \to X$ is continuous. Show that $f(Y)$ is analytic. (Hint: Since Y itself is the continuous image of $\mathbb{N}^{\mathbb{N}}$, it suffices to consider the case $Y = \mathbb{N}^{\mathbb{N}}$. Define $F_{n_1,\ldots,n_k}$ to be the closure of $\{f(\mathbf{n}) : \mathbf{n}|k = (n_1, \ldots, n_k)\}$.)

1AA. Assume again that the metric space (X, ρ) is separable and complete. A sequence $\{A_n\}$ of pairwise disjoint subsets of X is said to be *Borel separated* if there exists a sequence $\{B_n\} \subset \mathbf{B}_X$ of pairwise disjoint sets such that $A_n \subset B_n$ for all $n \in \mathbb{N}$.

(i) Assume that $A_j = \bigcup_{n=1}^{\infty} A_{j,n}$, and the family $\{A_j(n)\}_j$ is Borel separated for every j. Show that $\{A_j\}_j$ is Borel separated as well.

(ii) Consider two disjoint analytic sets A_1 and A_2. Show that they are Borel separated. (Hint: Let A_j be defined by the system $\{F^j_{n_1,\dots,n_k}\}$ of closed sets satisfying the conditions of Problem 1Z(i), and let $A^j_{m_1,\dots,m_p}$ be the system $\{F^j_{m_1,\dots,m_p,n_1,\dots,n_k}\}$, so $A^j_{m_1,\dots,m_p} = \bigcup_n A^j_{m_1,\dots,m_p,n}$. Assume that A_1 and A_2 are not Borel separated, and find integers $m_1(j), m_2(j),\dots$ such that the sets $A^1_{m_1(1),\dots,m_p(1)}$ and $A^{(2)}_{m_1(2),\dots,m_p(2)}$ are not Borel separated for any $p \in \mathbb{N}$. Deduce that the singletons $F^1_{\mathbf{m}(1)}$ and $F^2_{\mathbf{m}(2)}$ are not Borel separated, which is absurd.)

(iii) If $A \in \mathcal{A}_X$ and $X \setminus A \in \mathcal{A}_X$ then $A \in \mathbf{B}_X$.

1BB. Consider the space $C([0,1])$ of all continuous functions $f : [0,1] \to \mathbb{C}$. A complete metric on this space is defined by setting $\rho(f,g) = \sup_{t \in [0,1]} |f(t) - g(t)|$, $f, g \in C([0,1])$.

(i) Show that the set $\{f \in C([0,1]) : f'(1/2) \text{ exists}\}$ is Borel. Similarly, the linear manifold $C^1([0,1])$ of continuously differentiable functions on $[0,1]$ is a Borel set; we consider one-sided derivatives at the endpoints of the interval.

(ii) Given $\alpha \in (0,1]$, a function $f : [0,1] \to \mathbb{C}$ is said to be α-Hölder (Lipschitz when $\alpha = 1$) if $(f(s) - f(t))/|s - t|^\alpha$ is bounded on $\{(s,t) \in [0,1]^2 : s \neq t\}$. Denote by Λ_α the set of such functions. Show that Λ_α is a Borel set in $C([0,1])$ for every $\alpha \in (0,1]$. (Hint: a continuous function is in Λ_α if there exists a rational number $c > 0$ such that $|f(s) - f(t)| \leq c|s - t|^\alpha$ for all rational numbers $s, t \in [0,1]$.)

1CC. A number $x \in (0,1]$ has a unique binary expansion $x = \sum_{n=1}^\infty 2^{-n} x_n$ such that the infinitely many digits $x_n \in \{0,1\}$ are equal to 1. Such a number is said to be *normal* in base 2 if

$$\lim_{N \to \infty} \frac{\operatorname{card}\{n \in \{1,2,\dots N\} : x_n = 0\}}{N} = \frac{1}{2}.$$

Show that the set of normal numbers in base 2 is a Borel set in $(0,1]$.

1DD. Consider a metric space X. A set $E \subset X$ is said to be *meager* in X or *nowhere dense* in X if its closure $\overline{E}$ has no interior points. A set $E \subset X$ is said to be *of first Baire category* in X if it can be written as a countable union $E = \bigcup_{n \in \mathbb{N}} E_n$ such that each E_n is meager. (Sets which are not of the first category are said to be of the *second Baire category*. There is no third category.) Denote by $\mathbf{R}$ the collection of those sets $E \subset X$ for which there exists an open set $G \subset X$ such that the symmetric difference $E \triangle G$ is of the first category. Show that $\mathbf{R}$ is a σ-algebra.

Chapter 2
Measurability

Central to the discussion of measurability is the notion of a measurable space.

Definition 2.1. A *measurable space* is a set X together with a σ-algebra $\mathbf{S}$ of subsets of X.

Thus, strictly speaking, a measurable space is a pair $(X, \mathbf{S})$. Nevertheless, we frequently denote this space by the single symbol X and refer to X as measurable *with respect to* the σ-algebra $\mathbf{S}$. The elements of $\mathbf{S}$ are called the *measurable subsets* of X. (The requirement that the σ-ring $\mathbf{S}$ be complemented is equivalent to requiring the entire space X to be a measurable set.) Thus, as mentioned in Remark 1.3, in some texts a measurable space is defined to be a pair $(X, \mathbf{S})$ where $\mathbf{S}$ is a σ-ring of subsets of X.) In the event that any possibility of confusion exists, for instance, if X is simultaneously measurable with respect to some other σ-algebra in addition to $\mathbf{S}$, the elements of $\mathbf{S}$ will be said to be *measurable* $[\mathbf{S}]$.

Example 2.2. If X is a metric space or, more generally, a topological space then X is a measurable space with respect to the σ-algebra $\mathbf{B}_X$ of Borel sets in X. In the absence of any stipulation to the contrary, whenever, in the sequel, a metric space is regarded as a measurable space, it is the algebra $\mathbf{B}_X$ that is understood to be the σ-algebra of measurable sets.

Example 2.3. In the extended real number system $\mathbb{R}^\natural$ ([**I**, Chapter 2]) the closed rays $[a, +\infty] = \{t \in \mathbb{R}^\natural : t \geq a\}$, $a \in \mathbb{R}^\natural$, generate a σ-ring $\mathbf{B}^\natural$ containing the whole space $\mathbb{R}^\natural$, so $(\mathbb{R}^\natural, \mathbf{B}^\natural)$ is a measurable space. Whenever, in this book, we have occasion to regard $\mathbb{R}^\natural$ as a measurable space, it is this algebra $\mathbf{B}^\natural$ of "extended" Borel sets that we have in mind. (Since $\mathbf{B}^\natural$ contains the half-open interval $[a, b)$ for all finite real numbers a and b, it is clear that the trace of $\mathbf{B}^\natural$ on $\mathbb{R}$ coincides with the algebra $\mathbf{B}_\mathbb{R}$ of ordinary real Borel sets. Thus $\mathbf{B}^\natural$ is obtained from $\mathbf{B}_\mathbb{R}$ by adjoining the ideal numbers $\pm\infty$—either, neither or both—to the various Borel subsets of $\mathbb{R}$.)

© Springer International Publishing Switzerland 2016
H. Bercovici et al., *Measure and Integration*,
DOI 10.1007/978-3-319-29046-1_2

Example 2.4. If $\mathbf{S}$ is a σ-ring of subsets of a set X, and if $\mathbf{S}$ is *not* complemented, then there are various methods of turning X into a measurable space in such a way that the sets in $\mathbf{S}$ become measurable. The most economical of these procedures is, of course, the one set forth in Example 1.13; see also Problem 1T.

Example 2.5. If $(X, \mathbf{S})$ is a measurable space in which all of the singletons in X are measurable sets, then $\mathbf{S}$ must contain the collection $\mathbf{S}_0$ comprising all the countable subsets of X and their complements. But $\mathbf{S}_0$ is itself a σ-algebra. Thus the singletons in X are all measurable [$\mathbf{S}$] if and only if $\mathbf{S}$ refines $\mathbf{S}_0$.

Example 2.6. If $\{X_\gamma\}_{\gamma \in \Gamma}$ is an indexed partition of a set X, and if $\mathbf{S}_\gamma$ is a σ-algebra in X_γ for each index γ, then the full direct sum $\bigoplus_\gamma \mathbf{S}_\gamma$ of the family $\{\mathbf{S}_\gamma\}$ (Example 1.16) is also complemented in X, and therefore turns X into a measurable space, called the *full direct sum* of the family $\{(X_\gamma, \mathbf{S}_\gamma)\}$ and denoted by

$$\bigoplus_{\gamma \in \Gamma} (X_\gamma, \mathbf{S}_\gamma).$$

In particular, if $(X_1, \mathbf{S}_1)$ and $(X_2, \mathbf{S}_2)$ are measurable spaces with X_1 and X_2 disjoint, then the measurable space $(X_1 \cup X_2, \mathbf{S}_1 \oplus \mathbf{S}_2)$ can also be written as $(X_1, \mathbf{S}_1) \oplus (X_2, \mathbf{S}_2)$; see Example 1.15.

Let X be a measurable space with respect to a σ-algebra $\mathbf{S}$, and let A be an arbitrary subset of X. Then the trace $\mathbf{S}_A = \{E \cap A : E \in \mathbf{S}\}$ of $\mathbf{S}$ on A is a σ-algebra in A (Problem 1Q), and $(A, \mathbf{S}_A)$ is again a measurable space. Observe that in the event that A is itself measurable, the σ-algebra $\mathbf{S}_A$ simply consists of the measurable subsets of A: $\mathbf{S}_A = \{E \in \mathbf{S} : E \subset A\}$. In this case, and in this case only, we refer to $(A, \mathbf{S}_A)$ as a *subspace* of $(X, \mathbf{S})$.

Definition 2.7. If X is a measurable space with respect to a σ-algebra $\mathbf{S}$, then a *subspace* of X is a measurable set E (with respect to $\mathbf{S}$) equipped with the σ-algebra $\mathbf{S}_E$.

Example 2.8. If X is a metric space and A an arbitrary subset of X, then A itself becomes a metric subspace of X when equipped with its relative metric ([**I**, Chapter 6]), and the collection of all those subsets of A that are open with respect to this relative metric is precisely $\mathcal{G}_A$, the trace on A of the collection $\mathcal{G}$ of open sets in X ([**I**, Proposition 6.15]). The trace of the σ-algebra $\mathbf{B}_X$ of Borel sets in X is precisely the σ-algebra $\mathbf{B}_A$. Thus there is an unambiguous sense in which A is to be regarded as a measurable space in its own right. Nevertheless, according to the foregoing definition, $(A, \mathbf{B}_A)$ is not a subspace of the measurable space $(X, \mathbf{B}_X)$ unless A is a Borel set in X. Thus, for example, the measurable space $(\mathbb{R}, \mathbf{B}_\mathbb{R})$ of real numbers is a subspace of the measurable space $(\mathbb{R}^\natural, \mathbf{B}^\natural)$ of extended real numbers.

Given measurable spaces X and Y, there is a natural way of distinguishing a special class of mappings of X into Y, called the *measurable* mappings.

Definition 2.9. Let $(X, \mathbf{S})$ and $(Y, \mathbf{T})$ be measurable spaces. A mapping $\varphi : X \to Y$ is *measurable* (when necessary, *measurable* $[\mathbf{S}, \mathbf{T}]$) if the inverse image under φ of every set E in $\mathbf{T}$ is measurable $[\mathbf{S}]$, that is, if $\varphi^{-1}(E) \in \mathbf{S}$ for every set E in $\mathbf{T}$. More generally, if φ is defined only on some subset of X that contains the set A, then φ is said to be *measurable on* A if $\varphi|A$ is measurable on the measurable space $(A, \mathbf{S}_A)$. In the special case that Y is a metric space, in keeping with the general convention enunciated in Example 2.2, a mapping φ of X into Y is said simply to be *measurable* (or *measurable* $[\mathbf{S}]$) if it is measurable $[\mathbf{S}, \mathbf{B}_Y]$. Moreover, if both X and Y are metric spaces, the mapping φ is said to be *Borel measurable* if it is measurable $[\mathbf{B}_X, \mathbf{B}_Y]$.

Example 2.10. Any constant mapping of a measurable space $(X, \mathbf{S})$ into a measurable space $(Y, \mathbf{T})$ is measurable, since $\mathbf{S}$ always contains the whole space X. On the other hand, if $\mathbf{S} = \{\varnothing, X\}$, then it may happen that *only* the constant mappings of X into Y are measurable. Dually, if $\mathbf{S} = 2^X$, then *every* mapping of X into Y is measurable $[\mathbf{S}]$.

Example 2.11. Let X and Y be measurable spaces, let $\varphi : X \to Y$ be measurable, and suppose Y has the property that the singletons $\{y_0\}$ in Y are all measurable sets (see Example 2.5). Then all the level sets

$$\{x \in X : \varphi(x) = y_0\} = \varphi^{-1}(\{y_0\})$$

of φ are measurable sets in X. In particular, this is the case if Y is a metric space or, more generally, a Hausdorff topological space.

Example 2.12. Let $(X, \mathbf{S})$ and $(Y, \mathbf{T})$ be measurable spaces, let A and B be subsets of X such that $A \subset B$, and suppose $\varphi : X \to Y$ is measurable on B. Then for each set F in $\mathbf{T}$ there is a set E in $\mathbf{S}$ such that $(\varphi|B)^{-1}(F) = E \cap B$. But then

$$(\varphi|A)^{-1}(F) = \{x \in A : \varphi(x) \in F\} = E \cap A.$$

Thus φ is automatically measurable on A as well, so the collection $\mathcal{A}$ of all those subsets of X on which φ is measurable is closed with respect to the formation of subsets. But $\mathcal{A}$ is not closed with respect to unions in general. Indeed, if A is any *non*measurable subset of X, then the characteristic function χ_A is measurable on both A and $X \setminus A$—being constant on each set—but the real-valued function χ_A is not measurable on X.

On the other hand, the collection $\mathbf{M}$ of all those *measurable* subsets of X on which φ is measurable is clearly a σ-ring contained in $\mathbf{S}$. In particular, a mapping of X into Y that is measurable on each of a countable collection of measurable sets that covers X is itself measurable.

Proposition 2.13. *Let $(X, \mathbf{S})$ and $(Y, \mathbf{T})$ be measurable spaces, and let $\varphi : X \to Y$ be a map. Suppose given a collection $\mathcal{C}$ of subsets of Y such that $\mathbf{T} = \mathbf{S}(\mathcal{C})$. Then φ is measurable $[\mathbf{S}, \mathbf{T}]$ if (and only if) $\varphi^{-1}(C) \in \mathbf{S}$ for every C in $\mathcal{C}$.*

Proof. The collection $\{A \subset Y : \varphi^{-1}(A) \in \mathbf{S}\}$ is a σ-ring in Y and contains $\mathcal{C}$, so it must contain $\mathbf{T}$. $\Box$

Corollary 2.14. *Let X be a measurable space and let Y be a metric space. Then a mapping $\varphi \colon X \to Y$ is measurable if (and only if) the inverse image under φ of every open [closed] set in Y is measurable.*

In connection with these ideas, and in other contexts as well, the following notation is frequently useful.

Notation

For every real-valued function f on a set X and every real number a we write $E(f < a)$ and $E(f \leq a)$ for the sets $\{x \in X : f(x) < a\}$ and $\{x \in X : f(x) \leq a\}$, respectively. Similarly, $E(f > a)$ and $E(f \geq a)$ denote the sets $\{x \in X : f(x) > a\}$ and $\{x \in X : f(x) \geq a\}$, respectively. In the same vein, if a and b are two real numbers, we write $E(a < f < b)$ for the set

$$\{x \in X : a < f(x) < b\} = E(f > a) \cap E(f < b),$$

and similarly for symbols such as

$$E(a < f \leq b), \quad E(a \leq f \leq b) \quad \text{and} \quad E(f = a).$$

Example 2.15. All the sets mentioned above, associated with a measurable real-valued function f on a measurable space X, are clearly measurable. Conversely, if all of the sets $E(f < t), t \in \mathbb{R}$, are measurable, then f is measurable. Similar statements hold for the families $\{E(f \leq t), t \in \mathbb{R}\}$, $\{E(f > t) : t \in \mathbb{R}\}$, and $\{E(f \geq t) : t \in \mathbb{R}\}$. Exactly the same criteria ensure measurability when f is extended real-valued. See Problems 2A and 2B.

Example 2.16. Every semicontinuous extended real-valued function on a metric space is Borel measurable; see [**I**, Proposition 7.17].

Example 2.17. If X and Y are metric spaces, then every continuous mapping of X into Y is Borel measurable ([**I**, Theorem 7.4]). If, in particular, X is a discrete metric space (so every mapping of X into Y is continuous), then the continuous mappings of X into Y exhaust the Borel measurable mappings of X into Y. This case is exceptional; ordinarily one expects to find many Borel measurable mappings that are not continuous.

Example 2.18. A complex-valued function u on a measurable space X is measurable if and only if the real-valued functions $\Re u$ and $\Im u$ are both measurable. Indeed, if a, b, c, and d are any four real numbers such that $a \leq b$ and $c \leq d$, and if we write S_1 and S_2 for the closed strips

$$S_1 = \{\lambda \in \mathbb{C} : a \leq \Re\lambda \leq b\}, \quad S_2 = \{\lambda \in \mathbb{C} : c \leq \Im\lambda \leq d\},$$

then

$$u^{-1}(S_1) = (\Re u)^{-1}([a,b]) \quad \text{and} \quad u^{-1}(S_2) = (\Im u)^{-1}([c,d]).$$

This shows that if u is measurable, then $\Re u$ and $\Im u$ are also measurable. On the other hand, the rectangle $R = [a,b] \times [c,d]$ coincides with $S_1 \cap S_2$. If $\Re u$ and $\Im u$ are both measurable, then $u^{-1}(R)$ is a measurable set, and since the closed rectangles R generate the Borel sets in $\mathbb{C}$ as a σ-ring (Problem 1M), this shows that the complex-valued function u is measurable as well. (Note that this implies that u is measurable if and only if its complex conjugate $\bar{u}$ is measurable.) More generally, a mapping $x \mapsto (f_1(x), \ldots, f_d(x))$ of X into $\mathbb{R}^d$ is measurable if and only if the coordinate functions $f_i, i = 1, \ldots, n$, are all measurable, and similarly for $\mathbb{C}^d$ in place of $\mathbb{R}^d$. (See also Problem 2G.)

Example 2.19. Every monotone real-valued function f on $\mathbb{R}$ is Borel measurable. Indeed, if I is an interval of any type in $\mathbb{R}$, then $f^{-1}(I)$ is likewise an interval of some type. Similarly, every monotone extended real-valued function on $\mathbb{R}$ is Borel measurable. More generally, every monotone extended real-valued function defined on an arbitrary subset A of $\mathbb{R}$ is Borel measurable on A.

We turn next to the consideration of various ways of combining measurable mappings. Our first result is very simple but also very useful.

Proposition 2.20. *Let $(X, \mathbf{S}), (Y, \mathbf{T})$ and $(Z, \mathbf{U})$ be measurable spaces, and let $\varphi \colon X \to Y$ and $\psi \colon Y \to Z$ be measurable mappings. Then the composition $\psi \circ \varphi$ is measurable. In particular, if Y and Z are both metric spaces, and if ψ is Borel measurable, then $\psi \circ \varphi$ is measurable [$\mathbf{S}$] whenever φ is.*

Proof. If E belongs to $\mathbf{U}$, then $\psi^{-1}(E)$ belongs to $\mathbf{T}$, and thus $\varphi^{-1}(\psi^{-1}(E))$ belongs to $\mathbf{S}$. But $\varphi^{-1}(\psi^{-1}(E)) = (\psi \circ \varphi)^{-1}(E)$. $\qquad\square$

Example 2.21. If f is a measurable complex-valued function on a measurable space X, then the powers f^n of $f, n \in \mathbb{N}$, are all measurable as well. Similarly, the functions $|f|^r, r > 0$, are also all measurable. Moreover, the same is true of the functions f^{-n} and $|f|^{-r}$ on the (measurable) subspace $\{x \in X : f(x) \neq 0\}$.

Proposition 2.22. *If $f_1, \ldots, f_n$ are arbitrary measurable scalar-valued functions on a measurable space X, then $f_1 + f_2 + \cdots + f_n$ is also measurable.*

Proof. It suffices (by mathematical induction) to prove that the sum $f + g$ of two measurable functions is measurable. To this end, consider the mapping T of X into $\mathbb{C}^2$ defined by $T(x) = (f(x), g(x))$. If U and V are open sets in $\mathbb{C}$, then

$$T^{-1}(U \times V) = f^{-1}(U) \cap g^{-1}(V)$$

is a measurable set. But every open set in $\mathbb{C}^2$ is a countable union of sets of the form $U \times V$, since $\mathbb{C}$ is separable. Hence T is measurable by Proposition 2.13. Moreover, addition is continuous regarded as a mapping $s : \mathbb{C} \times \mathbb{C} \to \mathbb{C}$, whence it follows by Proposition 2.20 that $s \circ T$ is also measurable. Since $(s \circ T)(x) = f(x) + g(x)$, the proof is complete.   □

The modification in the proof of Proposition 2.22 required to establish the following result is obvious (see [**I**, Problem 7H]).

Proposition 2.23. *If* $f_1, \ldots, f_n$ *are measurable real-valued functions on a measurable space* X, *then the functions* $f_1 \vee \cdots \vee f_n$ *and* $f_1 \wedge \cdots \wedge f_n$ *are also measurable.*

Example 2.24. A real-valued function f on a measurable space X is measurable if and only if the positive and negative parts $f^+ = f \vee 0$ and $f^- = -(f \wedge 0)$ are both measurable on X.

Proposition 2.25. *If* $f_1, \ldots, f_n$ *are measurable scalar-valued functions on a measurable space* X, *and if* $\alpha_1, \ldots, \alpha_n$ *are scalars, then the linear combination*

$$\alpha_1 f_1 + \cdots + \alpha_n f_n \tag{2.1}$$

and the product

$$f_1 \cdots f_n \tag{2.2}$$

are also measurable.

Proof. The measurability of (2.1) is an immediate consequence of Proposition 2.22 and the second half of this result, in view of the fact that constant functions are measurable. As for the measurability of (2.2), it suffices as before to verify the measurability of the product fg of two measurable functions, and this may either be settled directly via another obvious modification of the proof of Proposition 2.22, or it may be derived from the identity $2fg = (f + g)^2 - (f^2 + g^2)$.   □

Corollary 2.26. *If* f *and* g *are measurable scalar-valued functions on a measurable space* X, *then* $\{x \in X : f(x) = g(x)\}$ *is a measurable set.*

Proof. This is the zero level set of the measurable function $f - g$.   □

Example 2.27. If $p(t)$ is a real polynomial and f is a measurable real-valued function on a measurable space X, then $p(f(x))$ is also measurable on X. Similarly, if $p(\lambda)$ is a complex polynomial and f is a measurable complex-valued function on X, then $p(f(x))$ is measurable. These facts may be viewed as consequences of Proposition 2.25, or may be derived directly from Proposition 2.20, since $p(f(x))$ coincides with the composition $p \circ f$, and the function $p(\lambda)$ is continuous.

A significant advantage that measurability enjoys over continuity is that it is readily preserved in passage to a limit. We begin our discussion of these matters with a detailed treatment of the important special case of a sequence of real-valued functions.

Proposition 2.28. *Let $\{f_n\}$ be a sequence of measurable, extended real-valued functions on a measurable space X. Then the functions*

$$\sup_n f_n \quad \text{and} \quad \inf_n f_n$$

are also measurable. In particular, the set on which $\{f_n\}$ is pointwise bounded above [below] in $\mathbb{R}$ is measurable.

Proof. It suffices to treat upper bounds because

$$\inf_n f_n = -\sup_n(-f_n).$$

Let A denote the set of points $x \in X$ for which the sequence $\{f_n(x)\}$ is bounded above in $\mathbb{R}$. Then $\sup_n f_n$ is constant $(= +\infty)$ on $X \setminus A$, so it is enough to verify that A is measurable and that $\sup_n f_n$ is measurable on A (see Problem 2B).

For each real number M set

$$E_M = \bigcap_{n=1}^{\infty} E(f_n \leq M).$$

Then E_M is clearly measurable, and we also have $E_M = E(\sup_n f_n \leq M)$. Thus, on the one hand,

$$A = \bigcup_{N=1}^{\infty} E_N$$

is measurable, while, on the other hand, $\sup_n f_n$ is measurable on A. $\quad\square$

Corollary 2.29. *If $\{f_n\}$ is a monotone sequence of measurable, extended real-valued functions on a measurable space X, then the pointwise limit $\lim_n f_n$ is also measurable.*

Proof. If $\{f_n\}$ is increasing [decreasing], $\lim_n f_n = \sup_n f_n$ $[= \inf_n f_n]$. $\quad\square$

Corollary 2.30. *For any sequence $\{f_n\}$ of measurable, extended real-valued functions on a measurable space X, the functions $\limsup_n f_n$ and $\liminf_n f_n$ are measurable.*

Proof. It suffices to treat the upper limit. For each positive integer m set

$$g_m = \sup_{n \geq m} f_n.$$

so that $\limsup_n f_n = \inf_m g_m$. The conclusion follows by two applications of Proposition 2.28. $\quad\square$

Theorem 2.31. *For an arbitrary sequence $\{f_n\}$ of measurable (finite) real-valued functions on a measurable space X, the set E of those points x for which the sequence $\{f_n(x)\}$ is convergent in $\mathbb{R}$ is measurable, and the pointwise limit*

$$\lim_n f_n(x), \quad x \in E,$$

is a measurable function on E.

Proof. This follows immediately from Corollary 2.30. $\qquad\qquad\qquad\square$

The result of Theorem 2.31 is true for complex-valued functions as well. (The proof of the following theorem can readily be reduced to the real case, but the argument given extends to more general situations; see Problem 2I.)

Theorem 2.32. *Let X be a measurable space, and let $\{f_n\}$ be a sequence of measurable complex-valued functions on X. Denote by E the set of all those points x of X for which the numerical sequence $\{f_n(x)\}$ is convergent in $\mathbb{C}$. Then E is a measurable set, and the function $f(x) = \lim_n f_n(x)$ is measurable on the subspace E.*

Proof. For each triple (k, m, n) of positive integers write

$$E_{k,m,n} = E(|f_k - f_m| < 1/n).$$

Since $|f_k - f_m|$ is a measurable function, each of the sets $E_{k,m,n}$ is measurable, and so therefore is the set

$$F_n = \bigcup_{p=1}^{\infty} \bigcap_{k=p}^{\infty} \bigcap_{m=p}^{\infty} E_{k,m,n}.$$

Now F_n is precisely the set of points x in X such that $|f_k(x) - f_m(x)| < 1/n$ eventually. The measurability of E follows from the equality $E = \bigcap_{n=1}^{\infty} F_n$, and this holds because a sequence of scalars converges if and only if it satisfies the Cauchy criterion.

In order to see that the function f is measurable on E, let U be an open set in the complex plane and for each $n \in \mathbb{N}$ set $G_n = E \cap f_n^{-1}(U)$. Then G_n is measurable, and so therefore is

$$H_U = \bigcup_{m=1}^{\infty} \bigcap_{n=m}^{\infty} G_n.$$

The set H consists precisely of those points x of E with the property that $\{f_n(x)\}$ is eventually in U. Next let V be an open set in $\mathbb{C}$ different from $\mathbb{C}$ itself, and let F denote the complement of V. For each positive integer j let U_j denote the set of points in $\mathbb{C}$ whose distance from the closed set F is greater than $1/j$:

$$U_j = \{\lambda \in \mathbb{C} \colon d(\lambda, F) > 1/j\}.$$

The set U_j is an open subset ([**I**, Problem 6J]) and we conclude that $f^{-1}(V) = \bigcup_{j=1}^{\infty} H_{U_j}$ is measurable, thus concluding the proof.   □

The following is a convenient summary of Propositions 2.23 and 2.25, Theorems 2.31 and 2.32, and Example 2.27. A linear manifold of scalar-valued functions that is closed with respect to multiplication is called a *function algebra*.

Theorem 2.33. *The collection $\mathcal{M}_{\mathbb{C}}$ of all measurable complex-valued functions on a measurable space X is a complex function algebra that contains all constant functions and is closed with respect to complex conjugation and the formation of limits of pointwise convergent sequences. The collection $\mathcal{M}_{\mathbb{R}}$ of all the real-valued functions in $\mathcal{M}_{\mathbb{C}}$ constitutes a real function algebra that is also a function lattice ([**I**, Problem 2M]).*

It is an interesting and useful fact that Theorem 2.33 has a valid converse. Before stating it, however, we introduce a concept that will play a major role in all that follows.

Definition 2.34. A scalar-valued function on a set X is said to be *simple* if it assumes only a finite number of distinct values.

The main facts about simple functions are summarized for convenience of reference in the following proposition, for which no proof need to be given.

Proposition 2.35. *A scalar-valued function on a set X is simple if and only if it can be expressed as a linear combination of characteristic functions of subsets of X. Among such representations of a given simple function s there is precisely one,*

$$s = \sum_{i=1}^{m} \alpha_i \chi_{E_i}, \tag{2.3}$$

in which the sets E_i are disjoint and nonempty and the coefficients α_i are distinct from one another and from zero. (If $s = 0$, the sum in (2.3) is empty.) If X is a measurable space, then a simple function s on X is measurable if and only if it can be expressed as a linear combination of characteristic functions of measurable subsets of X. Alternatively, s is measurable if and only if it assumes each of its values on a measurable set or, equivalently, if and only if the sets E_i in (2.3) are all measurable.

Simple functions provide an important link between measurable sets and measurable functions. Their usefulness stems largely from the following fact.

Proposition 2.36. *Let X be a measurable space and consider a function $f : X \to \mathbb{R}$. Then the following three conditions are equivalent.*

(1) *The function f is measurable.*
(2) *There exists a sequence $\{s_n\}$ of measurable simple real-valued functions on X that converges pointwise to f.*

(3) *There exists a sequence $\{s_n\}$ of measurable simple real-valued functions on X converging pointwise to f and satisfying the following additional conditions:*

   (a) *at each point x of X either $0 \leq s_1(x) \leq s_2(x) \leq \cdots$, or*
   $0 \geq s_1(x) \geq s_2(x) \geq \cdots$, *and*
   (b) *if $M \in \mathbb{N}$ and if $|f(x)| \leq M$, then $|f(x) - s_n(x)| \leq 1/2^n$ for every $n \geq M$.*

Before giving the proof, we note that property (3a) implies that $|s_n(x)| \leq |f(x)|$ for all x and all n. In particular, if $f(x) = 0$, then $s_n(x) = 0$ for all n. Note also that (3b) implies that $\{s_n\}$ converges uniformly to f on any set on which f is bounded.

Proof. It is obvious that (3) is a stronger condition than (2), while (2) implies (1) by Theorem 2.31. Hence it suffices to show that (1) implies (3). For each positive integer n consider the points

$$t_n = k/2^n, \qquad k = -n2^n, -n2^n + 1, \ldots, n2^n.$$

We define the simple function s_n as follows:

(α) If $f(x) \geq n$ or $f(x) < -n$ set $s_n(x) = 0$,
(β) If $-n \leq f(x) < n$, find k such that $t_k \leq f(x) < t_{k+1}$, and set $s_n(x) = t_k$ if $t_k \geq 0$ and $s_n(x) = t_{k+1}$ if $t_k < 0$.

A moment's reflection shows that the set on which the function s_n assumes each nonzero value t_k is the inverse image under f of a half-open interval, and since all intervals are Borel sets in $\mathbb{R}$, it follows that s_n is measurable. Moreover (3b) follows because $|f(x) - s_n(x)| \leq 1/2^n$ whenever $|f(x)| \leq n$. The verification of (3a) is left to the reader.

□

Remark 2.37. If $\{s_n\}$ is a sequence of measurable simple real-valued functions tending pointwise to a limit f as in (3a) of Proposition 2.36, then all three of the sequences $\{s_n^+\}, \{s_n^-\}, \{|s_n|\}$ are monotone increasing on X and tend pointwise to the limits f^+, f^- and $|f|$, respectively.

The analog of Proposition 2.36 for complex-valued functions is proved in a similar manner by partitioning the closed disk of radius $n2^n$ into parts of diameter $1/2^n$ and selecting for each part a point of minimum absolute value. We record the statement below.

Proposition 2.38. *Let X be a measurable space and consider a measurable function $f : X \to \mathbb{C}$. There exists a sequence $\{s_n\}$ of measurable simple complex-valued functions on X converging pointwise to f and satisfying the following additional conditions:*

(a) $0 \leq |s_1(x)| \leq |s_2(x)| \leq \cdots \leq |f(x)|$, $x \in X$, and
(b) if $M \in \mathbb{N}$ and if $|f(x)| \leq M$, then $|f(x) - s_n(x)| \leq 1/2^n$ for every $n \geq M$.

We can now prove the converse of Theorem 2.33.

Theorem 2.39. *Let X be a set and let $\mathcal{M}$ be a function algebra of real-valued functions on X that contains the constant functions and is closed with respect to the formation of limits of pointwise convergent sequences. Then there exists a unique σ-algebra $\mathbf{S}$ in X such that $\mathcal{M}$ is precisely the collection of all real-valued functions measurable $[\mathbf{S}]$. Similarly, a function algebra $\mathcal{M}$ of complex-valued functions on X that contains the constant functions and is closed with respect to complex conjugation and the formation of limits of pointwise convergent sequences is the collection of all complex-valued measurable functions with respect to a unique σ-algebra $\mathbf{S}$ in X.*

We only prove the theorem in the real case, leaving the complex case as an exercise (see Problem 2N). We need a lemma.

Lemma 2.40. *Let U be an open subset of the real line $\mathbb{R}$. Then there exists a sequence $\{p_n\}$ of real polynomials that converges pointwise to the characteristic function χ_U.*

Proof. First, there exists a sequence $\{f_n\}$ of continuous functions on $\mathbb{R}$ that converges pointwise to χ_U (see [**I**, Problem 7F]). By the Weierstrass approximation theorem, there exists, for each index n, a real polynomial p_n such that $|f_n(t) - p_n(t)| \leq 1/n$ for all $|t| \leq n$, and it is readily verified that the sequence $\{p_n\}$ satisfies the conclusion of the lemma. $\square$

The Weierstrass approximation theorem alluded to here, historically the simplest version of the theorem, may be stated as follows.

Theorem 2.41. (Weierstrass) *For any real-valued function f defined and continuous on a closed interval $[a, b]$ of real numbers, there exists a sequence $\{p_n\}$ of real polynomials converging uniformly to f on $[a, b]$.*

It suffices to prove the theorem when $[a, b] = [0, 1]$. The proof below is due to Bernstein and Korovkin.

Proof. Let $f : [0, 1] \to \mathbb{R}$ be continuous, and define for each $n \in \mathbb{N}$ the Bernstein polynomial

$$B_n(f, t) = \sum_{k=1}^{n} f\left(\frac{k}{n}\right) \binom{n}{k} t^k (1 - t)^{n-k},$$

where

$$\binom{n}{k} = \frac{n!}{k!(n-k)!}$$

is the usual binomial coefficient. It suffices to show that B_n converges uniformly to f on $[0, 1]$. In order to do this, we need to calculate B_n explicitly when f is a polynomial of degree at most two. Setting $p_j(t) = t^j$ for $j = 0, 1, 2$, we use the equations

$$B_n(p_0, t) = p_0(t), \quad B_n(p_1, t) = p_1(t), \quad B_n(p_2, t) = p_2(t) + \frac{t(1-t)}{n}$$

for $n \in \mathbb{N}$ and $t \in [0, 1]$. The relevant identities to be verified are

$$\sum_{k=0}^{n} \binom{n}{k} t^k (1-t)^{n-k} = 1, \tag{2.4}$$

which is simply the binomial theorem, and

$$\sum_{k=0}^{n} (k - nt)^2 \binom{n}{k} t^k (1-t)^{n-k} = nt(1-t) \tag{2.5}$$

(see Problem 2K). The fact that f is bounded and uniformly continuous on $[0, 1]$ ([**I**, Theorem 8.34 and Corollary 8.36]) easily implies that, given $\varepsilon > 0$, there exists a constant $M_\varepsilon > 0$ such that

$$|f(t) - f(s)| \leq \varepsilon + M_\varepsilon (t - s)^2, \quad s, t \in [0, 1].$$

This can be written as

$$-\varepsilon - M_\varepsilon (p_2(t) - 2sp_1(t) + s^2) \leq f(t) - f(s) \leq \varepsilon + M_\varepsilon (p_2(t) - 2sp_1(t) + s^2).$$

Using the fact that $B_n(g, t)$ is nonnegative when g is a nonnegative function, we deduce that

$$-\varepsilon - M_\varepsilon (B_n(p_2, t) - 2sB(p_1, t) + s^2) \leq B_n(f, t) - f(s)$$
$$\leq \varepsilon + M_\varepsilon (B_n(p_2, t) - 2sB_n(p_1, t) + s^2)$$

for $n \in \mathbb{N}$ and $t, s \in [0, 1]$. Set $t = s$ in this inequality and use the above formulas for $B_n(p_j, t)$ to obtain

$$|B_n(f, s) - f(s)| \leq \varepsilon + M_\varepsilon \frac{s(1-s)}{n} \leq \varepsilon + \frac{M_\varepsilon}{4n}, \quad n \in \mathbb{N}, s \in [0, 1].$$

Thus $|B_n(f, s) - f(s)| < 2\varepsilon$ for all $s \in [0, 1]$ provided that $n > M_\varepsilon / 4\varepsilon$, thereby concluding the proof.   □

We proceed now with the proof of Theorem 2.39 in the real case.

Proof. It is clear that on any measurable space the characteristic function of a set E is measurable if and only if E is measurable. Hence, if the theorem is to hold, the σ-algebra **S** must consist precisely of those subsets of X whose characteristic functions belong to $\mathcal{M}$. (Incidentally, this establishes the uniqueness of **S**.) Accordingly, we set $\mathbf{S} = \{E \subset X : \chi_E \in \mathcal{M}\}$, and proceed to verify:

(1) **S** is a σ-algebra.
(2) The functions in $\mathcal{M}$ are all measurable [**S**].
(3) Every real-valued function that is measurable [**S**] belongs to $\mathcal{M}$.

These arguments go as follows.

(1) Since $\chi_{E\setminus F} = \chi_E - \chi_E \chi_F$ and $\chi_{E \cup F} = \chi_E + \chi_F - \chi_E \chi_F$, it is clear that **S** is a ring of sets. To see that it is a σ-ring, let $\{E_n\}$ be a sequence of sets in **S** having union E. If $F_n = E_1 \cup \cdots \cup E_n$, then F_n belongs to **S** for each n and the sequence $\{\chi_{F_n}\}$ tends pointwise to χ_E. It follows that E also belongs to **S**. Moreover, it is obvious that **S** is complemented since the function identically equal to one is in $\mathcal{M}$.

(2) Consider a function $f \in \mathcal{M}$ and let U be an open subset of the real line $\mathbb{R}$. Since $\mathcal{M}$ is a function algebra containing the constants, it follows that $p(f(x))$ belongs to $\mathcal{M}$ for every real polynomial p. In particular, if $\{p_n\}$ is a sequence of real polynomials converging pointwise to χ_U as in Lemma 2.40, then the sequence of functions $h_n(x) = p_n(f(x))$ belongs to $\mathcal{M}$. But $h_n(x)$ converges pointwise to the characteristic function of $f^{-1}(U)$. Thus $f^{-1}(U)$ belongs to **S**, whence it follows that f is measurable [**S**] (Corollary 2.14).

(3) The simple real-valued functions on X that are measurable [**S**] are clearly in $\mathcal{M}$ (according to Proposition 2.35 they are linear combinations of the characteristic functions in $\mathcal{M}$), and every real-valued function on X that is measurable [**S**] is the pointwise limit of a sequence of such simple functions by Proposition 2.36.

$\square$

The taking of limits of pointwise convergent sequences of functions is an operation of great importance in the theory of measure and integration. We pay special attention to pointwise limits of sequences of continuous scalar-valued functions. (The concepts discussed below may also be found, in part, in [**I**]; see, in particular, [**I**, Problem 8K].)

Definition 2.42. A scalar-valued function f on a metric space X is of *Baire class one* on X if there exists a sequence of continuous scalar-valued functions on X that converges pointwise to f.

Example 2.43. All semicontinuous real-valued functions on a metric space X are of Baire class one ([**I**, Proposition 7.20]). In particular, the characteristic functions of all open subsets and all closed subsets of X are of Baire class one on X ([**I**, Problem 7F]).

Further Baire classes are defined in the same manner as the first. Thus one says that a scalar-valued function f on a metric space X is of *Baire class two* if f is the limit of some pointwise convergent sequence of functions of Baire class one, etc. We turn at once to the formal inductive definition toward which these initial constructions clearly point.

Definition 2.44. Let X be a metric space. Set $\mathcal{C}_0$ equal to the collection of all continuous scalar-valued functions on X, let α denote an arbitrary countable ordinal number, and suppose that $\mathcal{C}_\xi$ has already been defined for all ordinal numbers ξ such that $\xi < \alpha$. Then $\mathcal{C}_\alpha$ is defined to be the collection of all functions f with the property that there exists a sequence in $\bigcup_{\xi<\alpha}\mathcal{C}_\xi$ that converges pointwise to f. In this way we obtain by transfinite definition (see [**I**, Theorem 5.12]) a family $\{\mathcal{C}_\alpha\}_{\alpha<\Omega}$ of collections of scalar-valued functions on X indexed by the entire segment $W(\Omega)$ of countable ordinal numbers. The functions belonging to $\mathcal{C}_\alpha$ are said to be of *Baire class α* on X, a terminology obviously consistent with the earlier definitions of Baire classes one and two. (According to this terminology the continuous functions are exactly the functions of Baire class zero.) The functions belonging to the union

$$\mathcal{C}_\Omega = \bigcup_{\alpha<\Omega}\mathcal{C}_\alpha$$

are known as the *Baire functions* on X.

Theorem 2.45. *The class $\mathcal{C}_\Omega$ of Baire functions on a metric space X is the smallest collection of scalar-valued functions on X that contains the class $\mathcal{C}_0$ of all continuous scalar-valued functions on X and that is closed with respect to the formation of limits of pointwise convergent sequences.*

Proof. Assume first that $\{f_n\}$ is a sequence of Baire functions on X that converges pointwise to a limit f. Then each f_n is of Baire class α for some countable ordinal number α—say f_n is of Baire class α_n. According to [**I**, Example 5L] there exists a countable ordinal number η such that $\alpha_n < \eta$ for every n, and it follows that f is of Baire class η. Thus $\mathcal{C}_\Omega$ is indeed closed with respect to the formation of pointwise limits.

Assume next that $\mathcal{C}$ is a collection of scalar-valued functions on X that contains $\mathcal{C}_0$ and is closed with respect to the formation of pointwise limits. If all of the Baire classes $\mathcal{C}_\xi$ are contained in $\mathcal{C}$ for $\xi < \alpha$, where α denotes some countable ordinal number, then $\mathcal{C}_\alpha \subset \mathcal{C}$ as well. Indeed, every function f in $\mathcal{C}_\alpha$ is, by definition, the limit of some pointwise convergent sequence of functions belonging to $\mathcal{C}$. Thus every Baire class $\mathcal{C}_\alpha, \alpha < \Omega$, is contained in $\mathcal{C}$, and therefore $\mathcal{C}_\Omega \subset \mathcal{C}$. $\square$

The preceding result reveals a shorter route to the notion of a Baire function.

Proposition 2.46. *For any collection $\mathcal{F}$ of scalar-valued functions on a set X there is a smallest collection $\mathcal{B} = \mathcal{B}(\mathcal{F})$ of functions on X that contains $\mathcal{F}$ and is closed with respect to the formation of limits of pointwise convergent sequences. This collection will be called the* Baire class generated by $\mathcal{F}$.

Proof. The collection of all scalar-valued functions on X is closed with respect to the formation of limits of pointwise convergent sequences. If $\mathcal{D}$ denotes the intersection of any nonempty family of collections each of which is closed with respect to pointwise limits, then $\mathcal{D}$ is also closed with respect to pointwise limits. Thus $\mathcal{B}(\mathcal{F})$ is simply the intersection of the family of all collections of scalar-valued functions on X that contain $\mathcal{F}$ and are closed with respect to the formation of pointwise limits. □

Using this latter concept, we may paraphrase Theorem 2.45 by saying that the class $\mathcal{C}_\Omega$ of Baire functions on a metric space X is the Baire class generated by the class $\mathcal{C}_0$ of continuous functions on X. This simple characterization of the class of Baire functions stands to its initial transfinite definition exactly as the definition of Borel sets in Chapter 1 (as the σ-algebra $\mathbf{B}_X$ generated by the lattice of open sets) stands to the transfinite construction of $\mathbf{B}_X$ in Problem 1O. One definition has the advantage of brevity and avoids the use of transfinite numbers, but also, by that very token, fails to give any information about the grading of Baire functions into numbered classes. Interestingly enough, either definition can be used to prove theorems, as the following propositions demonstrate.

Proposition 2.47. *If f and g are Baire functions on a metric space X, then $f + g$ is also a Baire function on X.*

Proof. Consider first the collection $\mathcal{G}_0$ of all those scalar-valued functions g on X with the property that if f is an arbitrary continuous scalar-valued function on X, then $f + g$ is a Baire function. If $\{g_n\}$ is a sequence in $\mathcal{G}_0$ that converges pointwise to a limit h, and if f is some continuous scalar-valued function on X, then $\{f + g_n\}$ is a sequence of Baire functions converging pointwise to $f + h$ so that $f + h$ is also a Baire function. This shows that $\mathcal{G}_0$ is closed with respect to the formation of limits of pointwise convergent sequences. Since $\mathcal{G}_0$ obviously contains $\mathcal{C}_0$, it follows that $\mathcal{G}_0$ contains *all* Baire functions. Thus we have shown that if f is a continuous scalar-valued function on X, and g is an arbitrary Baire function on X, then $f + g$ is also a Baire function. To complete the proof, consider next the collection $\mathcal{G}$ of all those scalar-valued functions g on X with the property that if f is a Baire function on X, then $f + g$ is too. A repetition of the same argument shows that $\mathcal{G}$ is closed with respect to the formation of pointwise limits. On the other hand, we have just proved that $\mathcal{G}$ contains $\mathcal{C}_0$, and the proposition follows. □

Other results along the same line can be similarly obtained.

Proposition 2.48. *The class of complex-valued Baire functions on a metric space X is closed with respect to complex conjugation. Hence a complex-valued function f on X is a Baire function if and only if $\Re f$ and $\Im f$ are both Baire functions.*

Proof. Consider the collection $\overline{\mathcal{C}}_\Omega$ of all complex conjugates of complex-valued Baire functions on X. This class contains the continuous complex-valued functions on X (the complex conjugate of a continuous function is itself continuous) and is closed with respect to the formation of pointwise limits (if $\{\bar{f}_n\}$ converges pointwise to a limit f, then $\{f_n\}$ converges pointwise to $\bar{f}$, so $f \in \overline{\mathcal{C}}_\Omega$). Hence $\mathcal{C}_\Omega \subset \overline{\mathcal{C}}_\Omega$, and by symmetry $\mathcal{C}_\Omega = \overline{\mathcal{C}}_\Omega$. □

Continuing in this same vein, one easily establishes the following result, whose proof is omitted.

Theorem 2.49. *The real-valued Baire functions on a metric space X form a function algebra that contains the constant functions and is also a function lattice. The complex-valued Baire functions on X form a function algebra that contains the constant functions and is closed with respect to complex conjugation.*

To obtain these and other results concerning Baire functions directly from the original transfinite definition is rather more laborious, requiring as it does the machinery of transfinite induction, but then the end result is more informative. We begin with a pair of observations that clarify the later arguments to some degree.

Proposition 2.50. *On a metric space X the transfinite sequence $\{\mathcal{C}_\alpha\}_{\alpha<\Omega}$ of Baire classes is monotone increasing. For any countable ordinal number α, a function f is of Baire class $\alpha + 1$ on X if and only if it is the limit of a pointwise convergent sequence of functions of Baire class α on X. On the other hand, if λ is a countable limit number, then f is of Baire class λ on X if and only if it is the limit of a pointwise convergent sequence $\{f_n\}$ where each f_n is of Baire class η_n on X, and $\eta_1 < \eta_2 < \ldots < \eta_{n} < \ldots$ is a strictly increasing sequence in $W(\lambda)$.*

Proof. The stated conditions are obviously sufficient, and the asserted monotonicity follows from the simple fact that a constant sequence is convergent. Moreover, from monotonicity it is clear that $\bigcup_{\xi<\alpha+1} \mathcal{C}_\xi = \mathcal{C}_\alpha$, and hence that the functions in $\mathcal{C}_{\alpha+1}$ are limits of sequences in $\mathcal{C}_\alpha$. Finally, if $f \in \mathcal{C}_\lambda$ where λ is a limit number, then there is a sequence $\{f_n\}$ of functions in $\bigcup_{\xi<\lambda} \mathcal{C}_\xi$ converging pointwise to f. Thus each f_n belongs to some $\mathcal{C}_{\xi_n}, \xi_n < \lambda$, and we have but to define the strictly increasing sequence $\{\eta_n\}$ inductively, setting $\eta_1 = \xi_1$ and $\eta_{n+1} = (\eta_n + 1) \vee \xi_{n+1}$, for $n > 0$. □

Proposition 2.51. *Let X be a metric space, and let g be a continuous mapping of the scalar field into itself. Then for any countable ordinal number α and any function f of Baire class α on X, the composition $g \circ f$ is also of Baire class α.*

Proof. The result is obvious when $\alpha = 0$ because the composition of continuous mappings is continuous. Assume it holds for all $\xi < \alpha$, and $f \in \mathcal{C}_\alpha$. There is a sequence $\{f_n\}$ of functions converging pointwise to f with the property that each f_n is of some Baire class $\xi_n < \alpha$, and the sequence $\{g \circ f_n\}$ shares this property by the inductive hypothesis. In addition $\{g \circ f_n\}$ converges pointwise to $g \circ f$ because g is continuous, and therefore $g \circ f \in \mathcal{C}_\alpha$. □

The following result clearly contains Theorem 2.49 as a corollary.

Theorem 2.52. *Let X be a metric space, let α be a countable ordinal number, and let us write $\mathcal{C}_{\alpha,\mathbb{R}}$ and $\mathcal{C}_{\alpha,\mathbb{C}}$ for the collections of real and complex Baire functions of class α on X, respectively. Then $\mathcal{C}_{\alpha,\mathbb{C}}$ is a complex function algebra on X that contains the constant functions and is closed with respect to complex conjugation. Similarly, $\mathcal{C}_{\alpha,\mathbb{R}}$ is a real function algebra that is also a function lattice on X.*

We only sketch the proof.

Proof. The parts of the theorem that are not immediate consequences of Proposition 2.51 are all derived in the same way, by transfinite induction. As a typical example of such an argument, let us show that $\mathcal{C}_{\alpha,\mathbb{C}}$ is closed with respect to addition.

The result clearly holds for $\alpha = 0$. Suppose it is valid for all $\xi < \alpha$, and $f, g \in \mathcal{C}_{\alpha,\mathbb{C}}$. Choose sequences $\{f_n\}$ and $\{g_n\}$ of complex-valued functions on X converging pointwise to f and g, respectively, such that, for each index n, $f_n \in \mathcal{C}_{\xi_n,\mathbb{C}}$ and $g_n \in \mathcal{C}_{\eta_n,\mathbb{C}}$ for some $\xi_n < \alpha$ and $\eta_n < \alpha$. Then the functions f_n, g_n belong to $\mathcal{C}_{\zeta_n,\mathbb{C}}$, where $\zeta_n = \xi_n \vee \eta_n < \alpha$. The inductive hypothesis implies that $f_n + g_n \in \mathcal{C}_{\zeta_n,\mathbb{C}}$, and therefore $f + g = \lim_n (f_n + g_n) \in \mathcal{C}_{\alpha,\mathbb{C}}$. $\square$

According to Theorem 2.52 (or Theorem 2.49), the class $\mathcal{C}_\Omega$ of complex-valued Baire functions on a metric space X consists of all *measurable* functions with respect to some σ-algebra of subsets of X (Theorem 2.39). The following identifies the σ-algebra.

Theorem 2.53. *On any metric space X the class of scalar-valued Baire functions coincides with the class of Borel measurable scalar-valued functions.*

Proof. As seen earlier, $\mathcal{C}_\Omega$ consists of the scalar-valued functions which are measurable relative to the σ-field $\mathbf{S} = \{E \subset X : \chi_E \in \mathcal{C}_\Omega\}$. We must show that $\mathbf{S}$ coincides with the σ-algebra $\mathbf{B}_X$, and both parts of the proof are easy. On the one hand, $\mathbf{S}$ contains all open sets in X, so $\mathbf{S} \supset \mathbf{B}_X$. On the other hand, the collection of all Borel measurable real-valued functions on X contains the continuous real-valued functions and is closed with respect to the formation of pointwise limits, so every real-valued Baire function on X is Borel measurable. If $E \in \mathbf{S}$, so that χ_E is a Baire function, then χ_E is Borel measurable, and E is a Borel set. $\square$

Problems

2A. Fix a set M dense in $\mathbb{R}$, for example, the set of all rational numbers, or the set of all dyadic fractions. If f is a real-valued function on a measurable space X, then f is measurable if and only if $E(f \leq t)$ is measurable for every t in M. Similarly, f is measurable if and only if $E(f < t)[E(f \geq t), E(f > t)]$ is measurable for every $t \in M$.

2B. An extended real-valued function f on a measurable space X is measurable if and only if the two sets $E_{+\infty} = E(f = +\infty)$ and $E_{-\infty} = E(f = -\infty)$ are measurable and the restriction of f to the complement $X \setminus (E_{+\infty} \cup E_{-\infty})$ is a measurable (finite) real-valued function on that subspace.

2C. Let $\{t_n\}_{n=-\infty}^{+\infty}$ be a monotone increasing sequence of real numbers (indexed by the set $\mathbb{Z}$ of all integers), and set $a = \inf_n t_n, b = \sup_n t_n$ (the cases $a = -\infty$ and/or $b = +\infty$ are not excluded). Let f be a real-valued function defined on the interval (a, b), and suppose that f is monotone on each subinterval $(t_n, t_{n+1}), n \in \mathbb{Z}$. Show that f is a Borel measurable function.

2D. Let $\mathrm{mid}\{a, b, c\}$ denote that one of the three real numbers a, b, c that is bracketed by the other two. Show that if f, g and h are any three measurable real-valued functions on a measurable space X, then

$$\mathrm{mid}\{f(x), g(x), h(x)\}$$

is likewise a measurable function on X.

2E. (i) Let $\mathbf{S}_0$ be a σ-ring of subsets of a set X that is *not* complemented (so that $X \notin \mathbf{S}_0$), and let $\mathbf{S}_1$ denote the complemented σ-ring obtained by adjoining to $\mathbf{S}_0$ the complements of the sets in $\mathbf{S}_0$ (see Example 1.13). Describe the algebra of measurable scalar-valued functions on the measurable space $(X, \mathbf{S}_1)$. (Hint: Consider a countable partition of X into sets measurable $[\mathbf{S}_1]$.)

(ii) Let $X = X_1 \cup X_2$ be a partition of a set X, let $\mathbf{S} = \mathbf{S}_1 \oplus \mathbf{S}_2$, where $\mathbf{S}_i$ is a σ-algebra of subsets of $X_i, i = 1, 2$, and let $\mathcal{A}$ denote the algebra of measurable scalar-valued functions on $(X, \mathbf{S})$. If $\mathcal{J}_1$ denotes the set of those functions in $\mathcal{A}$ that vanish on X_2, then $\mathcal{J}_1$ is an ideal in $\mathcal{A}$ that is, in an obvious fashion, isomorphic as an algebra to the algebra $\mathcal{A}_1$ of measurable scalar-valued functions on $(X_1, \mathbf{S}_1)$. Similarly, the ideal $\mathcal{J}_2$ of those functions in $\mathcal{A}$ that vanish on X_1 is isomorphic to the algebra $\mathcal{A}_2$ of measurable scalar-valued functions on $(X_2, \mathbf{S}_2)$. The ideals $\mathcal{J}_1$ and $\mathcal{J}_2$ are complements as subspaces of (the vector space) $\mathcal{A}$; that is, $\mathcal{J}_1 \cap \mathcal{J}_2 = (0)$ while $\mathcal{J}_1 + \mathcal{J}_2 = \mathcal{A}$. Moreover, if functions f and g in $\mathcal{A}$ are written in the form $f = f_1 + f_2, g = g_1 + g_2$, where $f_i, g_i \in \mathcal{J}_i, i = 1, 2$, then $fg = f_1 g_1 + f_2 g_2$. (This situation is sometimes expressed by saying that $\mathcal{A}$ splits *internally* into the *direct sum* $\mathcal{A}_1 \oplus \mathcal{A}_2$.) Is the like true for more general notions of the direct sum of measurable spaces (Example 1.16)?

(iii) Let $\mathcal{P}$ be a countable partition of a set X, and let $\mathbf{S}$ be the σ-ring in X corresponding to $\mathcal{P}$ as in Example 1.7. Describe the measurable mappings of $(X, \mathbf{S})$ into $(Y, \mathbf{T})$, where Y is a set and $\mathbf{T}$ is some σ-algebra of subsets of Y containing all singletons (Example 2.5).

2F. (i) Let Y be a metric space, let $\mathcal{C}$ be a collection of subsets of a set X, and suppose given a sequence $\{\varphi_n\}_{n=1}^{\infty}$ of mappings of X into Y with the property that $\varphi_n^{-1}(U) \in \mathcal{C}$ for all n and all open sets U in Y. Show that if $\{\varphi_n\}$ converges pointwise on X to a limit ψ, then $\psi^{-1}(U) \in \mathcal{C}_{\delta\sigma}$ for every open set U in Y (see Problem 1J). In particular, if $(X, \mathbf{S})$ is a measurable space, then the limit of any pointwise convergent sequence of measurable mappings of X into Y is itself measurable.

(ii) Let X be an infinite measurable space, let Y be a metric space, and suppose given a mapping ψ of X into Y. Let $\mathcal{D}$ be the directed set of all finite subsets of X (directed by set inclusion $\subset$), let y_0 be a fixed point of Y, and for each Δ in $\mathcal{D}$ set

$$\varphi_\Delta(x) = \begin{cases} \psi(x), & x \in \Delta \\ y_0, & x \in X \setminus \Delta. \end{cases}$$

Then $\{\varphi_\Delta\}_{\Delta \in \mathcal{D}}$ is a net of mappings of X into Y. Show that this net converges pointwise on X to the given mapping ψ. Show also that the mappings φ_Δ are all measurable under the hypothesis that all singletons in X are measurable sets (see Example 2.5).

(iii) In the foregoing construction take for Y the real line $\mathbb{R}$, for ψ the characteristic function of some subset A of X, and set $y_0 = 0$. Show that in this case the net $\{\varphi_\Delta\}$ converges (pointwise) monotonically upward to χ_A no matter what A is.

Remark 2.54. These constructions clearly show that preservation of measurability under passage to a limit is essentially restricted to *sequential* convergence, and does not extend, in general, to convergent nets, not even monotone nets. This should come as no surprise; the very concept of measurability is rooted in the notion of a countable set.

2G. (Product metrics; see [**I**, Problem 6H]) Let $(X, \mathbf{S})$ be a measurable space, and let $Y_1, \ldots, Y_N$ be metric spaces.

   (i) If the product $Y = Y_1 \times \ldots \times Y_N$ is equipped with a product metric, and if $\varphi : X \to Y$ is measurable, then the coordinate mappings $\pi_i \circ \varphi, i = 1, \ldots, N$, are all measurable as well. (Here π_i denotes, as always, the projection of the product Y onto the ith factor Y_i.)
   (ii) Conversely, if the spaces Y_i are all separable, and if $\varphi : X \to Y$ has the property that the coordinate mappings $\pi_i \circ \varphi$ are all measurable, then φ is itself measurable. (Hint: See [**I**, Example 6W].)
   (iii) Generalize this discussion to the case of a (countably) infinite product $\prod_{n=1}^{\infty} Y_n$ of metric spaces equipped with a product metric.

2H. Let X be a measurable space, and let φ and ψ be measurable mappings of X into a separable metric space (Y, ρ). Show that the real-valued function $x \to \rho(\varphi(x), \psi(x))$ is measurable on X and that $\{x \in X : \varphi(x) = \psi(x)\}$ is a measurable set (see Corollary 2.26). In particular, if φ is a Borel measurable mapping of Y into itself, then the set $\{y \in Y : \varphi(y) = y\}$ of fixed points of φ is a Borel set.

2I. Let X be a measurable space and let $\{\varphi_n\}$ be a sequence of measurable mappings of X into a complete separable metric space Y. Show that the subset E of X consisting of those points x at which the sequence $\{\varphi_n(x)\}$ is convergent in Y is measurable and that $\varphi(x) = \lim_n \varphi_n(x)$ defines a measurable mapping $\varphi : E \to Y$.

2J. The complex version of (the essential part of) Proposition 2.36 goes as follows. For every measurable complex-valued function f on a measurable space X there exists a sequence $\{s_n\}$ of measurable simple complex-valued functions converging pointwise to f and satisfying the following additional conditions:

   (i) at each point $x \in X$ either $0 \leq \Re s_1(x) \leq \Re s_2(x) \leq \cdots$, or $0 \geq \Re s_1(x) \geq \Re s_2(x) \geq \cdots$, and likewise, either $0 \leq \Im s_1(x) \leq \Im s_2(x) \leq \cdots$, or $0 \geq \Im s_1(x) \geq \Im s_2(x) \geq \cdots$, and
   (ii) if $M \in \mathbb{N}$ and if $|f(x)| \leq M$, then $|f(x) - s_n(x)| \leq \sqrt{2}/2^n$ for every $n \geq M$.
   Show that this assertion is, in fact, valid. (Hint: The proof given in the text can be copied in the complex plane; alternatively, the complex case can be derived from the real case.)

2K. Use the definition of the combinatorial coefficient $\binom{n}{k}$ to derive the equivalent relations

$$\binom{n}{k} = \frac{n}{k}\binom{n-1}{k-1}, \quad \binom{n-1}{k-1} = \frac{k}{n}\binom{n}{k}, \tag{2.6}$$

for all $k = 1, \ldots, n$ (and all $n \in \mathbb{N}$; recall that $0! = 1$ by definition). Set $n = m - 1$ and $k = j - 1$ in the basic identity (2.4), and use (2.6) to show that

$$\sum_{j=1}^{m} \frac{j}{m}\binom{m}{j-1}x^j(1-x)^{m-j} = 1,$$

and hence that

$$\sum_{j=1}^{m} j\binom{m}{j}x^j(1-x)^{m-j} = mx \tag{2.7}$$

for $m = 2, 3, \ldots$. Then use the same trick over again to verify (2.5). The case $n = 1$ needs to be verified separately.

Remark 2.55. It need scarcely be said that these are not simply fortuitous calculations. In fact, in a Bernoulli process consisting of n independent repetitions of a simple trial with probability of success $p = x$ (and therefore with $q = 1 - x$) $\binom{n}{k}x^k(1-x)^{n-k}$ is the probability of precisely k successes. Thus (2.4) says merely that the sum of these probabilities is one, (2.7) states that the expected number of successes is np, and (2.5) gives the standard deviation of this number about its mean as npq—well-known facts of elementary probability theory.

2L. Prove the complex version of Theorem 2.39 by reducing it to the real case. (A direct proof of Theorem 2.39 in the complex case can be based on Problem 2J, along with a complex version of Lemma 2.40, but this would in turn necessitate the introduction of a complex version of the Weierstrass approximation theorem.)

2M. Let us call a mapping φ of a set X into an arbitrary set Y *simple* if it assumes only a finite number of distinct values in Y. (Thus the simple scalar-valued functions defined above in connection with Proposition 2.34 are just the simple mappings of X into the scalar field.)

(i) If $(X, \mathbf{S})$ is a measurable space and Y is a metric space, then a simple mapping φ of X into Y is measurable if and only if it assumes each of its values on a measurable set.

(ii) A metric space Y is said to be *σ-compact* if it is the union of some countable collection of compact sets. Show that if $(X, \mathbf{S})$ is a measurable space and Y is a σ-compact metric space, then every measurable mapping φ of X into Y is the limit of a pointwise convergent sequence $\{\varphi_n\}$ of measurable simple mappings of X into Y. (Hint: There exists an increasing sequence $\{K_n\}$ of compact subsets of Y such that $Y = \bigcup_n K_n$.)

2N. A mapping φ of a measurable space X into a set Y is said to be *elementary* if it assumes only a *countable* number of distinct values. If Y is a metric space, then an elementary mapping of X into Y is measurable if and only if it assumes each of its values on a measurable set. If Y is a separable metric space, then a mapping of X into Y is measurable if and only if it is the limit of a uniformly convergent sequence of measurable elementary mappings of X into Y.

2O. If $(X, \mathbf{S})$ and $(Y, \mathbf{T})$ are measurable spaces and $\varphi : X \to Y$ is measurable, then (as we have noted, see Example 2.12) φ is automatically measurable on any subset A of X (meaning that $\varphi | A$ is measurable as a mapping of $(A, \mathbf{S}_A)$ into Y).

(i) Show, in the converse direction, that if A is measurable, then any measurable mapping $\psi : A \to Y$ is the restriction to A of some measurable mapping $\varphi : X \to Y$.

(ii) Show also that if Y is a complete σ-compact metric space (Problem 2K), then the assumption in (i) that the set A is measurable can be dropped. In particular, then, this is the case if Y is the scalar field $\mathbb{R}$ or $\mathbb{C}$. (Hint: Consider first the case of a simple measurable mapping ψ; use Problem 2I.)

(iii) Let X be a separable metric space, and suppose A is some subset of X that is *not* a Borel set in X (see Problem 1I). Then the identity mapping $\iota : A \to A$ is Borel measurable on A, but does not extend to any Borel measurable mapping of X into A. (Hint: A Borel measurable mapping of X into A is also Borel measurable as a mapping of X into itself, and therefore admits a Borel set of fixed points; recall Problem 2H.)

2P. Let $\mathcal{A}$ denote the algebra of all measurable scalar-valued functions on a measurable space $(Y, \mathbf{T})$, and suppose given a mapping $\varphi : X \to Y$ of some set X onto Y. Show that $\tilde{\mathcal{A}} = \{f \circ \varphi : f \in \mathcal{A}\}$ is the algebra of measurable scalar-valued functions on X with respect to a unique σ-algebra $\mathbf{S}$ in X, and find $\mathbf{S}$. What becomes of this proposition if we drop the assumption that the mapping φ is onto?

2Q. (i) A scalar-valued function on a metric space X that differs from a continuous function at only finitely many points is of Baire class one on X. Give an example of a function on the real line differing from a continuous function at countably infinitely many points that is of Baire class one, and of another that is not. (Hint: See [**I**, Example 8Q].)

(ii) Every scalar-valued function on a metric space X differing from a continuous function at only a countable set of points is of Baire class two on X.

2R. If a function $f : X \to \mathbb{R}$ on a metric space X has the property that $a \leq f(x) \leq b$ on X, where $a \leq b$ are real numbers, and if f is of Baire class α on X, then f is the limit of a pointwise convergent sequence $\{f_n\}$ of real-valued functions of Baire class less than α on X such that $a \leq f_n(x) \leq b$ for all x and all n. In particular, if $|f| \leq M$ on X, then f is the pointwise limit of a sequence of functions each of Baire class less than α, each of which is similarly bounded. Show, in the same vein, that if f is a complex-valued function of Baire class α on X such that $|f| \leq M$ on X, then f is the pointwise limit of a sequence $\{f_n\}$ of complex-valued functions on X where each f_n is not only of Baire class less than α, but also satisfies the condition $|f_n| \leq M$ on X. (Hint: For the complex case construct a retraction of $\mathbb{C}$ onto the closed disc $D_M = \{\lambda \in \mathbb{C} : |\lambda| \leq M\}$; that is, a continuous mapping of $\mathbb{C}$ onto D_M that agrees with the identity mapping on D_M.)

2S. Let $g : \mathbb{R}^d \to \mathbb{C}$ be a continuous function, and suppose $f_1, \ldots, f_n$ are real-valued functions of Baire class α on a metric space X. Show that the function

$$h(x) = g(f_1(x), \ldots, f_d(x))$$

is also of Baire class α on X. Show that this result remains valid if g is merely defined and continuous on some closed cell containing the range of the mapping $x \mapsto (f_1(x), \ldots, f_d(x))$, and devise complex analogs of these results.

2T. Let $\{M_n\}_{n=1}^{\infty}$ be a sequence of positive real numbers with $\sum_n M_n < +\infty$, and suppose that $\{f_n\}$ is a sequence of scalar-valued functions on a metric space X such that $|f_n| \leq M_n$ on $X, n \in \mathbb{N}$. Show that if each of the functions f_n is of Baire class α on X (for some countable ordinal number α), then the sum $f = \sum_n f_n$ is also of Baire class α. (Hint: Each f_n is the limit of a pointwise convergent sequence $\{p_k^{(n)}\}_{k=1}^{\infty}$ of functions of Baire class less than α, where $|p_k^{(n)}| \leq M_n$ on X for all k and all n. Set

$$q_m = p_1^{(m)} + \cdots + p_m^{(m)},$$

and let m tend to infinity. For all m greater than or equal to a fixed positive integer k we can write $q_m = q_m' + q_m''$, where $\{q_m'\}$ tends pointwise to $f_1 + \cdots + f_k$, while $|q_m''| \leq M_{k+1} + \cdots + M_m$. Hence $\{q_m\}$ converges pointwise to f.) Use the foregoing fact to prove that each of the Baire classes $\mathcal{C}_\alpha$ on $X, \alpha < \Omega$, is closed with respect to the formation of limits of uniformly convergent sequences.

2U. A sequence $\{f_n\}$ of scalar-valued functions on a metric space X is said to be *locally uniformly bounded* if for every point x_0 of X there exist a neighborhood V of x and a constant $M > 0$ such that $|f_n(x)| \leq M$ for all $x \in V$ and all indices n. (Each of the functions f_n is, then, in particular, locally bounded; see [**I**, Problem 7D].) Prove that the class of locally bounded Baire functions on X is the smallest collection of scalar-valued functions on X that contains the class $\mathcal{C}_0$ of continuous scalar-valued functions and is closed with respect to the formation of limits of pointwise convergent and uniformly locally bounded sequences.

Remark 2.56. In view of Theorem 2.52, it is natural to look for connections between the classification of the Borel sets in a metric space X into numbered classes and the like classification of the Baire functions on X. Such connections do indeed exist, but they are not as tidy as one might hope. The following problem provides a small sampler of such results.

2V. Let X be a metric space.

(i) For each countable ordinal number α let $\mathcal{E}_\alpha$ denote the collection of subsets E of X such that χ_E belongs to the Baire class $\mathcal{C}_\alpha$ on X. Show that $\mathcal{E}_\alpha$ is a complemented lattice of subsets of X. Define inductively classes $\mathcal{G}^\alpha$ for $\alpha < \Omega$ by setting $\mathcal{G}^0 = \mathcal{G}$ (the collection of open sets), and $\mathcal{G}^\alpha = \left(\bigcup_{\beta < \alpha} \mathcal{G}\right)_\delta$ if α is odd and $\mathcal{G}^\alpha = \left(\bigcup_{\beta < \alpha} \mathcal{G}\right)_\sigma$ if $\alpha > 0$ is even. Show also that $\mathcal{G}^\alpha \subset \mathcal{E}_{\alpha+1}$ for every $\alpha < \Omega$.

(ii) Suppose each function of the sequence $\{f_n\}$ of complex-valued functions on X has the property that the inverse image $f_n^{-1}(U)$ belongs to the lattice $\mathcal{G}^\alpha$ for every open set U in $\mathbb{C}$, and suppose $\{f_n\}$ converges pointwise to a limit f. Verify that $f^{-1}(U)$ belongs to $\mathcal{G}^{\alpha+2}$ for every open set $U \subset \mathbb{C}$. (Thus, for example, a scalar-valued function f of Baire class one on X has the property that $f^{-1}(U)$ is a $G_{\delta\sigma}$ for every open set U of complex numbers; see Problem 2F.)

(iii) Use the foregoing observation to prove that if λ is a countable limit number, and if $f \in \mathcal{C}_{\lambda,\mathbb{C}}$, then $f^{-1}(U) \in \mathcal{G}^{\lambda+2}$ for every open set $U \subset \mathbb{C}$. Conclude by transfinite induction that for each countable ordinal number α there is a positive integer k (depending on α) such that if $f \in \mathcal{C}_{\alpha,\mathbb{C}}$, then $f^{-1}(U) \in \mathcal{G}^{\alpha+k}$ for every open set $U \subset \mathbb{C}$.

Remark 2.57. The basic idea of the construction of various Baire classes of mappings on a space X clearly makes sense in a broader context than that of scalar-valued functions. The following two problems are concerned with the classification of mappings taking values in arbitrary metric spaces.

2W. Let (Y, σ) be a metric space.

    (i) Show that for any subset $\mathcal{F}$ of Y^X there is a smallest set $\mathcal{B}(\mathcal{F}) \subset Y^X$ that contains $\mathcal{F}$ and is closed with the respect to the formation of pointwise limits. The collection $\mathcal{B}(\mathcal{F})$ is the *Baire class generated* by $\mathcal{F}$.

    (ii) What is $\mathcal{B}(\varnothing)$? That is, what collection of mappings of X into Y is generated in this way by the empty collection of mappings? Let A be a subset of Y, and let $\mathcal{F}_A$ denote the collection of constant mappings of X into A. Find $\mathcal{B}(\mathcal{F}_A)$.

    (iii) Let $\mathcal{F}$ be a collection of mappings of X into Y. Set $\mathcal{C}_0 = \mathcal{F}$, and supposing, for a countable ordinal number α, that $\mathcal{C}_\xi$ is already defined for all $\xi < \alpha$, define $\mathcal{C}_\alpha$ to consist of all limits of pointwise convergent sequences $\{\varphi_n\}$ of mappings belonging to $\bigcup_{\xi < \alpha} \mathcal{C}_\xi$. Show that $\{\mathcal{C}_\alpha\}_{\alpha < \Omega}$ is a monotone increasing family of subsets of Y^X and that $\mathcal{B}(\mathcal{F}) = \bigcup_{\alpha < \Omega} \mathcal{C}_\alpha$.

2X. Let (X, ρ) and (Y, σ) be metric spaces, and let $\mathcal{C}_0 = \mathcal{C}(X; Y)$ denote the collection of all continuous mappings of X into Y. The mappings in the Baire class $\mathcal{B}(\mathcal{C}_0)$ generated by $\mathcal{C}_0$ are known simply as *Baire mappings* of X into Y. More particularly, for each countable ordinal number α, the collection $\mathcal{C}_\alpha$ defined in the transfinite procedure indicated in part (iii) of the preceding problem (starting with $\mathcal{F} = \mathcal{C}_0$) constitutes the *Baire class* α of mappings of X into Y.

    (i) For every countable ordinal number α, a mapping $\psi : X \to Y$ is of Baire class $\alpha + 1$ if and only if it is the limit of a pointwise convergent sequence $\{\varphi_n\}$ of mappings belonging to $\mathcal{C}_\alpha$. On the other hand, if λ is a countable limit number, then ψ is of Baire class λ if and only if there exist a strictly increasing sequence $\{\eta_n\}$ in $W(\lambda)$ and, for each index n, a mapping φ_n in $\mathcal{C}_{\eta_n}$ such that the sequence $\{\varphi_n\}$ converges pointwise to ψ.

    (ii) Let (Z, τ) be a third metric space. Show that if $\varphi : X \to Y$ and $\psi : Y \to Z$ are both Baire mappings, then $\psi \circ \varphi : X \to Z$ is also a Baire mapping. Show, more particularly, that if φ is of Baire class α and ψ is of Baire class β, then $\psi \circ \varphi$ is of Baire class $\alpha + \beta$.

    (iii) Suppose Y is of the form $Y = Y_1 \times \ldots \times Y_N$, where $Y_1, \ldots, Y_N$ are metric spaces. Let Y be equipped with a product metric, and let α be a countable ordinal number. Show that a mapping $\varphi : X \to Y$ is of Baire class α if and only if the coordinate mappings $\pi_i \circ \varphi, i = 1, \ldots, N$, are all of Baire class α. (Thus, in particular, φ is a Baire mapping if and only if the mappings $\pi_i \circ \varphi$ are all Baire mappings.)

    (iv) Part (iii) goes through in the case of a countably infinite product

$$Y = \prod_{n=1}^{\infty} Y_n$$

as well (assuming, as always, that Y is equipped with a product metric). (Hint: Assume, as one may, that $Y \neq \varnothing$, and select a fixed point $y_{n,0}$ in each factor Y_n.)

    (v) The Baire mappings of $\mathbb{R}$ into the Cantor set C are just the constant functions $f : \mathbb{R} \to C$. Thus no counterpart of Theorem 2.39 holds in this generalized context.

2Y. Let $(X, \mathbf{S})$ be a measurable space and let (Y, σ) be a metric space. A mapping $\varphi : X \to Y$ will be said to have *property* (MSR) if φ is measurable and its range $\varphi(X)$ is separable (as a subspace of Y). If Y is separable, then every measurable mapping of X into Y has property (MSR); if X is a separable metric space, then every Baire mapping of X into Y has property (MSR).

(i) Prove that if $\varphi : X \to Y$ has property (MSR) and ε is a positive number, then there exists a measurable elementary mapping φ_ε of X into Y such that $\sigma(\varphi(x), \varphi_\varepsilon(x)) < \varepsilon$ for all $x \in X$ (recall Problem 2N). Show, conversely, that any mapping of X into Y that is thus approximable has property (MSR). (Hint: A subset of Y is separable as a subspace if and only if it is contained in the closure of a countable subset of Y.)

(ii) The collection of all mappings in Y^X with property (MSR) is closed under the formation of pointwise limits. (Hint: The separable subsets of Y form a σ-ideal in 2^Y.)

(iii) If $\varphi : X \to Y$ has property (MSR), and if ψ is a mapping of Y into some third metric space Z such that ψ has property (MSR), or such that ψ is a Baire mapping, then $\psi \circ \varphi$ has property (MSR).

(iv) Suppose Y is of the form $Y = Y_1 \times \ldots \times Y_N$, where (Y_i, σ_i) is a metric space, $i = 1, \ldots, N$, and suppose Y is equipped with a product metric. Show that a mapping φ of X into Y has property (MSR) if and only if the coordinate mappings $\pi_i \circ \varphi$ all have that property. Generalize this result to the case of an infinite product $Y = \prod_{n=1}^{\infty} Y_n$.

(v) If two mappings φ and ψ of X into Y both have property (MSR), then the function $x \mapsto \sigma(\varphi(x), \psi(x))$ is measurable.

2Z. Let $f : \mathbb{R} \to \mathbb{R}$ be an arbitrary function. Show that the set E of points of continuity of f and the set F of differentiability points of f are Borel sets in $\mathbb{R}$. Moreover, the function $f' : F \to \mathbb{R}$ is Borel measurable.

2AA. (Korovkin) Denote by $\mathcal{C}_{\mathbb{R}}([0, 1])$ the linear space of continuous functions $f : [0, 1] \to \mathbb{R}$. Let $T_n : \mathcal{C}_{\mathbb{R}}([0, 1]) \to \mathcal{C}_{\mathbb{R}}([0, 1])$, $n \in \mathbb{N}$, be a sequence of real-linear maps with the following properties.

(i) If $f \in \mathcal{C}_{\mathbb{R}}([0, 1])$ is a nonnegative function, then $T_n(f)$ is nonnegative for all $n \in \mathbb{N}$.

(ii) The sequence $\{T_n(f)\}_{n=1}^{\infty}$ converges uniformly to f when $f(t) = t^j$ for $j = 0, 1, 2$.

Show that $\{T_n(f)\}_{n=1}^{\infty}$ converges uniformly to f for every $f \in \mathcal{C}_{\mathbb{R}}([0, 1])$.

Chapter 3
Integrals and measures

In the language of modern integration theory the term *integral* refers to
a number of somewhat different concepts, arrived at through a variety of
constructions and definitions. About the only thing that can be said about
integration in reasonable generality is that an integral on a space X is a linear transformation that is defined on a vector space of functions on X and
satisfies certain continuity requirements. As regards the *Lebesgue* integral,
however, matters are in a much less chaotic state. Indeed, while a considerable number of different definitions and constructions can be found in the
literature, there is unanimous agreement on what a Lebesgue integral is.
We provide an axiomatic characterization.

Definition 3.1. Let $(X, \mathbf{S})$ be a measurable space. A linear functional L
defined on a real vector space $\mathcal{L}$ of measurable real-valued functions on X
is a *real Lebesgue integral* on X provided the following two conditions are
satisfied.

(L$_1$) If f belongs to $\mathcal{L}$ and if g is a measurable real-valued function on X such
that $|g| \leq |f|$, then

    (a) g also belongs to $\mathcal{L}$, and
    (b) $|L(g)| \leq L(|f|)$.

(L$_2$) If $\{f_n\}_{n=1}^{\infty}$ is a sequence of functions in $\mathcal{L}$, and if

    (a) $\lim_{m,n} L(|f_m - f_n|) = 0$, and
    (b) $\{f_n\}$ converges pointwise to g,

    then $g \in \mathcal{L}$ and $\lim_n L(f_n) = L(g)$.

Similarly, a *complex Lebesgue integral* on X is defined to be a complex
linear functional L on a vector space $\mathcal{L}$ of measurable complex-valued functions on X satisfying the corresponding conditions (L$_1$) and (L$_2$), but with
(L$_1$) rephrased in terms of complex-valued functions. The term "Lebesgue

© Springer International Publishing Switzerland 2016
H. Bercovici et al., *Measure and Integration*,
DOI 10.1007/978-3-319-29046-1_3

integral," written without qualification, is to be understood to mean either a real or complex Lebesgue integral. The functions belonging to the domain $\mathcal{L}$ of a Lebesgue integral L are said to be *integrable* with respect to L, and if f is such an integrable function, then $L(f)$ is called the *integral* of f with respect to L. A sequence $\{f_n\}$ of functions integrable with respect to L that satisfies condition ($\mathrm{L}_2 a$) will be said to be *Cauchy in the mean* with respect to L. In the same vein, if the functions f and f_n are all integrable with respect to L, and if $L(|f - f_n|) \to 0$, we say that the sequence $\{f_n\}$ *converges to f in the mean*.

Remark 3.2. The reader will observe that, strictly speaking, a Lebesgue integral is a triple $(X, \mathcal{L}, L)$, where X is a measurable space, $\mathcal{L}$ is a vector space of measurable functions on X, and L is a vector functional on $\mathcal{L}$. Note that the right member of ($\mathrm{L}_1 b$) is defined. Indeed, the first conclusion we draw from ($\mathrm{L}_1 a$) is that $|f|$ is integrable along with f, whence it follows that $|L(f)| \le L(|f|)$.

The following facts are also immediate consequences of axiom (L_1) only and are summarized here for convenience of reference.

Proposition 3.3. *If L is a complex Lebesgue integral on a measurable space X, and if f is integrable with respect to L, then so are $\overline{f}, \Re f$ and $\Im f$. (Thus the space $\mathcal{L}$ of integrable functions with respect to L is self-conjugate ([**I**, Example 3K]).) Likewise, if f is a real-valued function on X that is integrable with respect to L, then f^+ and f^-, the positive and negative parts of f, are also integrable. Hence the collection $\mathcal{L}_{\mathbb{R}}$ of real-valued functions integrable with respect to L is a function lattice as well as a real vector space ([**I**, Problem 2M]). Similarly, if L is a real Lebesgue integral on X, then f^+, f^- and $|f|$ are integrable with respect to L wherever f is, and the (real) vector space $\mathcal{L}$ is a function lattice.*

Example 3.4. Let $(X, \mathbf{S})$ be a measurable space, and let $\mathcal{M}$ denote the vector space of all measurable scalar-valued functions on X. Then the zero functional on $\mathcal{M}$ is a Lebesgue integral on X. Similarly, the zero functional on the trivial submanifold $(0) \subset \mathcal{M}$ is a Lebesgue integral on X. More generally, let $\mathbf{J}$ be an arbitrary σ-ideal in $\mathbf{S}$ (Problem 1W), and let $\mathcal{L}_{\mathbf{J}}$ denote the collection of all those scalar-valued functions f in $\mathcal{M}$ such that f vanishes outside of some set belonging to $\mathbf{J}$. Then the zero functional on $\mathcal{L}_{\mathbf{J}}$ is a Lebesgue integral on X. Indeed, the only points requiring verification are one, that $\mathcal{L}_{\mathbf{J}}$ is a vector space of functions, and two, that if the functions of a sequence $\{f_n\}$ all belong to $\mathcal{L}_{\mathbf{J}}$, and if $\{f_n\}$ converges pointwise to g, then g also belongs to $\mathcal{L}_{\mathbf{J}}$, and these are obviously true since $\mathbf{J}$ is a σ-ideal.

Example 3.5. If L is a Lebesgue integral on a measurable space X, if $\mathcal{L}$ denotes the vector space of integrable functions with respect to L, and if a is a positive real number, then the functional aL is also a Lebesgue integral on X having $\mathcal{L}$ for its space of integrable functions. (For $a \le 0$ this assertion is not valid in general. Indeed, it is clear that aL is not a Lebesgue integral for $a < 0$ unless $L = 0$. The case $a = 0$ is less obvious; see Problem 3E.) Likewise, if L' is some other Lebesgue integral on X and $\mathcal{L}'$ is the space of functions integrable with respect to L', then $L + L'$ acting on $\mathcal{L} \cap \mathcal{L}'$ is again a Lebesgue integral on X. (In each case the verification of the conditions of the definition is a trivial exercise which we leave to the reader.)

Example 3.6. Let $(X, \mathbf{S})$ be a measurable space, let $\mathcal{L}$ be the space of integrable functions with respect to a Lebesgue integral L on $(X, \mathbf{S})$, and suppose given a σ-algebra $\mathbf{S}_0$ in X such that $\mathbf{S}_0 \subset \mathbf{S}$. If $\mathcal{M}_0$ denotes the space of measurable scalar-valued functions on $(X, \mathbf{S}_0)$, and if we define $\mathcal{L}_0 = \mathcal{L} \cap \mathcal{M}_0$ and $L_0 = L|\mathcal{L}_0$, then it is easily verified that L_0 is a Lebesgue integral on $(X, \mathbf{S}_0)$. (The converse assertion is inexact. That is, if L_0 is a given Lebesgue integral on $(X, \mathbf{S}_0)$, there may or may not exist a Lebesgue integral L on $(X, \mathbf{S})$ such that $L_0 = L|\mathcal{L}_0$; see Example 3.32.)

Example 3.7. Let x_0 be a point in a measurable space $(X, \mathbf{S})$ and let $\mathcal{M}$ denote the vector space of all measurable complex-valued functions on X. The linear functional D defined on $\mathcal{M}$ by setting $D(f) = f(x_0)$ for each f in $\mathcal{M}$ is clearly a complex Lebesgue integral on X. Similarly, so is the functional $\widetilde{D}$ defined by setting

$$\widetilde{D}(f) = f(x_0) + \ldots + f(x_n), \quad f \in \mathcal{M},$$

where $\{x_0, \ldots, x_n\}$ is any fixed finite set of points in X.

Example 3.8. There is an interesting infinite analog of the foregoing construction. Suppose $\{x_n\}_{n=0}^{\infty}$ is a fixed sequence of distinct points in a measurable space X. Consider the collection $\mathcal{L}$ of all those measurable complex-valued functions f on X with the property that the infinite series $\sum_{n=0}^{\infty} |f(x_n)|$ is convergent (in $\mathbb{R}$), and set $S(f) = \sum_{n=0}^{\infty} f(x_n)$ for each f in $\mathcal{L}$. Then S is a complex Lebesgue integral on X. The nontrivial part of the proof of this fact is the verification of axiom (L_2). Let $\{f_k\}$ be a sequence of functions in $\mathcal{L}$ that converges pointwise on X to a function g and is Cauchy in the mean with respect to S, so that $\lim_{k,k'} S(|f_k - f_{k'}|) = 0$. Given $\varepsilon > 0$ there exists a positive integer K such that $S(|f_k - f_{k'}|) < \varepsilon$ when $k, k' \geq K$. In particular, $\sum_{n=0}^{N} |f_k(x_n) - f_{k'}(x_n)| < \varepsilon$ for every positive integer N and all $k, k' \geq K$. Passing to the limit as $k' \to \infty$, we obtain

$$\sum_{n=0}^{N} |f_k(x_n) - g(x_n)| \leq \varepsilon, \quad k \geq K,$$

for all N, and therefore

$$\sum_{n=0}^{\infty} |f_k(x_n) - g(x_n)| \leq \varepsilon, \quad k \geq K. \tag{3.1}$$

This shows, in the first place, that $f_k - g$ belongs to $\mathcal{L}$ for $k \geq K$, and hence that g does so too. In the second place, (3.1) shows that

$$|S(f_k) - S(g)| \leq S(|f_k - g|) \leq \varepsilon, \quad k \geq K,$$

and the verification of (L$_2$) is complete. The simplest and at the same time the most important instance of this construction is obtained by taking for X the set $\mathbb{N}_0$ of nonnegative integers, for $\mathbf{S}$ the power class on $\mathbb{N}_0$, and setting $x_n = n, n \in \mathbb{N}_0$. In this special case the vector space $\mathcal{L}$ is simply the collection of all complex sequences $\{\xi_n\}_{n=0}^{\infty}$ such that $\sum_{n=0}^{\infty} |\xi_n| < +\infty$, and the integral of such a sequence is $S(\{\xi_n\}) = \sum_{n=0}^{\infty} \xi_n$.

While Examples 3.4, 3.5, and 3.6 are somewhat trivial, in Examples 3.7 and 3.8 we are dealing with Lebesgue integrals that, while simply described, are of serious interest in both mathematics and physics. These examples are also of historical importance, pointing back, as they do, to the source of the whole concept of an integral, namely, the notion of a sum. As it turns out, both are special cases of the general idea of an *indexed sum*.

Example 3.9. (Finite Sums) Let Δ be a finite index set. Then for each indexed family of scalars $\{\lambda_\delta\}_{\delta \in \Delta}$ the sum

$$S = \sum_{\delta \in \Delta} \lambda_\delta$$

is a well-defined scalar. (For $\Delta = \varnothing$ this is the *empty sum*, that is, zero.) The collection of all indexed families $\{\lambda_\delta\}_{\delta \in \Delta}$ is a vector space (algebraically indistinguishable from $\mathbb{R}^N$ or $\mathbb{C}^N$ as the case may be, where $N = \operatorname{card} \Delta$), and the assignment

$$\{\lambda_\delta\} \mapsto S$$

is readily seen to be a Lebesgue integral on Δ (or, to be more precise, on the measurable space $(\Delta, 2^\Delta)$).

This example of an integral is about as elementary as can be imagined, and might well be regarded as not worth mentioning. It is however the primitive context from which all notions of integration ultimately derive.

In the real case there is a natural extension of the concept of finite indexed sum to families $\{t_\delta\}_{\delta \in \Delta}$ of *extended* real numbers. Here the sum $\sum_{\delta \in \Delta} t_\delta$ is a well-defined extended real number *provided not both $\pm\infty$ appear among the summands t_δ*. Of this extension, however, it should be explicitly noted that indexed families $\{t_\delta\}$ of extended real numbers do not form a real vector space, and the assignment $\{t_\delta\} \mapsto \sum_{\delta \in \Delta} t_\delta$ is accordingly not a linear functional. These matters are all touched on in [**I**]; see, in particular, [**I**, Problem 2T].

Example 3.10. (Indexed Sums) Let Γ be an index set (which may as well be taken to be infinite, the finite case having just been disposed of). The collection $\mathcal{D}$ of all finite subsets $\Delta \subset \Gamma$ is a directed set (under set inclusion

$\subset$), and for each family $\{\lambda_\gamma\}_{\gamma\in\Gamma}$ of scalars indexed by Γ we define the *net* $\{s_\Delta\}_{\Delta\in\mathcal{D}}$ *of finite sums*, setting

$$s_\Delta = \sum_{\gamma\in\Delta} \lambda_\gamma, \quad \Delta \in \mathcal{D}.$$

When this net is convergent, and only then, we declare the indexed family $\{\lambda_\gamma\}$ to be *summable*, and define its *sum* to be the limit of the convergent net $\{s_\Delta\}$:

$$\sum_{\gamma\in\Gamma} \lambda_\gamma = \lim_{\Delta\in\mathcal{D}} s_\Delta = \lim_{\Delta\in\mathcal{D}} \sum_{\gamma\in\Delta} \lambda_\gamma.$$

(See [**I**, p. 103] for terminology and notation.)

On the basis of this definition alone several things are apparent. For one thing, it is easily seen that the collection of all summable indexed families $\{\lambda_\gamma\}_{\gamma\in\Gamma}$ of scalars is a vector space $\mathcal{S}$, and likewise that the assignment of the sum S to each summable family is a linear functional on $\mathcal{S}$. Moreover, an indexed family $\{\lambda_\gamma\}$ of complex numbers is summable if and only if the real families $\{\Re\lambda_\gamma\}$ and $\{\Im\lambda_\gamma\}$ are, and, when this is the case, we have

$$\sum_{\gamma\in\Gamma} \lambda_\gamma = \sum_{\gamma\in\Gamma} \Re\lambda_\gamma + i \sum_{\gamma\in\Gamma} \Im\lambda_\gamma.$$

Thus $\mathcal{S}$ and S are both self-conjugate ([**I**, Example 3K]). In fact, we have here, once again, a Lebesgue integral. But this time matters are less transparent, and the details demand some elucidation. We begin with a form of the Cauchy criterion appropriate for nets of finite sums (see [**I**, Chapter 8]).

Let $\{\lambda_\gamma\}_{\gamma\in\Gamma}$ be an indexed family of scalars. *Then the corresponding net* $\{s_\Delta\}_{\Delta\in\mathcal{D}}$ *of finite sums is Cauchy if and only if, for given* $\varepsilon > 0$, *there exists a set* $\Delta_\varepsilon \in \mathcal{D}$ *with the property that* $|s_\Delta| < \varepsilon$ *for all* $\Delta \in \mathcal{D}$ *such that* $\Delta \cap \Delta_\varepsilon = \varnothing$. Indeed, if Δ_ε and Δ are any two disjoint finite sets of indices, then Δ_ε and $\Delta' = \Delta_\varepsilon \cup \Delta$ contain Δ_ε and, of course,

$$s_{\Delta'} - s_{\Delta_\varepsilon} = s_\Delta,$$

so the condition is clearly necessary. On the other hand, if it is satisfied, and $\varepsilon > 0$ is given, then there is a set Δ_ε such that $|s_\Delta| < \varepsilon/2$ for all Δ disjoint from Δ_ε. But then, for any two finite sets Δ' and Δ'' containing Δ_ε we have

$$s_{\Delta'} - s_{\Delta''} = s_{\Delta'\setminus\Delta''} - s_{\Delta''\setminus\Delta'}$$

and therefore $|s_{\Delta'} - s_{\Delta''}| < \varepsilon/2 + \varepsilon/2 = \varepsilon$. Thus the net $\{s_\Delta\}$ satisfies the Cauchy criterion.

Since the scalar fields $\mathbb{R}$ and $\mathbb{C}$ are complete, the Cauchy criterion is sufficient as well as necessary for the summability of an indexed family $\{\lambda_\gamma\}$. In order to exploit this fact, consider first a summable indexed family $\{t_\gamma\}$

of real numbers. Let ε be a positive number and let Δ_ε be chosen as in the Cauchy criterion. If Δ is a finite set disjoint from Δ_ε, and if Δ_+ and Δ_- denote the subsets of Δ where t_γ is positive and negative, respectively, then

$$\sum_{\gamma \in \Delta} t_\gamma^+ = \sum_{\gamma \in \Delta_+} t_\gamma < \varepsilon \quad \text{and} \quad \sum_{\gamma \in \Delta} t_\gamma^- = - \sum_{\gamma \in \Delta_-} t_\gamma < \varepsilon,$$

and therefore

$$\sum_{\gamma \in \Delta} |t_\gamma| = \sum_{\gamma \in \Delta} \left(t_\gamma^+ + t_\gamma^- \right) < 2\varepsilon.$$

Thus the indexed family $\{|t_\gamma|\}$ is also summable. It follows, of course, that for any summable family $\{\lambda_\gamma\}$ of *complex* numbers the family $\{|\lambda_\gamma|\}$ is summable as well, and since, in all cases,

$$\left| \sum_{\gamma \in \Delta} \lambda_\gamma \right| \leq \sum_{\gamma \in \Delta} |\lambda_\gamma|$$

for any finite set Δ of indices, we have the inequality

$$\left| \sum_{\gamma \in \Gamma} \lambda_\gamma \right| \leq \sum_{\gamma \in \Gamma} |\lambda_\gamma|,$$

valid for all summable indexed families of scalars.

Consider next an indexed family $\{u_\gamma\}$ of nonnegative real numbers. In this case, the net of partial sums $\{s_\Delta\}$ is monotone increasing, and is therefore convergent if and only if it is bounded above in $\mathbb{R}$, in which case we have

$$\sum_{\gamma \in \Gamma} u_\gamma = \sup_{\Delta \in \mathcal{D}} s_\Delta$$

([**I**, Example 6J]). Thus if $0 \leq u_\gamma \leq v_\gamma$ for each index γ, and if $\{v_\gamma\}$ belongs to $\mathcal{S}$, then so does $\{u_\gamma\}$ and we have

$$\sum_{\gamma \in \Gamma} u_\gamma \leq \sum_{\gamma \in \Gamma} v_\gamma.$$

This shows that the functional S satisfies axiom (L_1) for a Lebesgue integral (see Problem 3C). Let us also verify axiom (L_2).

Suppose $\{\lambda_\gamma^{(n)}\}_{n=1}^\infty$ is a sequence of summable indexed families that is Cauchy in the mean, that is, satisfies the condition

$$\lim_{m,n} \sum_{\gamma \in \Gamma} \left| \lambda_\gamma^{(m)} - \lambda_\gamma^{(n)} \right| = 0,$$

and that also converges indexwise to the family $\{\mu_\gamma\}$. Fix $\varepsilon > 0$ and choose N such that

$$\sum_{\gamma \in \Gamma} \left| \lambda_\gamma^{(m)} - \lambda_\gamma^{(n)} \right| < \varepsilon$$

for all $m, n \geq N$. Then for any finite set $\Delta \subset \Gamma$ we have

$$\sum_{\gamma \in \Delta} \left| \lambda_\gamma^{(m)} - \lambda_\gamma^{(n)} \right| < \varepsilon$$

when $m, n \geq N$. But now, as $n \to \infty$, the finite sum $\sum_{\gamma \in \Delta} |\lambda_\gamma^{(m)} - \lambda_\gamma^{(n)}|$ tends to $\sum_{\gamma \in \Delta} |\lambda_\gamma^{(m)} - \mu_\gamma|$, and it follows that

$$\sum_{\gamma \in \Delta} \left| \lambda_\gamma^{(m)} - \mu_\gamma \right| \leq \varepsilon \tag{3.2}$$

when $m \geq N$ and for all $\Delta \in \mathcal{D}$. From this we see first that the indexed family $\{|\lambda_\gamma^{(m)} - \mu_\gamma|\}$ is summable for $m \geq N$, so that $\{\mu_\gamma\}$ is summable as well. Moreover, it also follows from (3.2) that

$$\sum_{\gamma \in \Gamma} \left| \lambda_\gamma^{(m)} - \mu_\gamma \right| \leq \varepsilon$$

for $m \geq N$. Thus $\lim_m S(\{|\lambda_\gamma^{(m)} - \mu_\gamma|\}) = 0$, and therefore

$$\lim_m S(\{\lambda_\gamma^{(m)}\}) = S(\{\mu_\gamma\}),$$

as was to be proved.

We remark, once again, that the ideas discussed here extend, in part, to the case of indexed families $\{t_\gamma\}$ of *extended* real numbers. Indeed, if $\{t_\gamma\}$ is such a family in which *not both* $\pm\infty$ *appear*, then the net $\{s_\Delta\}$ is defined just as above. If for some index γ_0 we have $t_{\gamma_0} = +\infty$, then $s_\Delta = +\infty$ for every Δ containing γ_0, and

$$\sum_{\gamma \in \Gamma} t_\gamma = +\infty.$$

Similarly, $\sum_{\gamma \in \Gamma} t_\gamma = -\infty$ if $t_\gamma = -\infty$ for some index γ. Otherwise $\{s_\Delta\}$ is a net of finite real numbers. If it converges in $\mathbb{R}$, we are dealing with an ordinary summable family of real numbers. If it does not converge in $\mathbb{R}$, then it may happen that for every real number M (no matter how large) there exists a finite set Δ_M of indices such that $s_\Delta \geq M$ for all $\Delta \supset \Delta_M$, and we have $\sum_{\gamma \in \Gamma} t_\gamma = +\infty$. Dually, if for arbitrary m (no matter how small) there exists Δ_m such that $s_\Delta \leq m$ for $\Delta \supset \Delta_m$, then $\sum_{\gamma \in \Gamma} t_\gamma = -\infty$. (See [**I**, Example 6L].) In particular, for an indexed family $\{u_\gamma\}$ of *nonnegative* extended real numbers, $\sum_{\gamma \in \Gamma} u_\gamma = +\infty$ holds whenever the family is not summable. Thus for such a family we have $\sum_{\gamma \in \Gamma} u_\gamma < +\infty$ when and only when the family is summable.

It is, of course, no accident that the proof of axiom (L_2) for indexed sums is very like the corresponding argument in Example 3.8. It is apparent that an infinite series $\sum_{n=0}^{\infty} \lambda_n$ of scalars λ_n is absolutely convergent (that is, $\sum_{n=0}^{\infty} |\lambda_n| < +\infty$) if and only if the indexed family $\{\lambda_n\}_{n \in \mathbb{N}_0}$ is summable, and that $\sum_{n=0}^{\infty} \lambda_n = \sum_{n \in \mathbb{N}_0} \lambda_n$ when this is the case. Thus Example 3.8 may be viewed as a special case of this example.

The following result is elementary in that its proof uses only axiom (L_1) in the definition of the Lebesgue integral; it is nonetheless of profound importance.

Proposition 3.11. *If L is a complex Lebesgue integral on a measurable space X, and if f is a real-valued function on X that is integrable with respect to L, then the integral $L(f)$ is real. If L is either a real or a complex Lebesgue integral on X, and if f is a nonnegative function on X that is integrable with respect to L, then $L(f) \geq 0$.*

Proof. If f is a real-valued function on X, then $f = f^+ - f^-$, and these two functions are integrable along with f, as we have seen (Proposition 3.3). Thus it suffices to prove the second assertion of the proposition, and this follows at once upon setting $g = 0$ in axiom (L_1). $\qquad \square$

Corollary 3.12. *If L is a complex Lebesgue integral on a measurable space X, and $\mathcal{L}$ denotes the complex vector space of functions integrable with respect to L, then*

$$L(\Re f) = \Re L(f) \quad \text{and} \quad L(\Im f) = \Im L(f)$$

*for every $f \in \mathcal{L}$ (see [**I**, Problem 3I]). If L is either a real or complex Lebesgue integral, then L is monotone increasing on the real vector space of all real-valued functions integrable with respect to L.*

Example 3.13. Let L be a complex Lebesgue integral on a measurable space X, and let $\mathcal{L}$ denote the vector space of integrable functions with respect to L. Then $\mathcal{L}$ is self-conjugate and is therefore (identifiable with) the complexification $(\mathcal{L}_\mathbb{R})^+$ of the real vector space $\mathcal{L}_\mathbb{R}$ of real-valued functions in $\mathcal{L}$ (see [**I**, Example 3K]). Moreover it is clear that the linear functional $L_\mathbb{R} = L|\mathcal{L}_\mathbb{R}$ is a real Lebesgue integral, and the given integral L is its complexification. Conversely, if one starts with a real Lebesgue integral L defined on a real vector space $\mathcal{E}$ of real-valued measurable functions on a measurable space X, then the complexification L^+ (defined on the space $\mathcal{L}^+$) is a complex Lebesgue integral on X, but this fact is not immediate (Corollary 3.38); see also Problem 3C.

Example 3.14. The Cauchy integral of elementary calculus (say on the unit interval) is a linear functional defined on the real vector space $\mathcal{C}_\mathbb{R}([0,1])$ of continuous functions ([**I**, Example 3L]), but it is obviously not a real Lebesgue integral. Likewise, the Riemann integral R defined on the real vector space $\mathcal{R}$ of real-valued Riemann integrable functions on $[0, 1]$ fails to be a real Lebesgue integral. Indeed, if R were a real Lebesgue integral, then the functions in $\mathcal{R}$ would all be measurable with respect to some σ-algebra $\mathbf{S}$ of subsets of $[0, 1]$, and since $\mathcal{C}_{\mathbb{R}}([0, 1]) \subset \mathcal{R}, \mathbf{S}$ would contain all the Borel sets in $[0, 1]$. But then the characteristic function of the set of rational numbers in $[0, 1]$ would be Riemann integrable, contrary to fact.

The second basic result in the theory of Lebesgue integration is the following result.

Proposition 3.15. *Let L be a Lebesgue integral (real or complex) on a measurable space X, and let $\{f_n\}$ be a monotone increasing sequence of nonnegative functions on X, each of which is integrable with respect to L. Suppose that $\{f_n\}$ converges pointwise to a real-valued function g. Then g is integrable with respect to L if and only if the numerical sequence $\{L(f_n)\}$ is bounded in $\mathbb{R}$, and in this case we have $\lim_n L(f_n) = L(g)$.*

Proof. According to Proposition 3.11 and Corollary 3.12, the integrals $L(f_n)$ are nonnegative and the sequence $\{L(f_n)\}$ is monotone increasing. Moreover, if g is integrable, then $0 \leq L(f_n) \leq L(g)$ for every n. On the other hand, if the sequence $\{L(f_n)\}$ is bounded above, then it is convergent, and is therefore a Cauchy sequence:

$$\lim_{m,n} |L(f_m) - L(f_n)| = \lim_{m,n} |L(f_m - f_n)| = 0.$$

Since the sequence $\{f_n\}$ is monotone, we have $|L(f_m - f_n)| = L(|f_m - f_n|)$ for all indices m and n, so the sequence $\{f_n\}$ is Cauchy in the mean, and the result follows at once from axiom (L_2). $\quad\square$

Recall now (see Remark 3.2) that a (real or complex) Lebesgue integral is a triple $(X, \mathcal{L}, L)$ where $X = (X, \mathbf{S})$ is a measurable space, $\mathcal{L}$ is a vector space of measurable functions on $(X, \mathbf{S})$, and L is a (real or complex) linear functional defined on $\mathcal{L}$ subject to axioms (L_1) and (L_2) of Definition 3.1. An important characteristic of a Lebesgue integral is its close connection with a *measure space*, a concept which we introduce below in Definition 3.19. The remarkable fact, established in Theorems 3.29 and 3.37, is that the map $L \mapsto \mu$ given by (3.7) establishes a bijection between the set of all real Lebesgue integrals on a measurable space $(X, \mathbf{S})$ and the set of all measures on $(X, \mathbf{S})$. This connection between a (real or complex) Lebesgue integral and its associated measure is extremely important. Before giving the definition of a measure, however, it is desirable to define some ancillary concepts. To begin with, a function whose domain of definition is a collection of subsets of some set X is a *set function* on X. We are concerned with set functions that are either complex-valued or extended real-valued (or real-valued, of course, this being a special case of both).

Definition 3.16. Let φ be a complex-valued or extended real-valued set function defined on a collection $\mathcal{A}$ of subsets of a set X. Then φ is *finitely additive* on $\mathcal{A}$ if for every finite (pairwise) disjoint subcollection $\mathcal{C}$ of $\mathcal{A}$ such that the union $\bigcup \mathcal{C}$ also belongs to $\mathcal{A}$ we have

$$\varphi\left(\bigcup \mathcal{C}\right) = \sum_{A \in \mathcal{C}} \varphi(A). \tag{3.3}$$

Similarly, φ is *countably additive* on $\mathcal{A}$ if (3.3) holds for every countable disjoint subcollection $\mathcal{C}$ of $\mathcal{A}$ such that $\bigcup \mathcal{C} \in \mathcal{A}$. (Thus a countably additive set function is automatically finitely additive as well.)

Remark 3.17. Since the collection $\mathcal{C}$ in this definition is countable, it could, in every case, be arranged in a simple sequence, whereupon the scalar sum appearing in (3.3) would become an ordinary series, finite or infinite as the case might be. Moreover, it is in these terms that the definition of additivity is customarily given. This difference in formulation, however, is without real significance. Indeed, it clearly is of no consequence which way we define finite additivity, while if $\{A_n\}_{n=1}^{\infty}$ is some infinite sequence of sets in $\mathcal{A}$ such that $\varphi(\bigcup_n A_n) = \sum_n \varphi(A_n)$, the union $\bigcup_n A_n$ is clearly unaffected by a rearrangement of the sequence $\{A_n\}$, so the series $\sum_n \varphi(A_n)$ is also unconditionally, and therefore absolutely, convergent, and the indexed family $\{\varphi(A_n)\}_{n \in \mathbb{N}}$ is therefore necessarily summable. When the collection $\mathcal{C}$ is empty, the equality (3.3) implies that $\varphi(\varnothing) = 0$ if $\varnothing \in \mathcal{A}$ and φ is finitely additive.

Observe that for extended real-valued set functions this definition explicitly assumes the various sums involved to be defined in $\mathbb{R}^{\natural}$. Thus if φ is extended real-valued and finitely additive on $\mathcal{A}$, and if A and B are disjoint sets in $\mathcal{A}$ such that $A \cup B \in \mathcal{A}$, then $\varphi(A)$ and $\varphi(B)$ dare not be infinities of opposite sign. Moreover, in this case, the sum in (3.3) is to be interpreted in $\mathbb{R}^{\natural}$ as discussed in Example 3.10.

Another important concept with which we are concerned is defined only in the real case.

Definition 3.18. A nonnegative, extended real-valued set function φ defined on a collection $\mathcal{A}$ of subsets of a set X is *countably subadditive* on $\mathcal{A}$ if for every set B in $\mathcal{A}$ and every countable subcollection $\mathcal{C} \subset \mathcal{A}$ such that $B \subset \bigcup \mathcal{C}$ we have $\varphi(B) \leq \sum_{A \in \mathcal{C}} \varphi(A)$.

The most important kind of set function of all, from our point of view, is known as a *measure*.

Definition 3.19. If $(X, \mathbf{S})$ is a measurable space, then a *measure* on $(X, \mathbf{S})$ is a nonnegative, extended real-valued, countably additive set function μ defined on the σ-algebra $\mathbf{S}$ such that $\mu(\varnothing) = 0$. The value $\mu(E)$ of μ at a measurable set E is the *measure* of E. A measurable space $(X, \mathbf{S})$ together with a measure μ defined on $\mathbf{S}$ is called a *measure space*, and is denoted by $(X, \mathbf{S}, \mu)$, or, when no confusion is possible, simply by X. If the measure μ is finite-valued (or, what comes to the same thing, if $\mu(X) < +\infty$; see Proposition 3.27 below), then the measure μ and the measure space $(X, \mathbf{S}, \mu)$ are said to be finite. A *probability space* is a measure space X satisfying $\mu(X) = 1$.

Example 3.20. Let $(X, \mathbf{S})$ be a measurable space, let $\{w_x\}_{x \in X}$ be an arbitrary family of nonnegative extended real numbers indexed by X, and for each set $E \in \mathbf{S}$ set

$$\mu(E) = \sum_{x \in E} w_x$$

(see Example 3.10). Then μ is a measure on $(X, \mathbf{S})$. (Such a measure is called a *discrete* measure; the number w_x is the *weight* or *mass* of the point x. Recall that the sum of the empty indexed family of numbers is zero.) Indeed, if $\mathcal{C}$ is an *arbitrary* pairwise disjoint collection of sets in $\mathbf{S}$ with the property that $\bigcup \mathcal{C} \in \mathbf{S}$, then

$$\mu\left(\bigcup \mathcal{C}\right) = \sum_{E \in \mathcal{C}} \mu(E)$$

(see Problem 3F). Such a measure is said to be *completely additive*. In the converse direction it is not hard to see that if μ is a completely additive measure on a measurable space $(X, \mathbf{S})$, and if $\mathbf{S}$ is rich enough to contain all singletons in X, then μ is in fact a discrete measure with $w_x = \mu(\{x\})$ for each point x.

Example 3.21. If the weights w_x in the foregoing example are all equal to one, then the resulting measure γ on $(X, \mathbf{S})$ has the property that $\gamma(E) = +\infty$ for every infinite set $E \in \mathbf{S}$, while $\gamma(F)$ is the number of points in F whenever $F \in \mathbf{S}$ is finite. This measure is known as the *counting measure* on $(X, \mathbf{S})$.

In the preceding examples, there exist points $x \in X$ such that $\{x\} \in \mathbf{S}$ and $\mu(\{x\}) > 0$. One says that μ has a *point mass* at x when this happens. Point masses are particular cases of atoms.

Definition 3.22. Let $(X, \mathbf{S}, \mu)$ be a measure space. A set $A \in \mathbf{S}$ is called an *atom* of μ if $\mu(A) > 0$ and every set $E \in \mathbf{S}$ such that $E \subset A$ has measure 0 or $\mu(A)$. An atom A is *finite* if $\mu(A) < +\infty$ and *infinite* otherwise. The measure μ is said to be *semi-finite* if it has no infinite atoms.

Example 3.23. Consider an uncountable set X, and denote by $\mathbf{S}$ the σ-algebra generated by the finite subsets of X. The σ-algebra $\mathbf{S}$ consists of those sets $E \subset X$ such that either E or $X \setminus E$ is at most countable. Define a measure μ on $(X, \mathbf{S})$ by

$$\mu(E) = \begin{cases} 0 & \text{if } E \text{ is at most countable,} \\ a & \text{if } X \setminus E \text{ is at most countable,} \end{cases}$$

for some fixed $a \in (0, +\infty]$. Then X itself is an infinite atom, infinite if $a = +\infty$, but μ has no point masses.

Proposition 3.24. *Let $(X, \mathbf{S}, \mu)$ be a measure space. Then X is semi-finite if and only if the equality*

$$\mu(E) = \sup\{\mu(F) : F \subset E, F \in \mathbf{S}, \mu(E) < +\infty\}$$

holds for every $E \in \mathbf{S}$.

Proof. Assume that $E \in \mathbf{S}$ have infinite measure, and assume that the number

$$\alpha = \sup\{\mu(F) : F \subset E, F \in \mathbf{S}, \mu(E) < +\infty\}$$

is finite. There exist sets $F_n \in \mathbf{S}$ such that $\mu(F_n) > \alpha - 2^{-n}$. Set $F = \bigcup_{n=1}^{\infty} F_n$ and $A = E \setminus F$. Then $F \in \mathbf{S}$, $\mu(F) = \alpha$, and the definition of α implies that every set $B \in \mathbf{S}$ such that $B \subset A$ and $\mu(B) < +\infty$ satisfies $\mu(B) = 0$. Indeed, $F \cup B \subset E$ and

$$\alpha = \mu(F) \le \mu(F) + \mu(B) = \mu(F \cup B) \le \alpha.$$

Thus A is an infinite atom. We conclude that X is semi-finite precisely when the equality in the statement holds for sets E with infinite measures; for other sets, the equality is always satisfied. $\square$

Example 3.25. Let $\mathbf{H}_r$ denote the ring of all finite (pairwise disjoint) unions of half-open intervals $(a, b]$ in $\mathbb{R}$ (Problem 1B). If a set A in $\mathbf{H}_r$ is expressed as a disjoint union

$$A = \bigcup_{i=1}^{n}(a_i, b_i] \tag{3.4}$$

of such intervals, then it is easily seen that the sum

$$\nu(A) = \sum_{i=1}^{n}(b_i - a_i) \tag{3.5}$$

is determined solely by the set A and is independent of the representation (3.4). Indeed, if $A = (a, b]$ where a and b are real numbers such that $a < b$, then the intervals in (3.4) are (up to order) the intervals $(a_{i-1}, a_i]$ determined by some partition $\{a = a_0 < \ldots < a_n = b\}$ of $[a, b]$, so the sum in (3.5) reduces to $b - a$. But for an arbitrary element A of $\mathbf{H}_r$ there is a unique representation of A in the form (3.4) (up to order) in which the various *closed* intervals $[a_i, b_i]$ are pairwise disjoint, and the result follows.

Suppose now that A and B are disjoint elements of $\mathbf{H}_r$. If both A and B are expressed as in (3.4) as finite pairwise disjoint unions of half-open intervals, then the union of these two collections is likewise pairwise disjoint and has union $A \cup B$. Thus $\nu(A \cup B) = \nu(A) + \nu(B)$, from which it follows at once that ν is a finitely additive measure on $\mathbf{H}_r$.

Example 3.26. For each pair q, r of rational numbers such that $q < r$ let us write $(q, r]^{\sim}$ for the intersection $(q, r] \cap \mathbb{Q}$. The collection $\mathbf{H}_r^{\sim}$ of all finite unions of such intervals of rational numbers is obviously a ring of sets in the space $\mathbb{Q}$ of rational numbers. Moreover, the same argument as in Example 3.25 proves that there is a uniquely determined finitely additive measure $\tilde{\nu}$ on $\mathbf{H}_r^{\sim}$ with the property that $\tilde{\nu}((q,r]^{\sim}) = r - q$ for every pair q, r of rational numbers such that $q < r$.

Some of the elementary properties of measures are shared by all finitely additive set functions as well.

Proposition 3.27. *Let ν be a finitely additive set function defined on a ring of sets $\mathbf{R}$. Then ν is* monotone *and* subtractive *in the sense that if A and B belong to $\mathbf{R}$ and if $A \subset B$, then $\nu(A) \leq \nu(B)$ and, whenever $\nu(A) < +\infty, \nu(B \setminus A) = \nu(B) - \nu(A)$.*

Proof. Apply additivity to $B = A \cup (B \setminus A)$. $\qquad\qquad\qquad\qquad\qquad$ □

Proposition 3.28. *A measure μ on a measurable space $(X, \mathbf{S})$ is* countably subadditive *and also* semicontinuous *in the sense that*

$$\mu\left(\bigcup_n E_n \right) = \lim_n \mu(E_n) \qquad\qquad (3.6)$$

for every monotone increasing sequence $\{E_n\}$ of sets in $\mathbf{S}$. Conversely, if ν is a finitely additive measure defined on the σ-algebra $\mathbf{S}$ and if ν is either countably subadditive or semicontinuous in the above sense, then ν is countably additive and is therefore a measure.

Proof. Assume first that μ is a measure on $(X, \mathbf{S})$, and let $\{E_n\}_{n=1}^{\infty}$ be a sequence of measurable sets in X. We begin by disjointifying $\{E_n\}$, setting

$$F_n = E_n \setminus \left(\bigcup_{k<n} E_k \right), n \in \mathbb{N}$$

(see [**I**, Problem 1U]). Since $F_n \subset E_n$, we have $\mu(F_n) \leq \mu(E_n)$ for every n by the preceding proposition. Hence

$$\mu\left(\bigcup_{n=1}^{\infty} E_n \right) = \mu\left(\bigcup_{n=1}^{\infty} F_n \right) = \sum_{n=1}^{\infty} \mu(F_n) \leq \sum_{n=1}^{\infty} \mu(E_n),$$

which, together with the fact that μ is monotone, shows that μ is countably subadditive. Moreover, if the sequence $\{E_n\}$ is monotone increasing, we have $E_n = \bigcup_{k=1}^{n} F_k$, and therefore, $\mu(E_n) = \sum_{k=1}^{n} \mu(F_k)$ for every n. But then

$$\lim_n \mu(E_n) = \sum_{n=1}^{\infty} \mu(F_n) = \mu\left(\bigcup_{n=1}^{\infty} F_n \right) = \mu\left(\bigcup_{n=1}^{\infty} E_n \right),$$

as was to be proved.

Conversely, assume ν is a finitely additive measure defined on $\mathbf{S}$, let $\{F_n\}$ be a pairwise disjoint sequence of sets in $\mathbf{S}$, and let $F = \bigcup_n F_n$. If we define $E_n = \bigcup_{k=1}^n F_k$, $n \in \mathbb{N}$, then E_n is an increasing sequence whose union is also F, and since $\nu(E_n) = \sum_{k=1}^n \nu(F_k)$, it is clear that ν is countably additive if it is semicontinuous. On the other hand, $\sum_{k=1}^n \nu(F_k) \le \nu(F)$ for all n since ν is monotone, and therefore $\sum_{n=1}^\infty \nu(F_n) \le \nu(F)$. Hence ν is countably additive if and only if it is countably subadditive. $\qquad\square$

Thus far we have done little more than introduce the two concepts of Lebesgue integral and measure. It remains to study the relations between them. In one direction this is quite easy.

Theorem 3.29. *Let L be a (real or complex) Lebesgue integral on a measurable space $(X, \mathbf{S})$ and let $\mathcal{L}$ denote the vector space of functions integrable with respect to L. Given $E \in \mathbf{S}$, we define*

$$\mu(E) = \begin{cases} L(\chi_E) & \text{if } \chi_E \in \mathcal{L} \\ +\infty & \text{otherwise.} \end{cases} \tag{3.7}$$

Then μ is a measure on $(X, \mathbf{S})$. This measure, which we shall call the measure associated with L, determines L uniquely.

Proof. We first prove the uniqueness assertion, and then verify that the set function μ is indeed a measure on X. Suppose, accordingly, that L' is another Lebesgue integral on $(X, \mathbf{S})$ of the same type (real or complex) with vector space of integrable functions $\mathcal{L}'$, and suppose that the set function associated with L' as in (3.7) coincides with μ. We must show that $\mathcal{L}' = \mathcal{L}$ and that $L' = L$ on $\mathcal{L}$.

Clearly $\mathcal{L}$ and $\mathcal{L}'$ contain precisely the same characteristic functions. But then they contain exactly the same simple functions as well, since a measurable simple function s is integrable with respect to a Lebesgue integral if and only if the characteristic function of the *support* $N_s = E(s \ne 0)$ is integrable (see Problem 3K). Moreover, if s is a simple function belonging to $\mathcal{L}$ (and therefore to $\mathcal{L}'$), we clearly have $L(s) = L'(s)$.

Now let f be an arbitrary function belonging to $\mathcal{L}$. By symmetry it suffices to verify that f is also in $\mathcal{L}'$ and that $L(f) = L'(f)$. Suppose first that f is real-valued. Write $f = f^+ - f^-$, where f^+ and f^- denote as usual the positive and negative parts of f, and let $\{s_n\}$ be a monotone increasing sequence of nonnegative measurable simple functions converging pointwise to f^+ (Proposition 2.36). Since f^+ and the simple functions s_n all belong to $\mathcal{L}$ along with f, it follows from Proposition 3.15 that

$$L(f^+) = \lim_n L(s_n) = \lim_n L'(s_n) = L'(f^+).$$

Similarly we obtain $L(f^-) = L'(f^-)$, and therefore $L(f) = L'(f)$. Thus $\mathcal{L} \subset \mathcal{L}'$ and $L = L'|\mathcal{L}$ in the real case. To complete the proof in the complex case, it suffices to recall that both L and L' are self-conjugate.

It remains to verify that the set function μ defined in (3.7) is a measure on $(X, \mathbf{S})$. Clearly μ is a nonnegative, extended real-valued function defined on $\mathbf{S}$. Moreover, it is an easy consequence of the linearity of L that μ is finitely additive. Hence (Proposition 3.28) the proof will be complete if we verify that μ is semicontinuous. To this end let $\{E_n\}$ be an increasing sequence of sets in $\mathbf{S}$, and set $E = \bigcup_n E_n$. If $\mu(E_n) = +\infty$ for some n, then $\mu(E) = +\infty$ too (by axiom (L_1)), and (3.6) holds. On the other hand, if $\mu(E_n)$ is finite for all n, then $\{\chi_{E_n}\}$ is an increasing sequence of nonnegative functions in $\mathcal{L}$ that converges pointwise to χ_E and (3.6) follows immediately from Proposition 3.15.   □

In this book we chose, in the interest of clarity and brevity, to define the concept of a Lebesgue integral prior to defining the idea of a measure. The usual practice, however, is to begin with a measure μ on a measurable space $(X, \mathbf{S})$ and then to construct a Lebesgue integral $(\mathcal{L}, L)$ on $(X, \mathbf{S})$ that satisfies (3.7). This is done below in the proof of Theorem 3.37. Consequently, where we have referred to the measure μ *associated* with a Lebesgue integral L, the usual practice is to speak of L as *integration with respect to the measure μ*. Henceforth, we also employ this standard terminology. Moreover, in the same spirit, the value $L(f)$ of the integral L at a function f is called the *integral of f with respect* to μ, and is denoted by

$$\int_X f\, d\mu \quad \text{or} \quad \int_X f(x)\, d\mu(x)$$

when necessary to avoid confusion. Similarly, the functions we have called integrable with respect to L will be said to be *integrable* (*over X*) *with respect to μ*, or more briefly, *integrable* $[\mu]$. More generally, let E be a measurable subset of X and let f be a complex-valued function defined on some subset of X that contains E. If the function $\tilde{f}$ defined by setting

$$\tilde{f}(x) = \begin{cases} f(x), & x \in E, \\ 0, & x \notin E, \end{cases}$$

is integrable $[\mu]$, then f is said to be *integrable* $[\mu]$ *over E*, and $\int_X \tilde{f}\, d\mu$ is declared to be the *integral of f over E with respect to μ*, denoted by $\int_E f\, d\mu$. Thus, if f is integrable $[\mu]$ over X, then

$$L(\chi_E f) = \int_E f\, d\mu$$

for every measurable set E (since in this case $\chi_E f$ is integrable along with f).

Example 3.30. Let $(X, \mathbf{S})$ be a measurable space, and let $\mathbf{J}$ be a σ-ideal in $\mathbf{S}$. The set function $\mu_{\mathbf{J}}$ on X that assigns the value 0 to each set E in $\mathbf{J}$ and the value $+\infty$ to each set E in $\mathbf{S} \setminus \mathbf{J}$ is a measure on $(X, \mathbf{S})$, and this measure is, in fact, the measure associated with the Lebesgue integral defined in Example 3.4. (The special case $\mathbf{J} = \mathbf{S}$ yields the *zero measure* on $(X, \mathbf{S})$, associated with the zero functional on the space $\mathcal{M}$ of all measurable complex-valued functions on X. The special case $\mathbf{J} = \{\varnothing\}$ yields the measure on $(X, \mathbf{S})$ that assigns $+\infty$ to every nonempty set in $\mathbf{S}$, that is, the measure associated with the Lebesgue integral L for which $\mathcal{L}$ is the trivial linear manifold.)

Example 3.31. Let μ and μ' be two measures on the same measurable space $(X, \mathbf{S})$, and let a and b be positive numbers. Then it is easy to see that $a\mu + b\mu'$ is again a measure on $(X, \mathbf{S})$, and it is equally easy to see what integration with respect to this measure amounts to. A measurable complex-valued function f on X is integrable $[a\mu + b\mu']$ if and only if it is integrable over X with respect to both μ and μ', and, when this is the case,

$$\int_X f\, d(a\mu + b\mu') = a \int_X f\, d\mu + b \int_X f\, d\mu'.$$

Indeed, it was pointed out in Example 3.5 that this functional is a Lebesgue integral on X, and it is clear that its associated measure is precisely the measure $a\mu + b\mu'$. (Note that this description of the integral with respect to $a\mu + b\mu'$ need not hold when $a = b = 0$, since in this case the measure $a\mu + b\mu'$ is the zero measure and, as we have seen (Example 3.30), every measurable complex-valued function on X is integrable with respect to it.)

Example 3.32. Let $(X, \mathbf{S}, \mu)$ be a measure space, and suppose given a σ-algebra $\mathbf{S}_0$ such that $\mathbf{S}_0 \subset \mathbf{S}$. If L denotes Lebesgue integration with respect to μ, and L_0 is the restriction of L to those integrable functions that are measurable $[\mathbf{S}_0]$ (see Example 3.6), then by (3.7) the associated measure μ_0 on $\mathbf{S}_0$ is given by

$$\mu_0(E) = \begin{cases} L(\chi_E) = \mu(E), & \mu(E) < +\infty, \\ +\infty, & \mu(E) = +\infty. \end{cases}$$

In other words, L_0 is simply integration with respect to the measure $\mu_0 = \mu|\mathbf{S}_0$. This shows that the question whether a Lebesgue integral L_0 on a measurable space such as $(X, \mathbf{S}_0)$ is the restriction of some extended integral on the space $(X, \mathbf{S})$ is related to the possibility of extending the associated measure: *Given μ_0 on $(X, \mathbf{S}_0)$, does there exist a measure on $(X, \mathbf{S})$ such that $\mu_0 = \mu|\mathbf{S}_0$?* As mentioned before, the answer is not always in the affirmative.

Example 3.33. The Lebesgue integral D of Example 3.7 is readily seen to be integration with respect to the measure δ_{x_0} that assigns to each measurable set E the value $\delta_{x_0}(E) = \chi_E(x_0)$. This measure is known as the *point mass* or *Dirac measure* at x_0. (The measures associated with the other integrals $\widetilde{D}$ and S of Examples 3.7 and 3.8 are equally easy to describe; see Problem 3T.)

Example 3.34. Let $(X, \mathbf{S}, \mu)$ be a measure space, and let E be a measurable subset of X (so that $(E, \mathbf{S}_E)$ is a subspace of (X, S)). The set function that assigns the value $\mu(F)$ to each measurable subset F of E is clearly a measure on $(E, \mathbf{S}_E)$. This measure is called the *restriction* of μ to E and is denoted by $\mu|E$. A measure space of the form $(E, \mathbf{S}_E, \mu|E)$ is called a *subspace* of $(X, \mathbf{S}, \mu)$. Integration over such a subspace is easily described: A measurable complex-valued function f on E is integrable $[\mu|E]$ if and only if f is integrable $[\mu]$ over E, and, when this is the case,

$$\int_E f \, d(\mu|E) = \int_E f \, d\mu.$$

Indeed, this functional is readily seen to be a Lebesgue integral on E, and $\mu|E$ is its associated measure.

Once a Lebesgue integral is given, it is a simple matter, as noted in Theorem 3.29, to identify the measure associated with it. What of the converse? If a measure μ is given, does there exist a Lebesgue integral having μ as its associated measure? The answer is "yes" as we see in Theorem 3.37 The nontrivial proof of this fact requires, among other things, the following remarkable result.

Proposition 3.35. (Egorov's Theorem) *Let $(X, \mathbf{S}, \mu)$ be a measure space, let E be a measurable subset of X having finite measure, and let $\{f_n\}$ be a sequence of complex-valued functions defined and measurable on E and converging pointwise on E to a function f. Then for any $\varepsilon > 0$ there exists a subset F of E such that $\mu(F) < \varepsilon$ and such that $\{f_n\}$ converges uniformly to f on $E \setminus F$.*

Proof. There is no loss of generality in assuming that $E = X$, and hence that $\mu(X) < +\infty$, since we may simply replace X by the subspace E. For an arbitrary pair m, n of positive integers set

$$E_{m,n} = E(|f_n - f| < 1/m).$$

Since f_n and f are measurable functions (Theorem 2.33), the set $E_{m,n}$ is measurable, and so therefore are the sets

$$G_{m,n} = \bigcap_{k=n}^{\infty} E_{m,k} \quad \text{and} \quad F_{m,n} = X \setminus G_{m,n}.$$

The set $G_{m,n}$ is the set of those points x in X such that $|f_k(x) - f(x)| < 1/m$ for all $k \geq n$. Thus for each fixed index m, the sequence $\{G_{m,n}\}_{n=1}^{\infty}$ is monotone increasing with union X, and therefore $\lim_n \mu(G_{m,n}) = \mu(X)$ and $\lim_n \mu(F_{m,n}) = 0$ (Propositions 3.27 and 3.28). Hence for each m there is an index $n = n(m)$ such that

$$\mu(F_{m,n(m)}) < \varepsilon/2^m.$$

Set now $F = \bigcup_{m=1}^{\infty} F_{m,n(m)}$, so that

$$\mu(F) \le \sum_{m=1}^{\infty} \mu(F_{m,n(m)}) < \varepsilon.$$

Finally, observe that $|f_n - f| < 1/m$ on $G_{m,n(m)}$ for all $n \ge n(m)$, so that $\{f_n\} \to f$ uniformly on $X \setminus F = \bigcap_{m=1}^{\infty} G_{m,n(m)}$. $\qquad\square$

We also need the following version of Egorov's theorem. (Both versions of Egorov's theorem are special cases of a more general result; see Problem 3M.)

Proposition 3.36. *Let $(X, \mathbf{S}, \mu)$ be a measure space, let E be a measurable subset of X having finite measure, and let $\{f_n\}$ be a sequence of real-valued functions defined, measurable, and converging pointwise to $+\infty$ on E. Then for any positive number ε there exists a subset F of E such that $\mu(F) < \varepsilon$ and such that $\{f_n\}$ converges* uniformly *to $+\infty$ on $E \setminus F$; that is, such that for any positive real number M there is an index N such that $f_n \ge M$ on $E \setminus F$ for all $n \ge N$.*

Proof. Simply apply Proposition 3.35 to the sequence $\{1/(f_n \vee 1)\}$. $\qquad\square$

We are ready to state and prove the fundamental existence theorem of integration theory.

Theorem 3.37. *Let $(X, \mathbf{S}, \mu)$ be a measure space. Then there exists a unique real Lebesgue integral on X, as well as a unique complex Lebesgue integral on X with associated measure equal to μ. Consequently, in view of Theorem 3.29, given a measurable space $(X, \mathbf{S})$, there is a bijective mapping $\mu \leftrightarrow L$ between the set of all measures μ on $(X, \mathbf{S})$ and the set of all real [complex] Lebesgue integrals $(\mathcal{L}, L)$ on $(X, \mathbf{S})$ such that corresponding pairs (μ, L) satisfy (3.7).*

Proof. A Lebesgue integral consists of a linear manifold of functions together with a linear functional defined thereon. Thus, to prove the first statement of the theorem, it is necessary both to construct the linear manifold $\mathcal{L}$ of integrable functions on X and to define the integral $\int_X f \, d\mu$ of a function $f \in \mathcal{L}$. Uniqueness was already established in Theorem 3.29. We consider only the complex case since it subsumes the real case (recall Example 3.13).

Denote by $\mathcal{L}_0$ denote the collection of all simple functions s with the property that s assumes each of its nonzero values on a set of finite measure. A measurable simple function s belongs to $\mathcal{L}_0$ if and only if its support $N_s = E(s \ne 0)$ has finite measure. Since $N_s = N_{|s|}$, this observation makes it plain that if $s \in \mathcal{L}_0$, and if t is any measurable simple function on X such that $|t| \le |s|$, then t also belongs to $\mathcal{L}_0$. Likewise, it is clear that if s and t belong to $\mathcal{L}_0$, and if $\alpha, \beta \in \mathbb{C}$, then $\alpha s + \beta t \in \mathcal{L}_0$ so that $\mathcal{L}_0$ is a linear manifold of functions. Given a function $s \in \mathcal{L}_0$ written as

$$s = \sum_{i=1}^{n} \alpha_i \chi_{E_i}, \tag{3.8}$$

where $\mu(E_i) < +\infty, i = 1, \ldots, n$, we define the integral of s with respect to μ to be

$$\int_X s \, d\mu = \sum_{i=1}^{n} \alpha_i \mu(E_i). \tag{3.9}$$

The reader will verify with little difficulty that this number depends only on s, and not on the representation (3.8) (see Problem 3P). A moment's reflection discloses that the mapping $s \mapsto \int_X s \, d\mu$ is a linear functional on $\mathcal{L}_0$ that is self-conjugate and is real-valued and monotone increasing on the real-valued functions in $\mathcal{L}_0$. Likewise, if $s \in \mathcal{L}_0$, then $|\int_X s \, d\mu| \leq \int_X |s| \, d\mu$ (Problem 3P). Similarly, if $s \in \mathcal{L}_0$ and if E is a measurable set, then $\chi_E s$ also belongs to $\mathcal{L}_0$. In keeping with the notation introduced earlier we write $\int_E s \, d\mu$ for $\int_X \chi_E s \, d\mu$.

Next we define the space $\mathcal{L}$. A measurable function $f : X \to \mathbb{C}$ belongs to $\mathcal{L}$ if and only if there exists a sequence $\{s_n\} \subset \mathcal{L}_0$ which converges pointwise to f and $\lim_{m,n} \int_X |s_m - s_n| d\mu = 0$. It is routine to verify that the collection $\mathcal{L}$ is a vector space and $\mathcal{L}_0 \subset \mathcal{L}$. Assume that $f \in \mathcal{L}$ is the pointwise limit of the sequence $\{s_n\}$ such that $\lim_{m,n} \int_X |s_m - s_n| d\mu = 0$. Since

$$\left| \int_X s_n \, d\mu - \int_X s_m \, d\mu \right| \leq \int_X |s_n - s_n| d\mu,$$

the limit $\lim_n \int_X s_n \, d\mu$ exists, and we wish to set

$$\int_X f \, d\mu = \lim_n \int_X s_n \, d\mu. \tag{3.10}$$

We must therefore prove that the limit in (3.10) depends only on f and not on the particular sequence $\{s_n\}$. Let $\{t_n\} \subset \mathcal{L}_0$ be another sequence converging pointwise to f and satisfying the condition $\lim_{m,n} \int_X |t_m - t_n| d\mu = 0$, and define $u_n = s_n - t_n \in \mathcal{L}_0$. The sequence u_n converges pointwise to zero, and $\lim_{m,n} \int_X |u_m - u_n| d\mu = 0$. We show that this implies

$$\lim_n \int_X u_n \, d\mu = 0, \tag{3.11}$$

so that, indeed, $\lim_n \int_X s_n \, d\mu = \lim_n \int_X t_n \, d\mu$. To verify this statement, fix $\varepsilon > 0$, and choose n_0 such that $\int_X |u_n - u_m| d\mu < \varepsilon$ when $m, n \geq n_0$. Denote by $E \in \mathbf{S}$ the support of u_{n_0}, and note that $\mu(E) < +\infty$ and, since $u_{n_0} = 0$ on $X \setminus E$,

$$\int_{X \setminus E} |u_n| d\mu = \int_{X \setminus E} |u_n - u_{n_0}| d\mu \leq \int_X |u_n - u_{n_0}| d\mu < \varepsilon$$

when $n \geq n_0$. Set now $M = \max_x |s_{n_0}(x)|$, and apply Egorov's theorem on the subspace $E \subset X$ to find a measurable set $F \subset E$ satisfying $\mu(F) < \varepsilon/M$

and a number $n_1 > n_0$ such that $|u_n(x)| < \varepsilon/\mu(E)$ for $n \geq n_1$ and $x \in E \setminus F$. We have then

$$\int_X |u_n| d\mu = \int_{X \setminus E} |u_n| d\mu + \int_{E \setminus F} |u_n| d\mu + \int_F |u_n| d\mu$$

$$< \varepsilon + \frac{\varepsilon}{\mu(E)} \mu(E \setminus F) + \int_F |u_{n_0}| d\mu + \int_F |u_n - u_{n_0}| d\mu$$

$$\leq 4\varepsilon$$

provided that $n \geq n_1$. Since ε is arbitrary, we conclude that (3.11) is true, and hence the definition (3.10) makes sense.

It is immediate that the map $f \mapsto \int_X f \, d\mu$ is a linear functional on $\mathcal{L}$ which extends the notion of integration already defined on $\mathcal{L}_0$. Other direct consequences of the definitions are the facts that $\mathcal{L}$ is self-conjugate and that the functional $f \mapsto \int_X f \, d\mu$ is self-conjugate on $\mathcal{L}$ and order preserving on the real-valued functions in $\mathcal{L}$. Moreover if $f \in \mathcal{L}$, then $|f| \in \mathcal{L}$ and

$$\left| \int_X f \, d\mu \right| \leq \int_X |f| \, d\mu.$$

(If $\{s_n\}$ is a sequence in $\mathcal{L}_0$ that is Cauchy in the mean, then so is $\{|s_n|\}$, and if $\{s_n\}$ converges pointwise to f, then $\{|s_n|\}$ converges pointwise to $|f|$.) Finally, if $f \in \mathcal{L}$ and if E is a measurable set, then $\chi_E f \in \mathcal{L}$ as well. (If $\{s_n\}$ is a sequence in $\mathcal{L}_0$ that is Cauchy in the mean, then $\{\chi_E s_n\}$ is also Cauchy in the mean; in connection with these preliminary remarks see Problem 3Q.) Here too we write $\int_E f \, d\mu$ for $\int_X \chi_E f \, d\mu$.

It remains to verify the axioms defining a Lebesgue integral. As regards (L_1), in view of what has just been said, we need only verify it in the following simplified form: If f is a nonnegative function in $\mathcal{L}$ and if g is a measurable function on X such that $0 \leq g \leq f$, then $g \in \mathcal{L}$ and

$$0 \leq \int_X g \, d\mu \leq \int_X f \, d\mu$$

(see Problem 3A). Suppose, to begin with, that g is the characteristic function of some measurable set E. In this case what we need to show is that $\mu(E) \leq \int_X f \, d\mu$. To start with, if F is any measurable subset of E of finite measure, then $\chi_F \leq \chi_E \leq f$ so $\mu(F) \leq \int_X f \, d\mu$ (since in this case $\chi_F \in \mathcal{L}_0 \subset \mathcal{L}$). Moreover, since f is the pointwise limit of a sequence $\{s_n\}$ of simple functions each of which vanishes outside some set of finite measure, it is easy to see that there is an increasing sequence $\{F_n\}$ of measurable sets of finite measure such that $\bigcup_n F_n = E$. But then $\mu(F_n) \to \mu(E)$, and since $\mu(F_n) \leq \int_X f \, d\mu$ for each n, the result follows. (See Proposition 3.28. Note, in particular, that $\mu(E)$ is finite whenever $\chi_E \in \mathcal{L}$; the essence of this argument is explored in Problem 3R, to which the reader is referred.)

Suppose next that $g = s$ is a measurable simple function, let a be any nonzero value assumed by s, and let $E = E(s = a)$. Then $a\chi_E \leq \chi_E f$, and it follows at once from what has just been shown that $a\mu(E) \leq \int_E f\, d\mu$. But then, if $a_1, \ldots, a_n$ is an enumeration of the distinct nonzero values of s, and if $E_i = E(s = a_i), i = 1, \ldots, n$, and $E = E_1 \cup \cdots \cup E_n$, we have

$$\int_X s\, d\mu = \sum_{i=1}^n a_i \mu(E_i) \leq \sum_{i=1}^n \int_{E_i} f\, d\mu = \int_E f\, d\mu \leq \int_X f\, d\mu,$$

and this settles the case of a measurable simple function g.

Suppose finally that g is merely measurable ($0 \leq g \leq f$), and let $\{s_n\}$ be a monotone increasing sequence of measurable simple functions converging pointwise to g (Proposition 2.36). Then, as we have just seen, each function s_n is in $\mathcal{L}_0$ and

$$0 \leq \int_X s_1\, d\mu \leq \ldots \leq \int_X s_n\, d\mu \leq \ldots \leq \int_X f\, d\mu.$$

Hence the numerical sequence $\{\int_X s_n\, d\mu\}$ is convergent, and thus Cauchy. But this implies that $\{s_n\}$ is Cauchy in the mean, since

$$\int_X |s_m - s_n|\, d\mu = |\int_X (s_m - s_n)\, d\mu|$$

by virtue of the monotonicity. Thus, by definition, $g \in \mathcal{L}$ and

$$\int_X g\, d\mu = \lim_n \int_X s_n\, d\mu \leq \int_X f\, d\mu,$$

so that (L$_1$) is proved.

The fact that integration on $\mathcal{L}$ is a self-conjugate functional shows that it suffices to verify (L$_2$) when the functions f_n and g are real-valued. The inequality

$$\int_X |f_n^+ - f_m^+|d\mu \leq \int_X |f_n - f_m|d\mu$$

allows us to further restrict the argument to nonnegative functions f_n and g. Assume therefore that $\{f_n\} \subset \mathcal{L}$ is sequence of nonnegative functions satisfying conditions (L$_2$a) and (L$_2$b) with f_n in place of f and $\int$ in place of L. For each n, there exists a sequence $\{s_{n,k}\}_{k=1}^\infty$ of simple functions satisfying the conditions of 2.36. We have $s_{n,k} \leq s_{n,k+1}$ and therefore

$$0 \leq \int_X s_{n,k}\, d\mu \leq \int_X s_{n,k+1}\, d\mu \leq \int_X f\, d\mu.$$

This implies that the sequence $\left\{\int_X s_{n,k}\,d\mu\right\}_k$ is Cauchy, and since

$$\int_X |s_{n,\ell} - s_{n,k}|d\mu = \left| \int_x s_{n,\ell}\,d\mu - \int_X s_{n,k}\,d\mu \right|,$$

we conclude that $\lim_{k,\ell} \int_X |s_{n,\ell} - s_{n,k}|d\mu = 0$. The definition of the integral implies the existence for each $n \in \mathbb{N}$ of an integer $k_n \geq n$ such that

$$\int_X |f_n - s_{n,k_n}|\,d\mu = \int_X f_n\,d\mu - \int_X s_{n,k_n}\,d\mu < 1/n.$$

We set $s_n = s_{n,k_n}, n \in \mathbb{N}$. Then $\{s_n\} \subset \mathcal{L}_0$, and since

$$\int_X |s_m - s_n|\,d\mu < \int_X |f_m - f_n|\,d\mu + 1/m + 1/n,$$

the sequence $\{s_n\}$ is also Cauchy in the mean. Next, let x be a point of X and observe that, since $f_n(x) \to g(x)$, there exists a positive integer M such that $|f_n(x)| \leq M$ for all n. It follows that $|s_{n,k}(x) - f_n(x)| \leq 1/2^k$ for all integers $k \geq M$ and all positive integers n. In particular, if $n \geq M$, then $|s_n(x) - f_n(x)| \leq 1/2^n$. Thus $\{s_n\}$ converges pointwise to g. But then, by definition, the numerical sequence $\{\int_X s_n\,d\mu\}$ tends to $\int_X g\,d\mu$, and it follows at once that the equiconvergent sequence $\{\int_X f_n\,d\mu\}$ does so too ([**I**, Problem 6M]). This completes the proof of the fact that the functional $f \mapsto \int_X f\,d\mu$ on the vector space $\mathcal{L}$ is a Lebesgue integral on X. That the measure associated with this integral agrees with the given measure μ on all sets on which μ is finite follows directly from the above definitions. On the other hand, if $\mu(E) = +\infty$, then, as has already been seen, the function χ_E does not belong to $\mathcal{L}$. This completes the proof of the first statement in the proof of Theorem 3.37.

To prove the second statement, note that it is immediate from Theorem 3.29 that there is a unique mapping $\varphi : L = (\mathcal{L}, L) \mapsto \mu$ from the set of all real [complex] Lebesgue integrals on $(X, \mathbf{S})$ into the set of all measures on $(X, \mathbf{S})$ such that (3.7) holds. Moreover, the proof above of the first part of the theorem yields a mapping $\psi : \mu \mapsto L = (\mathcal{L}, L)$ from the set of all measures on $(X, \mathbf{S})$ into the set of all real [complex] Lebesgue integrals on $(X, \mathbf{S})$ that, in particular, satisfies (3.9). But (3.7) and (3.9) together show that $\psi = \varphi^{-1}$, which proves the second statement of the theorem. □

Corollary 3.38. *The complexification of a real Lebesgue integral on a measurable space $(X, \mathbf{S})$ is a complex Lebesgue integral on $(X, \mathbf{S})$.*

Proof. Let $(X, \mathcal{L}_0, L_0)$ be a real Lebesgue integral on $(X, \mathbf{S})$, let μ be the measure associated with L_0, and let L be the complex Lebesgue integral with associated measure μ. If $\mathcal{L}_{\mathbb{R}}$ denotes the collection of real-valued functions integrable with respect to L, then $L|\mathcal{L}_{\mathbb{R}}$ is another real Lebesgue integral on $(X, \mathbf{S})$, and it is obvious that its associated measure is μ. Uniqueness implies that $L|\mathcal{L}_{\mathbb{R}} = L_0$, and therefore $L = (L|\mathcal{L}_{\mathbb{R}})^+ = L_0^+$. □

Remark 3.39. This corollary shows that every real Lebesgue integral is just the restriction to real-valued functions of a complex Lebesgue integral, namely its own complexification. (For an alternative proof of Corollary 3.38 see Problem 3C.) Thus the theory of real Lebesgue integrals is completely subsumed by the complex case. In the sequel we continue to use the neutral term "Lebesgue integral," according to standing convention, to mean either a real or a complex integral, but henceforth, in giving proofs, the argument will usually be presented in the complex case, since that also covers the real case.

Example 3.40. If L and L' are Lebesgue integrals on the same measurable space $(X, \mathbf{S})$, and if for each measurable nonnegative function f that is integrable with respect to L' it is the case that f is also integrable with respect to L and $L(f) \leq L'(f)$, then it is clear that $\mu \leq \mu'$ setwise on $\mathbf{S}$, where μ and μ' denote the measures associated with L and L', respectively. Conversely, if μ and μ' are two measures on $(X, \mathbf{S})$ and if $\mu \leq \mu'$ setwise on $\mathbf{S}$, then the construction in the proof of Theorem 3.37 shows that every nonnegative measurable function f on X that is integrable $[\mu']$ is also integrable $[\mu]$ and that $\int_X f \, d\mu \leq \int_X f \, d\mu'$. This fact can be used to give another proof that the measure associated with a Lebesgue integral determines it uniquely.

Example 3.41. Let $(X, \mathbf{S})$ be a measurable space, let $\{w_x\}_{x \in X}$ be an indexed family of nonnegative extended real weights on X, and let μ denote the discrete measure on $(X, \mathbf{S})$ associated with this system of weights as in Example 3.20. The Lebesgue integral with respect to μ is once again readily described. A measurable complex-valued function f on X is integrable $[\mu]$ if and only if $\sum_{x \in X} |f(x)| w_x < +\infty$, and, when this is the case,

$$\int_X f \, d\mu = \sum_{x \in X} f(x) w_x.$$

(Indeed, it is an easy consequence of the properties of indexed sums that the functional $f \mapsto \sum_{x \in X} f(x) w_x$ is a Lebesgue integral, and μ is obviously the associated measure.) In particular, if γ is the counting measure on $(X, \mathbf{S})$ (Example 3.21), then integration with respect to γ of a function f simply reduces to the formation of the indexed sum $\sum_{x \in X} f(x)$.

The preceding examples show that Lebesgue integration with respect to a discrete measure reduces to a special case of the general idea of an indexed sum. On the other hand, every one of the measures introduced up to this point has been discrete (see Problem 3T). If that were all there were to it, the theory of the Lebesgue integral would be largely subsumed under the theory of indexed sums and there would be little more to discuss. But that is not the case. The interesting and useful measures are, by and large, *not* discrete. The following example is introduced here to assure the reader that further development of the Lebesgue integral is of greater significance than the elementary cases thus far introduced might suggest.

Example 3.42. There exists a unique measure λ_d on the σ-algebra $\mathbf{B}_{\mathbb{R}^d}$ that agrees with ordinary d-dimensional volume on the closed cells in $\mathbb{R}^d$. In other words, for an arbitrary closed cell $Z = [a_1, b_1] \times \ldots \times [a_d, b_d]$ in $\mathbb{R}^d$,

$$\lambda_d(Z) = (b_1 - a_1) \ldots (b_d - a_d). \tag{3.12}$$

This measure is known as d-*dimensional Lebesgue-Borel measure* (or simply as *Lebesgue-Borel measure* on $\mathbb{R}^d$). (The proof that such a measure exists requires some machinery that we develop later. For the time being, we take care not to base the proof of any proposition on the properties of Lebesgue-Borel measure, but, as will be seen, this presents no difficulty.) Note that if the cell Z in (3.12) is degenerate, that is, one or more of the edges $[a_i, b_i]$ has zero length, then $\lambda_d(Z) = 0$. In particular, λ_d vanishes on every singleton in $\mathbb{R}^d$, so it is neither discrete nor completely additive.

Remark 3.43. The terms *Lebesgue integral* and *Lebesgue integration* are ambiguous in that they are regularly used to refer to two quite different, albeit closely related, concepts. In the beginning, one-dimensional Lebesgue-Borel measure, painstakingly constructed by Borel, was used by Lebesgue to define Lebesgue integration with respect to λ_1 (naturally called *Lebesgue integration*). Soon however, it was observed that Lebesgue's construction worked equally well for integration with respect to d-dimensional Lebesgue-Borel measure, and integration with respect to this measure λ_d accordingly also became known as *Lebesgue integration*. Several years later, it was observed (the honors should probably be shared somehow between Kolmogorov and von Neumann) that the only properties of λ_1 or λ_d that were really needed to derive the useful properties of the Lebesgue integral were those common to all of the set functions we now call measures. Thus was born the notion of integration with respect to an arbitrary measure, and since all of the ideas needed to establish the properties of such an integral go back to Lebesgue himself, all such integrals are appropriately called *Lebesgue integrals* (as we have called them in this book), even though the integrals so designated are with respect to measures quite different from Lebesgue-Borel measure, and (in many cases) were unknown to Lebesgue.

Problems

3A. The space $\mathcal{L}$ of functions integrable with respect to a complex Lebesgue integral on a space X possesses, by definition, the following property:

(Γ) If $f \in \mathcal{L}$ and if g is a measurable complex-valued function on X such that $|g| \leq |f|$, then g also belongs to $\mathcal{L}$.

Show that a linear manifold $\mathcal{K}$ of measurable complex-valued functions on a measurable space X has property (Γ) (with $\mathcal{K}$ replacing $\mathcal{L}$) if and only if it possesses both of the following properties:}

(Γ') If f is a nonnegative function belonging to $\mathcal{K}$ and if g is measurable and satisfies $0 \leq g \leq f$, then g belongs to $\mathcal{K}$,

(Γ'') If f belongs to $\mathcal{K}$, so does $|f|$.

State and prove the analog of this result for real linear manifolds of measurable real-valued functions on X.

3B. Let X be a measurable space, let $\mathcal{K}$ be a linear manifold of measurable complex-valued functions on X possessing property (Γ) of Problem 3A, and let K be a linear functional on $\mathcal{K}$ such that $K(f) \geq 0$ whenever $f \geq 0$. (Such a functional is said to be *positive* (see [**I**, Problem 3I]); note that (Γ) ensures that $\mathcal{K}$ is self-conjugate.)

   (i) Verify that for any positive real number p the set $\mathcal{K}_p$ of all those measurable complex-valued functions f on X with the property that $|f|^p$ belongs to $\mathcal{K}$ is a vector space. (Hint: if u and v are any two nonnegative real numbers, then $(u+v)^p \leq 2^p(u^p \vee v^p)$.)

  (ii) Show that if f and g are any two functions belonging to the vector space $\mathcal{K}_2$, then the product $f\overline{g}$ belongs to $\mathcal{K}$. Verify also that

$$\psi(f,g) = K(f\overline{g}), \quad f, g \in \mathcal{K}_2,$$

defines a positive sesquilinear functional on $\mathcal{K}_2$ (see [**I**, Chapter 3] for definitions), and conclude that

$$|K(f\overline{g})|^2 \leq K(|f|^2)K(|g|^2)$$

for any two functions $f, g \in \mathcal{K}_2$ ([**I**, Problem 3Q]).

 (iii) Let u and v be real-valued functions belonging to $\mathcal{K}$. Prove that the functions $\sqrt{u^2 + v^2}, u^2/\sqrt{u^2 + v^2}$ and $v^2/\sqrt{u^2 + v^2}$ also belong to $\mathcal{K}$, where the latter two functions are to be given the value zero at those points of X where u and v both vanish. Conclude that

$$K(u)^2 + K(v)^2 \leq K\left(\sqrt{u^2 + v^2}\right)^2.$$

3C. (i) Let $\mathcal{K}$ be a complex vector space of measurable complex-valued functions on a measurable space X, and let K be a linear functional defined on $\mathcal{K}$. Prove that K satisfies axiom (L_1) in the definition of a complex Lebesgue integral on X if and only if $\mathcal{K}$ has property (Γ) (Problem 3A) and K is positive on $\mathcal{K}$.

  (ii) Show, analogously, that if $\mathcal{K}$ is a real vector space of measurable real-valued functions on X, and K is a linear functional on $\mathcal{K}$, then K satisfies axiom (L_1) in the definition of a real Lebesgue integral on X if and only if K is positive on $\mathcal{K}$ and $\mathcal{K}$ possesses the real analog of property (Γ) obtained by replacing "complex-valued" by "real-valued."

 (iii) Let L be a real Lebesgue integral on X, and for each pair u, v of functions integrable with respect to L set

$$L^+(u + iv) = L(u) + iL(v).$$

Verify that L^+ is a complex Lebesgue integral on X. (Thus, in brief, the complex-ification of a real Lebesgue integral on X is a complex Lebesgue integral on X.)

3D. Suppose given an indexed family $\{L_\gamma\}_{\gamma \in \Gamma}$ of complex Lebesgue integrals, all on the same measurable space $(X, \mathbf{S})$. Each integral L_γ has its own space $\mathcal{L}_\gamma$ of integrable functions. Let $\mathcal{L}$ denote the collection of all those complex-valued functions f on X in $\bigcap_{\gamma \in \Gamma} \mathcal{L}_\gamma$ with the property that $\sum_{\gamma \in \Gamma} L_\gamma(|f|) < +\infty$, and for each f in $\mathcal{L}$ set

$$L(f) = \sum_{\gamma \in \Gamma} L_\gamma(f).$$

(Since $|L_\gamma(f)| \le L_\gamma(|f|)$ for each f in $\mathcal{L}$ and each $\gamma \in \Gamma$, the indexed family $\{L_\gamma(f)\}_{\gamma \in \Gamma}$ is summable; see Example 3.10.) Verify that $\mathcal{L}$ is a vector space of measurable complex-valued functions on X and that L is a Lebesgue integral on X. (Hint: The only difficulty lies in verifying axiom (L2). If $\{f_k\}$ is a sequence in $\mathcal{L}$ that converges pointwise on X to a function g, and if $\{f_k\}$ is Cauchy in the mean with respect to L, then $\{f_k\}$ is also Cauchy in the mean with respect to each L_γ, so g belongs to $\mathcal{L}_\gamma$ for each γ and therefore belongs to $\bigcap_\gamma \mathcal{L}_\gamma$. Follow the method used in Example 3.10.)

3E. Show that if $(X, \mathcal{L}, 0)$ is a Lebesgue integral, then the space $\mathcal{L}$ of integrable functions is closed with respect to the formation of pointwise limits. Conclude that every zero Lebesgue integral is one of those introduced in Example 3.4. (Hint: If $f \in \mathcal{L}$, then $\chi_{N_f} \in \mathcal{L}$. This result can also be obtained via Theorem 3.29; see Example 3.30.)

3F. Let Γ be an index set and let $\{\lambda_\gamma\}_{\gamma \in \Gamma}$ be a summable indexed family of complex numbers.

  (i) Prove that $\lambda_\gamma = 0$ except for some countable subcollection of indices γ (that may vary with the family $\{\lambda_\gamma\}$).
 (ii) Show that $\{\lambda_\gamma\}_{\gamma \in \Gamma_0}$ is also summable for any subcollection Γ_0 of Γ.
(iii) Conclude that if $\mathcal{P}$ is an arbitrary partition of Γ, then

$$(*) \qquad \sum_{\gamma \in \Gamma} \lambda_\gamma = \sum_{E \in \mathcal{P}} \sum_{\gamma \in E} \lambda_\gamma.$$

Give an example of an indexed family and a partition of the index set such that the right side of $(*)$ is defined but the family is not summable.
 (iv) Show that if $\{u_\gamma\}_{\gamma \in \Gamma}$ is an indexed family of nonnegative extended real numbers, then

$$\sum_{\gamma \in \Gamma} u_\gamma = \sum_{E \in \mathcal{P}} \sum_{\gamma \in E} u_\gamma$$

holds generally for any partition $\mathcal{P}$ of Γ.
  (v) Discuss the extension of the foregoing results to the case of indexed families $\{t_\gamma\}_{\gamma \in \Gamma}$ of extended real numbers (positive or negative). (Hint: Treat first the case of a family $\{t_\gamma\}$ of ordinary finite real numbers. If such a family fails to satisfy the Cauchy criterion of Example 3.10, then either (1) there exists a positive number ε_0 such that for every finite subset Δ_0 of Γ there is a finite set Δ disjoint from Δ_0 such that $s_\Delta > \varepsilon_0$ (this is the case $\sum_\gamma t_\gamma^+ = +\infty$), or (2) there exists a positive number ε_0 such that for every Δ_0 there is a Δ disjoint from Δ_0 such that $s_\Delta < -\varepsilon_0$ (this is the case $\sum_\gamma t_\gamma^- = +\infty$). Consider all possibilities.)
 (vi) Verify that

$$\sum_{\gamma \in \Gamma} t_\gamma = \sum_{\gamma \in \Gamma} t_\gamma^+ - \sum_{\gamma \in \Gamma} t_\gamma^-$$

holds for an indexed family of finite real numbers when (and only when) the right side is defined. How are matters altered if the summands t_γ are allowed to assume infinite values? (For the purposes of this exercise let us agree to write $+\infty^+ = -\infty^- = +\infty$ and $+\infty^- = -\infty^+ = 0$.)

3G. Let X be a measurable space and let K be a linear functional defined on a linear manifold $\mathcal{K}$ of measurable complex-valued functions on X such that K satisfies axiom (L_1) in the definition of a Lebesgue integral. Verify that (3.7), with L and $\mathcal{L}$ replaced by K and $\mathcal{K}$, respectively, defines a finitely additive set function on the σ-algebra of measurable sets in X.

3H. Let $\mathcal{C}$ be a collection of subsets of a space X such that $\varnothing \in \mathcal{C}$, and let φ be a set function defined on $\mathcal{C}$ such that $\varphi(A \cup B) = \varphi(A) + \varphi(B)$ whenever A, B and $A \cup B$ belong to $\mathcal{C}$. (This condition is slightly weaker than finite additivity as introduced in Definition 3.16.) Show that $\varphi(\varnothing)$ must be either 0 or $\pm\infty$. In particular, $\varphi(\varnothing) = 0$ if φ is complex-valued.

3I. Propositions 3.27 and 3.28 are also valid for complex-valued set functions in so far as they make sense. Thus a finitely additive complex-valued set function defined on a ring of sets is subtractive. Likewise, a finitely additive complex-valued set function φ defined on the σ-algebra $\mathbf{S}$ of measurable subsets of a measurable space $(X, \mathbf{S})$ is countably additive if and only if φ is *semicontinuous* in the sense that $\varphi\left(\bigcup_n E_n\right) = \lim_n \varphi(E_n)$ for every monotone increasing sequence of sets $\{E_n\}$ in $\mathbf{S}$.

3J. Let μ be a measure on a measurable space $(X, \mathbf{S})$.

(i) It is shown in the text (Proposition 3.28) that if $\{E_n\}_{n=1}^{\infty}$ is an arbitrary increasing sequence of measurable sets in X, then the numerical sequence $\{\mu(E_n)\}$ tends upward to $\mu\left(\bigcup_n E_n\right)$. Show by example, on the other hand, that if $\{E_n\}_{n=1}^{\infty}$ is a decreasing sequence of measurable sets in X, then, while the sequence $\{\mu(E_n)\}$ is also decreasing, it need not tend to $\mu(\bigcap_n E_n)$. Show, however, that we *do* have $\mu\left(\bigcap_n E_n\right) = \lim_n \mu(E_n)$ for a decreasing sequence $\{E_n\}$ of measurable sets provided $\mu(E_n) < +\infty$ for at least one value of n.

(ii) Recall that the *limit superior* and *limit inferior* of an arbitrary sequence of subsets of a set X are defined by the formulas

$$\limsup_n E_n = \bigcap_{n=1}^{\infty} \bigcup_{k=n}^{\infty} E_k \text{ and } \liminf_n E_n = \bigcup_{n=1}^{\infty} \bigcap_{k=n}^{\infty} E_k.$$

(See [**I**, Example 1U] for these definitions and also for alternative characterizations.) Suppose $\{E_n\}$ is a sequence of measurable subsets of X. Prove that

$$\mu(\liminf_n E_n) \le \liminf_n \mu(E_n)$$

in general, and also that

$$\mu(\limsup_n E_n) \ge \limsup_n \mu(E_n)$$

in the event that some one of the sets $\bigcup_{k=n}^{\infty} E_k$ has finite measure. Can this latter condition be replaced by the weaker assumption that some one of the sets E_n has finite measure?

(iii) Suppose the sequence $\{E_n\}$ in (ii) is convergent, that is, that $\limsup_n E_n = \liminf_n E_n (= \lim_n E_n)$. Give a condition ensuring that

$$\mu(\lim_n E_n) = \lim_n \mu(E_n).$$

Give also an example of a convergent sequence of measurable sets for which this equation fails.

3K. A bounded measurable complex-valued function f on a measure space $(X, \mathbf{S}, \mu)$ is integrable $[\mu]$ if its support $N_f = E(f \neq 0)$ has finite measure. Moreover, if f is bounded away from zero on N_f, then this sufficient condition is also necessary. (Thus, for example, a measurable simple function s on X is integrable $[\mu]$ if and only if $\mu(N_s) < +\infty$.) Is the condition necessary in general?

3L. Let $(X, \mathbf{S}, \mu)$ be a measure space, and let f be a complex-valued function defined on an arbitrary subset of X. Show that if f is integrable $[\mu]$ over some measurable subset E of X, and if F is a measurable subset of E, then f is also integrable $[\mu]$ over F. Let $\mathbf{J}$ denote the collection of those measurable subsets of X over which f is integrable $[\mu]$. Conclude that $\mathbf{J}$ is an ideal in $\mathbf{S}$ (and therefore a ring) and verify that if we set $\nu(E) = \int_E f \, d\mu$ for every set E in $\mathbf{J}$, then ν is a finitely additive set function on $\mathbf{J}$.

3M. Let $(X, \mathbf{S}, \mu)$ be a finite measure space, let (Y, ρ) be a separable metric space, and let $\{\varphi_n\}_{n=1}^\infty$ be a sequence of measurable mappings of X into Y that converges pointwise on X to a mapping φ. Show that for an arbitrarily given positive number ε there exists a measurable set F in X such that $\mu(F) < \varepsilon$ and such that $\{\varphi_n\}$ converges to φ uniformly on $X \setminus F$. (Hint: See Problems 2F and 2H.) Show by example that the hypothesis $\mu(X) < +\infty$ cannot be omitted.

3N. (Problems 3N and 3O anticipate results from Chapter 4. The relevant techniques come from the proof of Theorem 3.37.) Let $(X, \mathbf{S}, \mu)$ be a measure space, and let $f : X \to \mathbb{C}$ be integrable $[\mu]$.

(i) Show that
$$\mu(\{x : |f(x)| \geq \alpha\}) \leq \frac{1}{\alpha} \int_X |f| \, d\mu$$
for every $\alpha > 0$. (Hint: $\alpha \chi_{E(|f| \geq \alpha)} \leq |f|$.)

(ii) Show that $\lim_{\alpha \to \infty} \int_{\{x : |f(x)| > \alpha\}} |f| \, d\mu = 0$. (Hint: Use (L2).)

3O. Let $(X, \mathbf{S}, \mu)$ be a measure space, and let $\{f_n\}$ be a sequence of integrable functions which is Cauchy in the mean.

(i) Given $\varepsilon > 0$, show that there exists a set $E \in \mathbf{S}$ such that $\mu(E) < \infty$ and $\int_{X \setminus E} |f_n| \, d\mu < \varepsilon$ for all $n \in \mathbb{N}$. (Hint: Adapt the argument following (3.11).)

(ii) Given $\varepsilon > 0$, show that there exists $\delta > 0$ such that $\mu(A) < \delta$ implies $\int_A |f_n| \, d\mu < \varepsilon$ for every $n \in \mathbb{N}$. (Hint: Use the Cauchy property to reduce to the case of a finite set $f_1, \ldots, f_n$, then use Problem 3N to further reduce to the case where these functions are bounded.)

3P. Let $(X, \mathbf{S}, \mu)$ be a measure space, and let s be a measurable simple complex-valued function on X. Let us call a representation of s of the form
$$(\dagger) \qquad s = \sum_{i=1}^n \alpha_i \chi_{E_i}$$

standard if the sets $E_i, i = 1, \ldots, n$, are all measurable and of finite measure and are also pairwise disjoint. Show that s admits a standard representation if and only if its support has finite measure, that is, if and only if s belongs to the vector space

$\mathcal{L}_0$ introduced in the proof of Theorem 3.37. Show also that if $s = \sum_{k=1}^{m} \beta_k \chi_{F_k}$ is another standard representation of s along with (†), then

$$\sum_{i=1}^{n} \alpha_i \mu(E_i) = \sum_{k=1}^{m} \beta_k \mu(F_k),$$

so that the number $\sum_{i=1}^{n} \alpha_i \mu(E_i)$ associated with s by means of a standard representation (†) depends only on s and not on the representation. Use this fact to prove that the functional thus defined on $\mathcal{L}_0$ is linear, and hence coincides with the functional $s \mapsto \int_X s\, d\mu$ introduced in the proof of Theorem 3.37. Show also that this functional is self-conjugate and that

$$\left| \int_X s\, d\mu \right| \leq \int_X |s|\, d\mu$$

for any $s \in \mathcal{L}_0$.

3Q. Let $(X, \mathbf{S}, \mu)$ be a measure space, and consider the functional $f \to \int_X f\, d\mu$ defined on the vector space $\mathcal{L}$ in the proof of Theorem 3.37. Show directly from the definition of this linear functional that it is self-conjugate and that if $f \in \mathcal{L}$, then $|f| \in \mathcal{L}$ and

$$\left| \int_X f\, d\mu \right| \leq \int_X |f|\, d\mu.$$

Show also that if $f \in \mathcal{L}$ and if $E \in \mathbf{S}$, then $f\chi_E \in \mathcal{L}$. Verify that if we write $\int_E f\, d\mu$ for $\int_X f\chi_E\, d\mu$, then the set function $\nu_f(E) = \int_E f\, d\mu$ is finitely additive on $\mathbf{S}$.

3R. If $(X, \mathbf{S}, \mu)$ is a measure space, then a measurable set E in X is said to be σ-*finite* with respect to μ if there exists a countable sequence $\{E_n\}$ of sets of finite measure with respect to μ such that $E = \bigcup_n E_n$. The collection of all sets in $\mathbf{S}$ that are σ-finite with respect to μ is a σ-ideal in $\mathbf{S}$ (Problem 1W). So is the collection $\mathbf{Z}$ of all sets of measure zero with respect to μ. (The sets of finite measure constitute an ideal in $\mathbf{S}$ between these two σ-ideals.)

(i) Verify that if f is a function that is integrable $[\mu]$, then the support of f is σ-finite with respect to μ.
(ii) Assume that E is a σ-finite set in $\mathbf{S}$ such that $\mu(F) \leq a$ for every measurable subset F of E having finite measure. Show that $\mu(E) \leq a$. Show by example that the assumption that E be σ-finite cannot be omitted.
(iii) Let f be a complex-valued function defined and measurable on X with σ-finite support. Show that the approximating functions in Problem 2J can be taken to be integrable $[\mu]$.

3S. Let $(X, \mathbf{S}, \mu)$ be a measure space such that X is σ-finite. Show that there exists a finite measure ν on $(X, \mathbf{S})$ such that $\{E \in \mathbf{S} : \mu(E) = 0\} = \{E \in \mathbf{S} : \nu(E) = 0\}$.

3T. Verify that the measures associated with the Lebesgue integrals introduced in Examples 3.4, 3.7, and 3.8 are discrete, by calculating in each case an appropriate weight for each point of X. (Prove also that these weights are uniquely determined if the σ-algebra $\mathbf{S}$ contains all singletons in X.) Show, in the same vein, that if the measures associated with the integrals L_γ in Problem 3D are all discrete, then the same is true of the integral L of that problem. How are matters changed if, in Example 3.8, we drop the assumption that the points of the sequence $\{x_n\}$ are all distinct?

3U. Let $(X, \mathbf{S}, \mu)$ be a measure space, and let f be a nonnegative function defined and measurable on X and having σ-finite support. Consider the supremum M of the set of integrals $\int_X s \, d\mu$ of all integrable simple functions s such that $0 \le s \le f$. Show that f is integrable $[\mu]$ if and only if $M < +\infty$, and show also that, if this is the case, then $M = \int_X f \, d\mu$. (This fact can be exploited to obtain an alternative construction of the Lebesgue integral with respect to μ. That the requirement of σ-finiteness of the support of f is really needed in this problem may be seen by considering the function $f = 1$ on the measure spaces in Example 3.30.)

Remark 3.44. These last results show clearly that if f is a nonnegative measurable function on a measure space $(X, \mathbf{S}, \mu)$, then the only way for f *not* to be integrable $[\mu]$ is for f to be too large. Indeed, if the support of f is not σ-finite, then f is too large to be integrable $[\mu]$ (Problem 3R). On the other hand, if f has σ-finite support, then there exists a monotone increasing sequence $\{f_n\}$ of nonnegative integrable functions that converges pointwise to f, and either the sequence $\{\int_X f_n \, d\mu\}$ is bounded, in which case f is integrable, or $\lim_n \int_X f_n \, d\mu = +\infty$. For this reason it is customary to define

$$\int_X f \, d\mu = +\infty$$

whenever f is nonnegative and measurable on X but not integrable over X. It should be noted that the custom of writing $\sum_{\gamma \in \Gamma} t_\gamma = +\infty$ when $\{t_\gamma\}_{\gamma \in \Gamma}$ is an indexed family of nonnegative real numbers that is not summable (see Example 3.10) is a special case of this convention.

3V. Let $(X, \mathbf{S}, \mu)$ be a finite measure space, and let f be a measurable real-valued function defined on X.

(i) Suppose first that f is bounded and that the range of f is contained in the half-open interval $(a, b]$. Let

$$\{a = t_0 < \cdots < t_n = b\}$$

be a partition of $[a, b]$, and for $i = 1, \ldots, n$, let $E_i = E(t_{i-1} < f \le t_i)$. Let ε be a positive number and suppose that $t_i - t_{i-1} \le \varepsilon$ for all i. Show that

$$0 \le \sum_{i=1}^{n} t_i \mu(E_i) - \int_X f \, d\mu \le \varepsilon \mu(X).$$

(ii) In the general case fix $\varepsilon > 0$ and let $\{t_n\}_{n=-\infty}^{+\infty}$ be a two-way infinite sequence of real numbers such that $0 \le t_n - t_{n-1} \le \varepsilon$ for every integer n and such that $t_n \to \pm\infty$ as $n \to \pm\infty$. Then f is integrable $[\mu]$ if and only if the series

$$\sum_{n=-\infty}^{+\infty} t_n \mu(E_n)$$

is absolutely convergent, and, if this is the case, then

$$0 \le \sum_{n=-\infty}^{+\infty} t_n \mu(E_n) - \int_X f \, d\mu \le \varepsilon \mu(X).$$

Remark 3.45. These observations can also be made the basis for an alternative construction of the Lebesgue integral with respect to μ when $\mu(X) < +\infty$.}

3W. Let X be a set and let $\mathbf{S}_0$ denote the σ-ring generated by X and the singletons in X, that is, the σ-algebra consisting of all countable subsets of X and their complements.

Let γ denote the counting measure on the measurable space $(X, 2^X)$, and let γ_0 be the restriction of γ to $\mathbf{S}_0$. Show that the functions integrable $[\gamma_0]$ over X are precisely the functions integrable $[\gamma]$. (See Example 3.32. The like observation may also be made, of course, for any σ-algebra between $\mathbf{S}_0$ and $\mathbf{S}$.)

Remark 3.46. The point of this problem is that *measurability* is not a matter of consequence in the context of *discrete* measures. The integration theory for a discrete measure is completely independent of the σ-algebra $\mathbf{S}$ of measurable sets provided $\mathbf{S}$ contains all singletons. It is appropriate to recall, however, that a Lebesgue integral is a triple $(X, \mathcal{L}, L)$, where X is a measurable space, which presupposes, of course, that *some* particular σ-algebra of measurable sets has been specified.

3X. (This problem and the following one refer to Lebesgue-Borel measure in Euclidean space; see Example 3.42.) Prove that if E is a Borel set in $\mathbb{R}^d$ and if ε is a positive number, then there exists an open set U in $\mathbb{R}^d$ such that $E \subset U$ and such that $\lambda_d(U \setminus E) < \varepsilon$. Use this fact to show that for any Borel set E in $\mathbb{R}^d$ there is a decreasing sequence $\{U_m\}$ of open sets containing E and having the property that $\lim_m \lambda_d(U_m \setminus E) = 0$, as well as an increasing sequence $\{F_m\}$ of closed sets contained in E such that $\lim_m \lambda_d(E \setminus F_m) = 0$. Conclude that there are sets H and K, where H is a G_δ and K an F_σ, such that $K \subset E \subset H$ and such that $\lambda_d(H \setminus E) = \lambda_d(E \setminus K) = 0$. (Hint: If $\mathbf{L}$ denotes the collection of all Borel subsets A of $\mathbb{R}^d$ with the property that for each positive number ε there exists an open set U such that $A \subset U$ and $\lambda(U \setminus A) < \varepsilon$, then $\mathbf{L}$ is a σ-lattice (Problem 1N). Verify directly that every closed set belongs to $\mathbf{L}$.)

3Y. Let $\{\theta_n\}_{n=0}^{\infty}$ be a sequence of numbers such that $0 < \theta_n < 1$ for each n, and let $C_{\{\theta_n\}}$ be the generalized Cantor set associated with this sequence ([**I**, Example 6P]). Show that $\lambda_1(C_{\{\theta_n\}}) > 0$ when and only when $\sum_{n=0}^{\infty} \theta_n < +\infty$. (In particular, the Lebesgue-Borel measure of the Cantor set C itself is zero.) Show too that, while $\mu(C_{\{\theta_n\}})$ is always strictly less than one, if r is any real number such that $r < 1$, then there are sequences $\{\theta_n\}$ for which $r < \mu(C_{\{\theta_n\}}) < 1$. (Hint: Let F_n be the nth closed set defined in the inductive construction of $C_{\{\theta_n\}}$. Direct calculation shows that

$$\lambda_1(F_n) = (1 - \theta_0)\dots(1 - \theta_{n-1}), \quad n \in \mathbb{N}.$$

Hence $\lambda(C_{\{\theta_n\}})$ is given by the infinite product $\prod_{n=0}^{\infty}(1 - \theta_n)$.)

3Z. Consider an algebra $\mathbf{R}$ of subsets of a set X, and a (finite) finitely additive measure $\mu : \mathbf{R} \to [0, +\infty)$. Denote by $\mathcal{L}_\infty$ the vector space of all bounded functions $f : X \to \mathbb{C}$ which are measurable $[\mathbf{R}]$, that is, $f^{-1}(A) \in \mathbf{R}$ for every Borel set $A \subset \mathbb{C}$.

(i) Show that for every $f \in \mathcal{L}_\infty$ there exist simple functions $s_1, s_2, \dots \in \mathcal{L}_\infty$ such that $|s_n| \leq |f|$ and $|f - s_n| \leq 1/n$ for every $n \in \mathbb{N}$.

(ii) Show that there exists a linear functional $L : \mathcal{L}_\infty \to \mathbb{C}$ such that $L(\chi_A) = \mu(A)$ for every $A \in \mathbf{R}$ and $|L(f)| \leq \mu(X) \sup_X |f|$ for every $f \in \mathcal{L}_\infty$. This functional is positive, that is $L(f) \geq 0$ if $f \geq 0$.

3AA. Consider an algebra $\mathbf{R}$ of subsets of a set X, and a positive linear functional $L : \mathcal{L}_\infty \to \mathbb{C}$ satisfying the inequality $|L(f)| \leq \sup_X |f|$ for every $f \in \mathcal{L}_\infty$. Show that the formula $\mu(A) = L(\chi_A)$ defines a finitely additive set function on $\mathbf{R}$.

Remark 3.47. The preceding two problems provide an analog of the developments in this chapter for (finite) finitely additive set functions defined on a ring. The relationship between L and μ is also indicated by the notation $L(f) = \int_X f \, d\mu$. Note that this functional satisfies axiom (L$_1$), while (L$_2$) must be replaced by the condition that $\lim_n L(f_n) = L(g)$ if $f_n \to g$ uniformly.

3BB. Let $(X, \mathbf{S}, \mu)$ be a measure space, let $A \in \mathbf{S}$ be an infinite atom for μ, and let $f : X \to \mathbb{C}$ be integrable $[\mu]$. Show that $f = 0$ almost everywhere $[\mu]$ on A.

Chapter 4
Convergence theorems for Lebesgue integrals

Lebesgue integration is a powerful tool principally on account of several *convergence theorems* (Theorems 4.24, 4.29, 4.31, and 4.35), and these are the main focus of this chapter. There are, however, several other things to be established. We begin by introducing the signed and complex counterparts of a measure. We recall that finite, finitely additive set functions φ defined on a ring always satisfy $\varphi(\varnothing) = 0$.

Definition 4.1. Let $(X, \mathbf{S})$ be a measurable space. Then a *signed measure* on $(X, \mathbf{S})$ is a countably additive set function $\nu : \mathbf{S} \to \mathbb{R}^{\natural}$ such that $\nu(\varnothing) = 0$. The value $\nu(E)$ of ν at a measurable set E is called the *measure* of E, and the measurable space $(X, \mathbf{S})$ together with a signed measure ν on $(X, \mathbf{S})$ is called a *signed measure space*, denoted by $(X, \mathbf{S}, \nu)$. Similarly, a complex measure on $(X, \mathbf{S})$ is a countably additive set function $\varphi : \mathbf{S} \to \mathbb{C}$. The value $\varphi(E)$ of φ at a measurable set E will be called the *measure* of E, and the measurable space $(X, \mathbf{S})$ together with a complex measure φ on $(X, \mathbf{S})$ is a *complex measure space*, denoted by $(X, \mathbf{S}, \varphi)$. Finally, when no confusion is possible, both the signed measure space $(X, \mathbf{S}, \nu)$ and the complex measure space $(X, \mathbf{S}, \varphi)$ will be denoted simply by X.

This terminology, while standard in the literature, is less logical than it might be. Thus a complex measure is not a measure in general, nor is a signed measure necessarily a measure. Moreover, a measure is not, in general, a complex measure, since it may assume the value $+\infty$. It is true, however, that every measure is a signed measure, and a finite measure (or, more generally, any finite-valued signed measure) is also a complex measure.

Definition 4.2. Let $(X, \mathbf{S}, \mu)$ be a measure space, and let φ be a set function defined on $\mathbf{S}$ that is either complex-valued or extended real-valued. We say that φ is *absolutely continuous with respect to* μ or, more simply, *absolutely continuous* $[\mu]$ if the condition $\mu(E) = 0$ implies $\varphi(E) = 0$ for every $E \in \mathbf{S}$. We write

© Springer International Publishing Switzerland 2016
H. Bercovici et al., *Measure and Integration*,
DOI 10.1007/978-3-319-29046-1_4

$$\varphi \ll \mu$$

to indicate that φ is absolutely continuous $[\mu]$.

In this chapter, the above definition is used primarily when φ is a complex measure. In this case absolute continuity takes on a stronger aspect. For the following proposition, we only consider positive measures, though the result is true for complex measures. Indeed, complex measures are linear combinations of finite measures as seen in Chapter 6.

Proposition 4.3. *Let $(X, \mathbf{S}, \mu)$ be a measure space, and let $\varphi : \mathbf{S} \to [0, +\infty)$ be a measure such that $\varphi \ll \mu$. Then, for every positive number ε there exists a positive number δ such that $\mu(E) < \delta$ implies $\varphi(E) < \varepsilon$ for every $E \in \mathbf{S}$.*

Proof. Assume, to get a contradiction, that the conclusion of the proposition is not true for some $\varepsilon_0 > 0$. Then there exist sets $E_n \in \mathbf{S}$ such that $\mu(E_n) < 2^{-n}$ and $\varphi(E_n) \geq \varepsilon_0$ for $n \in \mathbb{N}$. Define $F_N = \bigcup_{n \geq N} E_n$ and $F = \bigcap_{N=1}^{\infty} F_N$. We have

$$\mu(F) \leq \mu(F_N) \leq \sum_{n=N}^{\infty} \mu(E_n) < \sum_{n=N}^{\infty} 2^{-N} = 2^{-N+1}$$

for every $N \in \mathbb{N}$, and therefore $\mu(F) = 0$. On the other hand,

$$\varphi(F_N) \geq \varphi(E_N) \geq \varepsilon_0, \quad N \in \mathbb{N}.$$

We show that $\varphi(F) \geq \varepsilon_0$, thus contradicting absolute continuity. Indeed, the countable additivity of φ implies

$$\varphi(F) = \varphi(F_N) - \sum_{n=N}^{\infty} \varphi(F_{n+1} \setminus F_n).$$

Since the series $\sum_{n=1}^{\infty} \varphi(F_{n+1} \setminus F_n)$ converges, we can let $N \to +\infty$ to conclude that indeed $\varphi(F) \geq \varepsilon_0$, as claimed. $\qquad\square$

Example 4.4. The notion of absolute continuity was originally applied to functions of one real variable where it is seen to be a stronger property than uniform continuity. Consider indeed $\mu = \lambda_1$ on the Borel sets of the real line, and let φ be a finite measure defined on the Borel subsets of $\mathbb{R}$ such that $\varphi \ll \lambda_1$. Define a monotone increasing function $h : \mathbb{R} \to [0, +\infty)$ by setting $h(t) = \varphi((-\infty, t])$ for $t \in \mathbb{R}$. Thus, we have $\varphi((a, b]) = h(b) - h(a)$ for $a, b \in \mathbb{R}$, $a \leq b$. Proposition 4.3 implies the following statement: for every $\varepsilon > 0$ there exists $\delta > 0$ such that, for any collection of pairwise disjoint intervals $\{(a_i, b_i]\}_{i=1}^{n}$ satisfying $\sum_{i=1}^{n}(b_i - a_i) < \delta$, we have $\sum_{i=1}^{n} |h(b_i) - h(a_i)| < \varepsilon$. In particular, the function h is uniformly continuous, as can be seen by looking at a single interval. More generally, a function $h : \mathbb{R} \to \mathbb{C}$ is said to be *absolutely continuous* when the preceding statement is true. This terminology is also used for functions h defined on a proper interval.

Definition 4.5. Let $(X, \mathbf{S}, \mu)$ be a measure space, and let $\{\varphi_\gamma\}_{\gamma \in \Gamma}$ be an indexed family of set functions defined on $\mathbf{S}$ which are either complex-valued or extended real-valued.. We say that the family $\{\varphi_\gamma\}_{\gamma \in \Gamma}$ is *uniformly absolutely continuous* $[\mu]$ if for every positive number ε there exists a positive number δ such that

$$\mu(E) < \delta \text{ implies } |\varphi_\gamma(E)| < \varepsilon \text{ for every } \gamma \in \Gamma.$$

We need one more important property which a set function may possess.

Definition 4.6. Let $(X, \mathbf{S}, \mu)$ be a measure space, and let φ be a set function defined on $\mathbf{S}$ that is either complex-valued or extended real-valued. We say that φ is *concentrated on sets of finite measure* with respect to $[\mu]$ if for every $\varepsilon > 0$ there exists a set $E_\varepsilon \in \mathbf{S}$ such that

$$\mu(E_\varepsilon) < +\infty \text{ and } |\varphi(F)| < \varepsilon \text{ for every } F \in \mathbf{S} \text{ such that } F \subset X \setminus E_\varepsilon. \quad (4.1)$$

An indexed family $\{\varphi_\gamma\}_{\gamma \in \Gamma}$ is *uniformly* concentrated on sets of finite measure with respect to μ if the set E_ε in (4.1) can be chosen independently of γ.

An important kind of set function is the *indefinite integral* of an integrable function. Readers who recall the definition of indefinite integral given in elementary calculus may find this terminology puzzling. The connection between point functions and set functions on $\mathbb{R}$ is discussed later.

Definition 4.7. Let $(X, \mathbf{S}, \mu)$ be a measure space, and let $f : X \to \mathbb{C}$ be integrable $[\mu]$. Then the set function ν_f defined by setting

$$\nu_f(E) = \int_E f \, d\mu, \qquad E \in \mathbf{S},$$

is called the *indefinite integral* of f. Let $\mathcal{F}$ be a family of measurable complex valued functions on X all of which are integrable $[\mu]$. We say that the family $\mathcal{F}$ is *uniformly integrable* $[\mu]$ if the family of measures $\{\nu_{|f|} : f \in \mathcal{F}\}$ is uniformly absolutely continuous $[\mu]$ and uniformly concentrated on sets of finite measure with respect to μ.

Indefinite integrals have appeared before, but not by name, and only in a special case. The following result summarizes their main properties.

Proposition 4.8. *Let $(X, \mathbf{S}, \mu)$ be a measure space and let f be a complex-valued function that is integrable $[\mu]$ over X. Then the indefinite integral ν_f is a complex measure on X that is absolutely continuous and concentrated on sets of finite measure with respect to μ.*

Proof. If $E = E_1 \cup \cdots \cup E_N$ is any partition of a (measurable) set E into disjoint measurable sets $E_1, \ldots, E_N$, then $f\chi_E = f\chi_{E_1} + \cdots + f\chi_{E_N}$ and therefore

$$\nu_f(E) = \nu_f(E_1) + \cdots + \nu_f(E_N).$$

Thus ν_f is finitely additive by virtue of the linearity of the integral. To complete the proof that ν_f is a complex measure it suffices to verify semicontinuity (Problem 3I). Let $\{F_n\}$ be a monotone increasing sequence of measurable sets having union E. We must show that $\nu_f(F_n) \to \nu_f(E)$, or what comes to the same thing, that $\nu_f(R_n) \to 0$, where $R_n = E \setminus F_n, n \in \mathbb{N}$. Since $\left| \int_{R_n} f \, d\mu \right| \leq \int_{R_n} |f| \, d\mu$, it clearly suffices to prove this latter fact for $|f|$ in place of f, and since it also suffices to prove the other two assertions of the proposition for $|f|$ in place of f, we assume henceforth that f itself is nonnegative. Consider the sequence $\{f_n\}$ of nonnegative functions defined by $f_n = f\chi_{F_n}, n \in \mathbb{N}$. Clearly $\{f_n\}$ is monotone increasing and converges pointwise to $f\chi_E$. But then $\nu_f(F_n) \to \nu_f(E)$ by Proposition 3.15, and the countable additivity follows.

The absolute continuity $\nu_f \prec \mu$ is obvious. To complete the proof, we continue assuming that $f \geq 0$. We introduce the sequence of measurable sets $\{H_n\}_{n=1}^\infty$, where $H_n = E(f \geq 1/n)$, and the corresponding sequence $\{h_n = f\chi_{H_n}\}_{n=1}^\infty$ of measurable functions. The sequence $\{H_n\}$ is monotone increasing with union N_f, so $\{h_n\}$ is monotone increasing and converges pointwise to $f = f\chi_{N_f}$. Hence $\nu_f(H_n) \to \nu_f(N_f) = \nu_f(X)$ (Proposition 3.15). Thus it suffices to show that the sets H_n are all of finite measure with respect to μ. This follows from the fact that $\chi_{H_n} \leq nf$. $\qquad\square$

As noted in the preceding proof, the measure ν_f is a linear combination of finite positive measures, and therefore its absolute continuity implies the stronger property described in Proposition 4.3.

Example 4.9. If φ is an arbitrary complex measure on a measurable space $(X, \mathbf{S})$, then it is obvious that the set functions

$$\alpha(E) = \Re\varphi(E) \quad \text{and} \quad \beta(E) = \Im\varphi(E), \quad E \in \mathbf{S},$$

are finite-valued signed measures on $(X, \mathbf{S})$. Conversely, if α and β are any two finite-valued signed measures on $(X, \mathbf{S})$, then setting

$$\varphi(E) = \alpha(E) + i\beta(E), \quad E \in \mathbf{S},$$

defines a complex measure on $(X, \mathbf{S})$. If, in particular, $\varphi = \nu_f$ for some integrable function f on X, then $\alpha = \Re\varphi$ and $\beta = \Im\varphi$ are the indefinite integrals of $\Re f$ and $\Im f$, respectively.

Example 4.10. If a real-valued function f is integrable $[\mu]$ on a measure space $(X, \mathbf{S}, \mu)$, then the indefinite integral ν_f is a finite-valued signed measure on X. Let $A = E(f \geq 0)$ and $B = E(f < 0)$. This partition $X = A \cup B$ of X into two measurable sets has the property that $\nu_f(E) \geq 0$ for every measurable subset E of A, while $\nu_f(E) \leq 0$ for every measurable subset E

of B. (As a matter of fact, such a partition of a signed measure space always exists, but this is a relatively deep result. Signed measures are studied in their own right in Chapter 6.)

Example 4.11. Let f be an integrable real-valued function on a measure space $(X, \mathbf{S}, \mu)$. Then the functions f^+ and f^- are also integrable $[\mu]$ (Proposition 3.3) and it is obvious that

$$\nu_f = \nu_{f^+} - \nu_{f^-}$$

setwise on $\mathbf{S}$. Thus the signed measure ν_f is expressed as the difference of two ordinary finite measures. This construction is generalized in Chapter 6.

An important element in the theory of Lebesgue integrals, and one that helps to give that theory its characteristic flavor, is the prominent role played by sets of measure zero, or *null* sets as they are sometimes called.

Example 4.12. Let $(X, \mathbf{S})$ be a measurable space, and let $\{w_x\}_{x \in X}$ be a family of nonnegative weights. Then a subset Z of X is a null set for the associated discrete measure on $(X, \mathbf{S})$ (Example 3.20) if and only if Z belongs to $\mathbf{S}$ and the weight w_x of every point x of Z is zero. In particular, the only null set with respect to the counting measure on $(X, \mathbf{S})$ (Example 3.21) is the empty set.

Example 4.13. A singleton in $\mathbb{R}$ has measure zero with respect to Lebesgue-Borel measure λ_1 (Example 3.42). Hence every countable set in $\mathbb{R}$ is a null set with respect to λ_1. In particular, the set $\mathbb{Q}$ has Lebesgue-Borel measure zero.

Example 4.14. A subset $Z \subset \mathbb{R}$ satisfies $\lambda_1(Z) = 0$ if and only if for each $\varepsilon > 0$ there is a countable covering $\{(a_n, b_n)\}_{n=1}^{\infty}$ of Z by intervals (a_n, b_n) such that

$$\sum_{n=1}^{\infty} (b_n - a_n) < \varepsilon$$

(see Problem 3X). The Cantor set C ([**I**, Example 6O]) is covered by the system $\mathcal{F}_n$ of 2^n intervals for each index n, and since each of the intervals in $\mathcal{F}_n$ has length $1/3^n$, the sum of the lengths of the covering intervals is $(2/3)^n$. Thus C is a noncountable null set in $\mathbb{R}$. (In this context, recall Problem 3Y).

Example 4.15. Given an arbitrary measurable subset Z of a measure space $(X, \mathbf{S}, \mu)$, there is a simple way to define a measure on X that makes Z into a null set and that leaves the measure μ unchanged on the complement $E = X \setminus Z$. Indeed, if we write $\mu_E(A) = \mu(A \cap E)$ for each measurable set A in X, then μ_E is a measure on $(X, \mathbf{S})$ with the required properties. The measure μ_E, which we shall call the *concentration* of μ on E, may also be described as the indefinite integral of χ_E with respect to μ whenever $\mu(E) < +\infty$. It is instructive to compare Lebesgue integration with respect to μ_E with Lebesgue integration on the subspace $(E, \mathbf{S} \cap E, \mu|E)$ (Example 3.34).

Example 4.16. It is clear from the monotonicity and countable additivity of every measure that the collection $\mathbf{Z}$ of all null sets in an arbitrary measure space $(X, \mathbf{S}, \mu)$ is a σ-ideal in $\mathbf{S}$. (Indeed, this fact is implicit in several of the foregoing examples.) Conversely, if $\mathbf{Z}$ is an arbitrary σ-ideal in the σ-algebra $\mathbf{S}$ of measurable sets in an arbitrary measurable space $(X, \mathbf{S})$, then there are measures on $(X, \mathbf{S})$ having $\mathbf{Z}$ as the σ-ideal of null sets; see Example 3.4.

Given a property $p(x)$ which a point x in a measure space $(X, \mathbf{S}, \mu)$ may or may not have, we say that the property holds *almost everywhere* with respect to μ, or *almost everywhere* $[\mu]$, if there exists a null set Z with respect to μ such that $p(x)$ holds for every $x \in X \setminus Z$. (If there can be no doubt as to which measure μ is intended, we may simply say that the property holds *almost everywhere*. When μ is a probability measure, we may also say *almost surely* instead of *almost everywhere*.) For instance, two functions f and g on X are *equal almost everywhere* $[\mu]$ ($f = g$ *almost everywhere* $[\mu]$) if there exists a null set Z such that $\{x \in X : f(x) \neq g(x)\} \subset Z$, and two subsets E and F of X are *almost equal* $[\mu]$ if $\chi_E = \chi_F$ almost everywhere $[\mu]$. Likewise, a sequence $\{f_n\}$ of complex-valued functions on $(X, \mathbf{S}, \mu)$ *converges almost everywhere* to a limit g if there exists a null set Z with respect to μ such that $\{f_n(x)\}$ converges to $g(x)$ for $x \in X \setminus Z$. It is important to note that if each of a countable sequence $\{p_n\}$ of properties holds almost everywhere $[\mu]$, then all of the properties p_n hold simultaneously almost everywhere $[\mu]$. This is true because a countable union of null sets is again a null set.

The reason sets of measure zero are so important is that, for most purposes, they are totally negligible. The next two results illuminate this apparent paradox.

Proposition 4.17. *Let X be a measure space equipped with a measure μ, and let f be a measurable complex-valued function on X such that $f = 0$ almost everywhere $[\mu]$. Then f is integrable $[\mu]$ and $\int_X f \, d\mu = 0$.*

Proof. It suffices to treat the case $f \geq 0$. By hypothesis the support $N_f = E(f > 0)$ of f has measure zero. Consider the sequence of truncations $g_n = f \wedge n, n \in \mathbb{N}$. Since g_n is measurable and $0 \leq g_n \leq n\chi_N$, it follows that g_n is integrable $[\mu]$ and that $0 \leq \int_X g_n \, d\mu \leq n \int_X \chi_N \, d\mu = n\mu(N) = 0$ for every n. But then, by Proposition 3.15 it follows that f is also integrable $[\mu]$ and that $\int_X g_n \, d\mu$ tends to $\int_X f \, d\mu = 0$. $\qquad\qquad\square$

Corollary 4.18. *Let X be a measure space equipped with a measure μ and let f and g be measurable complex-valued functions on X such that $f = g$ almost everywhere $[\mu]$. Then f is integrable $[\mu]$ if and only if g is, and when both are integrable, $\int_X f \, d\mu = \int_X g \, d\mu$.*

Proof. Assume that f is integrable $[\mu]$. Proposition 4.17 implies that $g - f$ is integrable $[\mu]$, and therefore $g = f + g - f$ is integrable as well. Moreover, $\int_X (g - f) \, d\mu = 0$ and this yields the desired identity. The corollary follows by symmetry. $\qquad\qquad\square$

One can paraphrase these last two results in the language of quotient spaces (see [**I**, Chapter 3] for definitions). Let $(X, \mathbf{S}, \mu)$ be a measure space and let $\mathcal{L}$ and $\mathcal{M}$ denote, respectively, the (complex) vector space of all complex-valued functions integrable $[\mu]$ on X, and the (complex) vector space of all measurable complex-valued functions on X. Then according to Proposition 4.17 the set $\mathcal{M}_0$ of those functions in $\mathcal{M}$ that vanish almost everywhere $[\mu]$ is a linear submanifold of $\mathcal{L}$ as well as of $\mathcal{M}$. Let f be a function in $\mathcal{M}$ and let $\dot{f}$ denote the coset of f in the quotient space $\mathcal{M}/\mathcal{M}_0$. Then a function g in $\mathcal{M}$ belongs to $\dot{f}$ if and only if $\int_X |f - g| \, d\mu = 0$. In particular, if f is integrable $[\mu]$ and if g belongs to $\dot{f}$, then $\int_X g \, d\mu = \int_X f \, d\mu$. Thus it makes sense to define $\int_X \dot{f} \, d\mu = \int_X f \, d\mu$. Indeed, once a measure μ is fixed, it is frequently advantageous to think of integration with respect to μ as a linear functional on the quotient space $\mathcal{L}/\mathcal{M}_0$ rather than on $\mathcal{L}$ itself.

In this same connection there is also a modest expansion of the concept of measurability that is sometimes useful.

Definition 4.19. If $(X, \mathbf{S}, \mu)$ is a measure space, and if f is either a complex-valued function or an extended real-valued function whose domain of definition is a measurable subset D of X such that $\mu(X \setminus D) = 0$, then f will be said to be *measurable with respect to* $[\mu]$, or *measurable* $[\mu]$, on X if f is measurable on the subspace D.

The point of this usage is that such a function certainly possesses measurable extensions to all of X. Corollary 4.18 indicates that it is immaterial (for purposes of integration theory) which extension is chosen, and often there is no reason to choose one at all. In the same spirit we shall say of a complex-valued function f that is measurable $[\mu]$ on X that f is *integrable* $[\mu]$ *on* X if the measurable extensions of f to X are integrable $[\mu]$ (if one is, then all are), and we declare the common integral of these extensions to be the *integral* $\int_X f \, d\mu$ of f with respect to μ.

With this extension of the notion of integration with respect to a measure μ on a measurable space X we admit as integrable $[\mu]$ functions that are defined on various subsets of X, and such functions, of course, do not form a vector space. Thus it becomes slightly inaccurate to call integration with respect to μ a linear functional. It is clear, however, that if f and g are any two complex-valued functions that are measurable $[\mu]$ and if α and β are any complex numbers, then $\alpha f + \beta g$, defined pointwise almost everywhere on X, is also measurable $[\mu]$, and likewise that if f and g are both integrable $[\mu]$, then

$$\int_X (\alpha f + \beta g) \, d\mu = \alpha \int_X f \, d\mu + \beta \int_X g \, d\mu.$$

Thus integration retains the property of linearity for all practical purposes. (Another way to view the linearity of this slightly generalized version of the integral with respect to μ is to note that each function f that is integrable $[\mu]$ "almost belongs" to a unique element $\dot{f}$ of the quotient space $\mathcal{L}/\mathcal{M}_0$

introduced above, and to identify $\int_X \dot{f} \, d\mu$ with $\int_X f \, d\mu$.) We observe also that for the notion of integrable function thus extended it is true that if f is a function that is integrable $[\mu]$ on X, and if g is a complex-valued function that is measurable $[\mu]$ on X such that $|g| \leq |f|$ almost everywhere $[\mu]$, then g is also integrable $[\mu]$, and $\left| \int_X g \, d\mu \right| \leq \int_X |f| \, d\mu$. In other words, axiom (L_1) continues to hold in a suitably generalized form, and the same is clearly true of axiom (L_2) as well.

There are still other conventions concerning the integral of a function f on a measure space $(X, \mathbf{S}, \mu)$ that are customarily observed when the integrand f is real-valued or extended real-valued. Suppose, first, that f is an extended real-valued function that is measurable and nonnegative almost everywhere $[\mu]$ on X. If the (measurable) set $E_{+\infty} = E(f = +\infty)$ has measure zero, then there is a nonnegative, finite real-valued and measurable function $\widetilde{f}$ on X equal almost everywhere to f, and it is natural in this case to call f itself integrable $[\mu]$ if the function $\widetilde{f}$ is integrable $[\mu]$, and to set $\int_X f \, d\mu = \int_X \widetilde{f} \, d\mu$. (Once again, if one such function $\widetilde{f}$ is integrable $[\mu]$, then all are and all have the same integral.) If, on the other hand, f is *not* integrable $[\mu]$, we write

$$\int_X f \, d\mu = +\infty.$$

In the event that $\mu(E_{+\infty}) > 0$, we also write $\int_X f \, d\mu = +\infty$. (Note that we do not say that f is "integrable $[\mu]$" in these latter two cases.) Thus the symbol $\int_X f \, d\mu$ is defined as an extended real number for any extended real-valued function f that is defined and nonnegative almost everywhere $[\mu]$ and measurable $[\mu]$ on X. (All of this could have been said somewhat differently if the support of f were σ-finite; a consideration of cases shows that for such a function f we have

$$\int_X f \, d\mu < +\infty \quad \text{or} \quad \int_X f \, d\mu = +\infty$$

according as the set of integrals $\int_X g \, d\mu$ of the nonnegative integrable (simple) functions g on X such that $g \leq f$ almost everywhere $[\mu]$ is bounded or not (in $\mathbb{R}$), and also that, in any case, $\int_X f \, d\mu$ is simply the supremum (in $\mathbb{R}^{\natural}$) of this set of integrals.)

Finally, let f be an arbitrary extended real-valued function that is measurable $[\mu]$ on X. Then, according to the foregoing agreements, both $\int_X f^+ \, d\mu$ and $\int_X f^- \, d\mu$ are defined as nonnegative extended real numbers. If the difference

$$\int_X f^+ \, d\mu - \int_X f^- \, d\mu \tag{4.2}$$

is also defined as an extended real number (that is, if either $\int_X f^+ \, d\mu$ or $\int_X f^- \, d\mu$ is finite), then we write $\int_X f \, d\mu$ for the difference in (4.2).

If both $\int_X f^+ \, d\mu$ and $\int_X f^- \, d\mu$ are finite, then there is a finite real-valued function $\tilde{f}$ defined on all of X such that $\tilde{f}$ is integrable over X and equal to f almost everywhere, and we have $\int_X f \, d\mu = \int_X \tilde{f} \, d\mu$, so that in this case our last extension of the notion of integration with respect to μ does not really present us with anything new. Hence it is only the extended real-valued functions f for which $\int_X f \, d\mu = \pm\infty$ that require further attention at this time. In this connection we note, in the first place, that assigning values $\pm\infty$ to such integrals is the exact counterpart (indeed, a generalization) of the common and useful practice of assigning $\pm\infty$ to various divergent series and indexed sums of extended real numbers. Likewise, a consideration of cases discloses that if f and g are any two extended real-valued functions that are measurable $[\mu]$ on X and for which the symbols $\int_X f \, d\mu$ and $\int_X g \, d\mu$ are both defined, and if $f \leq g$ almost everywhere, then $\int_X f \, d\mu \leq \int_X g \, d\mu$. Moreover, it is easily seen that in all cases, given a measurable set E, if the integral $\int_X f \, d\mu$ is defined then $\int_X f\chi_E \, d\mu$ is too, and we continue the practice of writing $\int_E f \, d\mu$ for this latter integral. Finally, it is important to note that integrals taking values $\pm\infty$ are here introduced only for *extended real-valued integrands* f; if a function f assumes any nonreal complex values, then the integral of f, if it exists, must be an ordinary complex number. The extended real integrals just introduced obey the usual rules of algebra when the operations are defined.

Proposition 4.20. *Let f be an extended real-valued function that is measurable $[\mu]$ on a measure space $(X, \mathbf{S}, \mu)$ and for which the integral $\int_X f \, d\mu$ is defined as an extended real number, and let a be a (finite) real number. Then $\int_X af \, d\mu = a \int_X f \, d\mu$. Likewise, if g is another extended real-valued function that is measurable $[\mu]$ on X and for which $\int_X g \, d\mu$ is defined, and if the sum $\int_X f \, d\mu + \int_X g \, d\mu$ is also defined as an extended real number, then $\int_X (f + g) \, d\mu = \int_X f \, d\mu + \int_X g \, d\mu$. Consequently, if f and g are extended real-valued functions such that $\int_X f \, d\mu$ and $\int_X g \, d\mu$ are both defined, and if a and b are real numbers for which the combination $a \int_X f \, d\mu + b \int_X g \, d\mu$ is defined as an extended real number, then*

$$\int_X (af + bg) \, d\mu = a \int_X f \, d\mu + b \int_X g \, d\mu.$$

Proof. Consider first a single function f and a single real number a. The case in which $\int_X f \, d\mu$ is finite has already been covered, and by symmetry it suffices to treat the case $\int_X f \, d\mu = +\infty$. If $a = 0$, then $af = 0$ almost everywhere so $\int_X af \, d\mu = 0 = a \int_X f \, d\mu$. If $a > 0$, then $(af)^+ = af^+$ and $(af)^- = af^-$, whence it follows that $\int_X (af)^+ = +\infty$ while $\int_X (af)^- \, d\mu = a \int_X f^- \, d\mu < +\infty$, and hence that $\int_X af \, d\mu = +\infty = a \int_X f \, d\mu$. Finally, if $a < 0$, then $(af)^+ = (-a)f^-$ and $(af)^- = (-a)f^+$. But then $\int_X (af)^+ \, d\mu = (-a) \int_X f^- \, d\mu < +\infty$, while $\int_X (af)^- \, d\mu = (-a) \int_X f^+ \, d\mu = +\infty$, so that $\int_X af \, d\mu = -\infty = a \int_X f \, d\mu$.

Next, as regards the sum $f+g$ of two such functions we note once again that if $\int_X f\,d\mu$ and $\int_X g\,d\mu$ are both finite, then the desired result is already known. Hence, by symmetry, it suffices to treat the following two cases: (i) $\int_X f\,d\mu = +\infty$ with $\int_X g\,d\mu$ finite, and (ii) $\int_X f\,d\mu = \int_X g\,d\mu = +\infty$. Moreover, in both of these cases there is a set Z_1 of measure zero such that f is defined and measurable and satisfies the inequality $-\infty < f \leq +\infty$ on $X \setminus Z_1$ and likewise a set Z_2 of measure zero such that g is defined and measurable and such that $-\infty < g \leq +\infty$ on $X \setminus Z_2$. But then $f + g$ is defined and measurable on $X \setminus (Z_1 \cup Z_2)$, so $f + g$ is measurable $[\mu]$. Likewise, in both cases, we have $(f + g)^- \leq f^- + g^-$, and therefore

$$\int_X (f + g)^-\,d\mu < +\infty.$$

Thus in both cases everything comes down to showing that $\int_X (f+g)^+ = +\infty$. Suppose this were not the case. Then both $\int_X (f+g)^+\,d\mu$ and $\int_X (f+g)^-\,d\mu$ are finite, which implies that $f + g$ is integrable $[\mu]$. But then, in case (i), we find that $f = (f + g) - g$ is also integrable $[\mu]$, contrary to hypothesis, and this contradiction proves the result in case (i). Finally, if $f + g$ is integrable $[\mu]$, then, in case (ii), we obtain

$$\int_X f\,d\mu = \int_X [(f + g) - g]\,d\mu = -\infty$$

by case (i), contrary to hypothesis. Thus the desired result is also valid in case (ii).

Finally, to complete the proof, set $\widetilde{f} = af$ and $\widetilde{g} = bg$, so that $\int_X \widetilde{f}\,d\mu = a \int_X f\,d\mu$ and $\int_X \widetilde{g}\,d\mu = b \int_X g\,d\mu$ by the first part of the proof. Since, by hypothesis, the combination $a \int_X f\,d\mu + b \int_X g\,d\mu$ is defined as an extended real number, we see that the same is true of the sum $\int_X \widetilde{f}\,d\mu + \int_X \widetilde{g}\,d\mu$. But then

$$\int_X (af + bg)\,d\mu = \int_X (\widetilde{f} + \widetilde{g})\,d\mu = a \int_X f\,d\mu + b \int_X g\,d\mu. \qquad \square$$

We conclude this discussion with the following partial converse of Proposition 4.17.

Proposition 4.21. *Let X be a measure space equipped with a measure μ, let f be an extended real-valued function that is defined and nonnegative almost everywhere $[\mu]$ on X, and suppose $\int_X f\,d\mu = 0$. Then $f = 0$ almost everywhere $[\mu]$.*

Proof. According to the foregoing conventions and definitions there is a subspace X_0 of X such that $\mu(X \setminus X_0) = 0$ and such that f is defined, measurable, and nonnegative on X_0, and we have $\int_{X_0} f\,d\mu = 0$. Moreover, it is clear, as before, that we may assume without loss of generality that $X = X_0$. If, for some positive integer n, F is a measurable set of finite measure in X such that $f(x) \geq 1/n$ for every point x of F, then

$$0 = \int_X f \, d\mu \geq \int_F f \, d\mu \geq \mu(F)/n \geq 0,$$

and therefore $\mu(F) = 0$ by virtue of the monotonicity of integration with respect to μ. Since the set $E_n = E(f \geq 1/n)$ is σ-finite as a subset of the σ-finite support N_f of f, this shows that $\mu(E_n) = 0$. But $N_f = \bigcup_n E_n$, so $\mu(N) = \lim_n \mu(E_n)$, and the result follows. $\qquad\square$

Example 4.22. Let $(\mathbb{R}, \mathbf{B}_{\mathbb{R}}, \lambda_1)$ be the real numbers equipped with Lebesgue-Borel measure (Example 3.42), and let f be the function on $\mathbb{R}$ that is equal to $+\infty$ at each rational number and equal to zero at each irrational number ($f = +\infty \chi_{\mathbb{Q}}$). Then $\int_E f \, d\lambda_1 = 0$ for every Borel set E. On the other hand, the function identically equal to $+\infty$ on $\mathbb{R}$ has integral $+\infty$ over every subset of $\mathbb{R}$ having positive measure with respect to λ_1, and integral zero over every null set in $\mathbb{R}$.

Example 4.23. Let $(\mathbb{R}, \mathbf{B}_{\mathbb{R}}, \lambda_1)$ be as in Example 4.22, and let $g(t) = 1/t$, $t \neq 0$. Then g is measurable $[\mu]$ on $\mathbb{R}$, and if E is an arbitrary Borel subset of $\mathbb{R}_+ = \{t \in \mathbb{R} : t > 0\}$, then $\int_E g \, d\lambda_1 \geq 0$. In particular, $\int_{(1,+\infty)} g \, d\lambda_1 = +\infty$. (The function $g\chi_{(1,+\infty)}$ dominates the simple function

$$s_n = (1/2)\chi_{(1,2)} + \cdots + (1/(n+1))\chi_{(n,n+1))}$$

for every n, and the sequence $\left\{ \int_{\mathbb{R}} s_n \, d\lambda_1 \right\}$ is not bounded above in $\mathbb{R}$.) Similarly, $\int_{(0,1)} g \, d\lambda_1 = +\infty$. If E is an arbitrary Borel subset of $\mathbb{R}$ that meets either $\mathbb{R}_+$ or $\mathbb{R}_- = \{t \in \mathbb{R} : t < 0\}$ in a set of measure zero, then $\int_E g \, d\lambda_1$ is defined. On the other hand, $\int_{(-1,+1)} g \, d\lambda_1$, for instance, is undefined. If h is the function that is identically $+\infty$ on $\mathbb{R}_+$ and identically $-\infty$ on $\mathbb{R}_-$, then $\int_E h \, d\lambda_1$ is defined if and only if either $E \cap \mathbb{R}_+$ or $E \cap \mathbb{R}_-$ has measure zero.

The fact that null sets are negligible in integration theory opens up, as we have seen, the possibility of improving many of the results of Chapter 3. Thus, for example, it is obvious that if f is a function that is measurable $[\mu]$ on a measure space $(X, \mathbf{S}, \mu)$, and if f is merely nonnegative almost everywhere $[\mu]$, then $\int_X f \, d\mu \geq 0$. For the most part, to be sure, sprinkling "almost everywhere" into the hypotheses of theorems results in only token improvement, and we generally leave to the reader the task of reformulating (and reproving) in appropriate fashion the results of Chapter 3. In one important case, however, namely in Proposition 3.15, it is possible to effect a very significant improvement by allowing for exceptional null sets. The resulting theorem, which we now prove, is the first in the array of convergence theorems in this chapter.

Theorem 4.24. (Monotone Convergence Theorem). *Let $(X, \mathbf{S}, \mu)$ be a measure space and let $\{f_n\}$ be a sequence of extended real-valued functions defined and nonnegative almost everywhere $[\mu]$ and measurable $[\mu]$ on X. Suppose also that the sequence $\{f_n\}$ is monotone increasing almost everywhere $[\mu]$. Then*

$$\int_X \lim_n f_n \, d\mu = \lim_n \int_X f_n \, d\mu, \tag{4.3}$$

where $\lim_n f_n$ *is formed pointwise in* $\mathbb{R}^\natural$.

Proof. According to the hypotheses there exists $X_0 \in \mathbf{S}$ such that $\mu(X \setminus X_0)$ $= 0$, the functions f_n are all defined, nonnegative and measurable on X_0, and the sequence $\{f_n(x)\}$ is increasing for each $x \in X_0$. This fact implies, to begin with, that the function $\lim_n f_n$ is measurable $[\mu]$ on X, and hence that both sides of (4.3) are defined as extended real numbers. Moreover, it clearly suffices to prove the theorem on the subspace X_0, so that, without loss of generality, we may and do assume that $X = X_0$.

Let F denote the (measurable) set of those points x of X at which $\lim_n f_n(x)$ is finite, and let Y denote the complement $E(\lim_n f_n = +\infty)$. If the function $\lim_n f_n$ is integrable $[\mu]$, so that $\int_X \lim_n f_n \, d\mu < +\infty$, then $\mu(Y) = 0$ by convention, and $\int_X \lim_n f_n \, d\mu = \int_F \lim_n f_n \, d\mu$. But then $\int_X f_n \, d\mu = \int_F f_n \, d\mu$ for each index n, and the desired result follows at once from Proposition 3.15 (applied on the subspace F).

Suppose, on the other hand, that $\int_X \lim_n f_n \, d\mu = +\infty$. To complete the proof we must show that in this event $\lim_n \int_X f_n \, d\mu = +\infty$ as well. But if this is not true, then the increasing sequence $\{\int_X f_n \, d\mu\}$ is bounded above—say by M, and another application of Proposition 3.15 shows that $\int_F \lim_n f_n \, d\mu \leq M < +\infty$. Thus Y must have positive measure.

We derive a contradiction, and thus complete the proof, by showing that $\mu(Y) = 0$. The first step in this direction is modest but essential. The set Y is at least σ-finite with respect to μ. Indeed, Y is clearly contained in the union of the supports of the functions f_n, and each f_n, being integrable, has σ-finite support (see Problem 3R(i)). Hence it is enough to show that every measurable subset W of Y having finite measure actually has measure zero. But suppose W is a subset of Y such that $\mu(W) = a$ where $0 < a < +\infty$. Then by Proposition 3.36 there also exist a subset G of W such that $\mu(G) < a/2$ and an index N such that $f_N(x) \geq 2M/a$ at every point x of $W \setminus G$. But then

$$\int_X f_N \, d\mu \geq \int_{W \setminus G} f_N \, d\mu \geq (2M/a)\mu(W \setminus G) > M,$$

which is contrary to assumption. We conclude that $\mu(Y) = 0$, and the proof of the theorem is complete. $\qquad\square$

Corollary 4.25. (Theorem of Beppo-Levi) *Let* $(X, \mathbf{S}, \mu)$ *be a measure space, let* $\{p_n\}$ *be a sequence of extended real-valued functions defined and nonnegative almost everywhere* $[\mu]$ *and measurable* $[\mathbf{S}]$ *on* X, *and let* p *denote the (almost everywhere defined) pointwise sum* $p(x) = \sum_n p_n(x)$. *Then*

$$\int_X p \, d\mu = \sum_n \int_X p_n \, d\mu.$$

In particular, p is integrable [μ] if and only if all of the functions p_n are integrable [μ] and the numerical series $\sum_n \int_X p_n \, d\mu$ is convergent in $\mathbb{R}$. Hence $\sum_n \int_X p_n \, d\mu < +\infty$ implies $\sum_n p_n(x) < +\infty$ at almost every point x.

Proof. Just as in the preceding proof it is easy to see that it is enough to treat the case in which the functions p_n are defined and nonnegative everywhere on X. Set $f_n = p_1 + p_2 + \cdots + p_n, n = 1, 2, \ldots$, and apply the monotone convergence theorem.   □

Example 4.26. Let $(X, \mathbf{S}, \mu)$ be a measure space, and let f be an extended real-valued function that is measurable [**S**] on X and has the property that $\int_X f \, d\mu$ is defined as an extended real number. As has been noted, for any set E in **S** the integral $\int_E f \, d\mu$ is also defined, so that setting

$$\nu_f(E) = \int_E f \, d\mu, \qquad E \in \mathbf{S},$$

defines a set function ν_f on **S**, called, as before, the *indefinite integral* of f with respect to μ. This notion of indefinite integral differs from the one introduced initially in that it may assume one or the other of the values $\pm\infty$ (but not both). It is an easy consequence of all that has been said up to now that ν_f is absolutely continuous [μ].

Example 4.27. Let $(X, \mathbf{S}, \mu)$ be a measure space and let f be a nonnegative extended real-valued function defined almost everywhere [μ] and measurable [**S**] on X. Then the sequence $\{g_n\}_{n=1}^{\infty}$ of truncates $g_n = f \wedge n$ always has the property that

$$\int_X f \, d\mu = \lim_n \int g_n \, d\mu.$$

Example 4.28. Let $(X, \mathbf{S}, \mu)$ be a measure space and $\{E_n\}_{n=1}^{\infty} \subset \mathbf{S}$. For each $x \in X$ denote by $N(x)$ the number of sets E_n in this sequence such that $x \in E_n$, with the understanding that $N(x) = +\infty$ whenever the latter number is $\aleph_0$. Then

$$\int_{\mathbb{R}} N \, d\mu = \sum_{n=1}^{\infty} \mu(E_n).$$

Indeed, N is simply the sum $\sum_{n=1}^{\infty} \chi_{E_n}$ (in $\mathbb{R}^{\natural}$) so

$$\int_{\mathbb{R}} N \, d\mu = \sum_{n=1}^{\infty} \int_{\mathbb{R}} \chi_{E_n} \, d\mu = \sum_{n=1}^{\infty} \mu(E_n)$$

by the theorem of Beppo-Levi. When $\sum_{n=1}^{\infty} \mu(E_n) < +\infty$, it follows that N is integrable [μ], so N must be finite almost everywhere [μ]. This last observation is one half of the Borel-Cantelli lemma, frequently used in probability arguments. (The other half is a converse which holds when μ is a probability measure and the sets E_n are statistically independent; see Problem 4II.)

Theorem 4.29. (Fatou's Lemma). *Let $(X, \mathbf{S}, \mu)$ be a measure space, and let $\{f_n\}$ be a sequence of functions defined and nonnegative almost everywhere $[\mu]$ and integrable $[\mu]$ on X. Suppose that there exists a real number M such that $\int_X f_n \, d\mu \leq M$ for every n, and suppose also that $\{f_n\}$ converges pointwise almost everywhere $[\mu]$ to a limit f. Then f is integrable $[\mu]$ over X and $\int_X f \, d\mu \leq M$.*

Proof. For each positive integer n the function $g_n = \inf_{k \geq n} f_k$ is defined and nonnegative almost everywhere. Moreover, g_n is measurable $[\mu]$ and, since $0 \leq g_n \leq f_n$ almost everywhere, it follows that g_n is integrable $[\mu]$. Finally, the sequence $\{g_n\}$ is monotone increasing and convergent almost everywhere to f. The desired conclusion follows at once from the monotone convergence theorem.   □

Example 4.30. Let $\{f_n\}_{n=1}^{\infty}$ be a sequence of integrable complex-valued functions on a measure space $(X, \mathbf{S}, \mu)$ such that

$$\sum_{n=1}^{\infty} \int_X |f_n - f_{n+1}| \, d\mu < +\infty.$$

(Such a sequence might be said to be of *bounded variation in the mean*.) Then, by the theorem of Beppo-Levi (Corollary 4.25), the numerical series

$$\sum_{n=1}^{\infty} (f_n(x) - f_{n+1}(x))$$

converges absolutely almost everywhere $[\mu]$. Since this series telescopes, we see that the sequence $\{f_n\}$ converges almost everywhere to some limit—say f. Since $\{f_n\}$ is Cauchy in the mean, one sees, by applying Fatou's lemma to the sequence $\{|f_n - f_m|\}_{m=1}^{\infty}$, that $\{f_n\}$ also tends to f in the mean with respect to μ. (This final conclusion could also be obtained directly via an application of axiom (L_2) suitably generalized, as discussed above.)

The convergence theorems proved apply only to sequences of (extended) real-valued functions. We turn now to the standard convergence theorems for complex-valued functions. The sufficient conditions given in this result are also necessary; see Problems 4Q, 3N, and 3O.

Theorem 4.31. *Let $(X, \mathbf{S}, \mu)$ be a measure space, let $\{f_n\}$ be a sequence of complex-valued functions integrable $[\mu]$ over X, and suppose that $\{f_n\}$ is uniformly integrable $[\mu]$ and converges almost everywhere $[\mu]$ to a limit f. Then f is integrable $[\mu]$ and $\{f_n\}$ converges to f in the mean.*

Proof. There exists a measurable subset X_0 of X such that $\mu(X \setminus X_0) = 0$ and the functions f_n are everywhere defined and pointwise convergent to f on X_0. The limit f is measurable on X_0 and therefore measurable $[\mu]$ on X. If f is integrable $[\mu]$ over X_0, then it is also integrable $[\mu]$ over X. Similarly, by

Corollary 4.18 it suffices to verify that $\int_{X_0} |f - f_n|\, d\mu \to 0$. Thus, as before, we may and do assume without loss of generality that $X = X_0$. For each positive integer n let ν_n denote the indefinite integral of $|f_n|$. Given $\varepsilon > 0$, the hypothesis implies the existence of a set E_0 of finite measure such that

$$\nu_n(X \setminus E_0) = \int_{X \setminus E_0} |f_n|\, d\mu < \varepsilon/5, \quad n \in \mathbb{N}, \tag{4.4}$$

and of a number $\delta > 0$ with the property that if E is any measurable set such that $\mu(E) < \delta$, then $\nu_n(E) = \int_E |f_n|\, d\mu < \varepsilon/5$ for every $n \in \mathbb{N}$. By Egorov's theorem (Proposition 3.35) there exists a subset F of E_0 such that $\mu(F) < \delta$ and such that $\{f_n\}$ converges to f uniformly on $E_0 \setminus F$, and it follows that

$$\nu_n(F) = \int_F |f_n|\, d\mu < \varepsilon/5, \quad n \in \mathbb{N}. \tag{4.5}$$

Furthermore, the uniform convergence of $\{f_n\}$ on $E_0 \setminus F$ implies the existence of a positive integer N such that

$$|f(x) - f_n(x)| \leq (\mu(E_0) + 1)\varepsilon/5$$

for all $x \in E_0 \setminus F$ and all $n \geq N$. But then $\int_{E_0 \setminus F} |f - f_n|\, d\mu < \varepsilon/5$ for all $n \geq N$. Thus f is integrable $[\mu]$ over $E_0 \setminus F$. Moreover, (4.4) together with an application of Fatou's lemma (Theorem 4.29) to the sequence $\{|f_n|\}$ shows that f is also integrable $[\mu]$ over $X \setminus E_0$, and that, in fact,

$$\int_{X \setminus E_0} |f|\, d\mu \leq \varepsilon/5.$$

Similarly, by (4.5), f is also integrable over F and

$$\int_F |f|\, d\mu \leq \varepsilon/5.$$

Hence f is integrable $[\mu]$ over X and we have

$$\int_X |f - f_n|\, d\mu \leq \int_{X \setminus E_0} (|f| + |f_n|)\, d\mu + \int_{E_0 \setminus F} |f - f_n|\, d\mu + \int_F (|f| + |f_n|)\, d\mu < \varepsilon$$

for all $n \geq N$. Thus the sequence $\left\{ \int_X |f - f_n|\, d\mu \right\}$ tends to zero. $\qquad\square$

Corollary 4.32. *If f and the sequence $\{f_n\}$ satisfy the hypotheses of Theorem 4.31, then $\lim_n \int_X f_n\, d\mu = \int_X f\, d\mu$ in $\mathbb{C}$.*

Proof. Use the inequality $\left| \int_X f\, d\mu - \int_X f_n\, d\mu \right| \leq \int_X |f - f_n|\, d\mu$. $\qquad\square$

Remark 4.33. While the conclusion of Corollary 4.32 is certainly weaker than that of Theorem 4.31, it is nonetheless true that Theorem 4.31 can be derived from Corollary 4.32. Indeed, if the hypotheses of Theorem 4.31 hold for a sequence $\{f_n\}$, and if it is known for any reason that the limit f is integrable, then it is easy to verify that all of the hypotheses of the theorem also hold for the sequence $\{|f - f_n|\}$, which converges to zero almost everywhere

Remark 4.34. Consider once again the sequence $\{\nu_n\}$ of indefinite integrals in Theorem 4.31. If $\{F_k\}$ is a monotone decreasing sequence of measurable sets with limit F, then $\lim_k \nu_n(F_k) = \nu_n(F)$ for each index n since ν_n is a finite measure (Problem 3J). It is sometimes said that the sequence $\{\nu_n\}$ is *equicontinuous from above* at F if this convergence is uniform in n, that is, if for every monotone decreasing sequence $\{F_k\}$ converging to F, and for every positive number ε there exists K such that $\nu_n(F_k) - \nu_n(F) < \varepsilon$ for every $k \geq K$ and every n. It turns out that Theorem 4.31 remains valid if we assume of the sequence $\{\nu_n\}$ simply that it is *equicontinuous from above* at $\varnothing$. Indeed, it can be shown that this single condition actually implies that $\{\nu_n\}$ is both uniformly absolutely continuous and uniformly concentrated on sets of finite measure with respect to μ.

The following two results are simple consequences of Theorem 4.31, but are of considerable importance in their own right.

Theorem 4.35. (Dominated Convergence Theorem). *Let $(X, \mathbf{S}, \mu)$ be a measure space, let $\{f_n\}_{n=1}^{\infty}$ be a sequence of complex-valued functions measurable $[\mu]$ on X, and suppose that $\{f_n\}$ converges almost everywhere $[\mu]$ to a limit f. Suppose also that there exists a nonnegative function g such that g is integrable $[\mu]$ on X and $|f_n| \leq g$ almost everywhere $[\mu], n \in \mathbb{N}$. Then f is integrable $[\mu]$, and $\{f_n\}$ converges to f in the mean.*

Proof. The functions f_n are clearly integrable $[\mu]$ on X. Moreover, the indefinite integrals of the functions $|f_n|$ are all dominated setwise by the indefinite integral of g, and it follows at once (Proposition 4.8) that they are uniformly absolutely continuous and uniformly concentrated on sets of finite measure with respect to μ. The theorem follows immediately from Theorem 4.31. $\square$

Theorem 4.36. (Bounded Convergence Theorem) *Let $(X, \mathbf{S}, \mu)$ be a finite measure space, and let $\{f_n\}_{n=1}^{\infty}$ be a sequence of complex-valued functions measurable $[\mu]$ on X. Suppose that there exists a real number M such that $|f_n| \leq M$ almost everywhere $[\mu]$ for every n and that $\{f_n\}$ converges almost everywhere $[\mu]$ to a limit f. Then the functions f_n and f are integrable $[\mu]$ on X and $\{f_n\}$ converges to f in the mean.*

Proof. The constant function $g \equiv M$ is integrable $[\mu]$ since $\mu(X) < +\infty$. $\square$

Example 4.37. (The Darboux Integral) Let f be a real-valued function defined and bounded on a real interval $I = [a, b]$, and let

$$\mathcal{P} = \{a = t_0 < \cdots < t_N = b\}$$

be a partition of I. For each index $i = 1, \ldots, n$ we set

$$M_i = \sup \{f(t) : t_{i-1} \leq t \leq t_i\}, \quad m_i = \inf \{f(t) : t_{i-1} \leq t \leq t_i\},$$

and define

$$D_{\mathcal{P}}(f) = \sum_{i=1}^{n} M_i(t_i - t_{i-1}), \quad d_{\mathcal{P}}(f) = \sum_{i=1}^{n} m_i(t_i - t_{i-1}),$$

the *upper* and *lower Darboux sums*, respectively, of f *based on* the partition $\mathcal{P}$. Concerning these sums it is evident that for any $\mathcal{P}$ we have

$$m_0(b - a) \leq d_{\mathcal{P}}(f) \leq D_{\mathcal{P}}(f) \leq M_0(b - a),$$

where m_0 and M_0 denote, respectively, the infimum and supremum of f over I. More generally, if $\mathcal{P}'$ is another partition of I that refines $\mathcal{P}$, then

$$d_{\mathcal{P}}(f) \leq d_{\mathcal{P}'}(f) \leq D_{\mathcal{P}'}(f) \leq D_{\mathcal{P}}(f). \tag{4.6}$$

Thus $\{d_{\mathcal{P}}(f)\}$ and $\{D_{\mathcal{P}}(f)\}$ are bounded monotone nets (increasing and decreasing, respectively) indexed by the directed set of all partitions of the interval I, and it follows that both nets converge in $\mathbb{R}$ ([**I**, Example 6J]). The two limits

$$\underline{J}(f) = \lim_{\mathcal{P}} d_{\mathcal{P}}(f) \quad \text{and} \quad \overline{J}(f) = \lim_{\mathcal{P}} D_{\mathcal{P}}(f)$$

are the *upper* and *lower Darboux integrals*, respectively, of f over I.

If $\mathcal{P}_1$ and $\mathcal{P}_2$ are any two partitions of I, and if $\mathcal{P}_+$ is a common refinement of $\mathcal{P}_1$ and $\mathcal{P}_2$, then from (4.6) we obtain

$$d_{\mathcal{P}_1}(f) \leq d_{\mathcal{P}_+}(f) \leq D_{\mathcal{P}_+}(f) \leq D_{\mathcal{P}_2}(f).$$

This shows that any one lower Darboux sum of f is a lower bound of the entire net of all upper Darboux sums, and hence that $\underline{J}(f) \leq \overline{J}(f)$. (The same calculation shows that the difference $\overline{J}(f) - \underline{J}(f)$ is given by

$$\overline{J}(f) - \underline{J}(f) = \lim_{\mathcal{P}}(D_{\mathcal{P}}(f) - d_{\mathcal{P}}(f)) = \lim_{\mathcal{P}} \sum_{i=1}^{n} (M_i - m_i)(t_i - t_{i-1}),$$

where $M_i - m_i = \omega(f; [t_{i-1}, t_i])$ is the oscillation of f over the subinterval $[t_{i-1}, t_i]$ ([**I**, Problem 7X]).) When $\underline{J}(f) = \overline{J}(f)$ (and only then) f is said to be *Darboux integrable* over I, and the common value $J(f) = \underline{J}(f) = \overline{J}(f)$ is the *Darboux integral* of f over I.

Thus far we have said little that is new. (The reader is referred to [**I**, Problem 2S] for basic definitions.) Our main purpose here is to relate these notions to the Lebesgue integral. As regards the Darboux sums this is easily done. Given f and the partition $\mathcal{P}$ as above, we simply define two auxiliary functions by setting

$$g_{\mathcal{P}}(t) = m_i \quad \text{and} \quad G_{\mathcal{P}}(t) = M_i, \quad t_{i-1} < t < t_i,$$

for each index $i = 1, \ldots, n$. Then $g_{\mathcal{P}}$ and $G_{\mathcal{P}}$ are simple Borel measurable functions defined almost everywhere on I with respect to Lebesgue-Borel measure λ_1 (Example 3.42), and it is clear that

$$d_{\mathcal{P}}(f) = \int_a^b g_{\mathcal{P}} \, d\lambda_1, \quad D_{\mathcal{P}}(f) = \int_a^b G_{\mathcal{P}} \, d\lambda_1.$$

To express the upper and lower Darboux integrals as Lebesgue integrals is more difficult. We need to replace the limits appearing in their definition by ordinary sequential limits, and this requires some work.

Example 4.38. To begin with, let $\mathcal{P}$ be a partition of I, and suppose $\mathcal{P}'$ is another partition obtained from $\mathcal{P}$ be adjoining one or more partition points in some *single* subinterval of $\mathcal{P}$—say the subinterval $[t_{i-1}, t_i]$. To fix ideas, let us consider lower sums. If we write

$$\Sigma' = \sum_{j \neq i} m_j(t_j - t_{j-1}),$$

then, of course, $d_{\mathcal{P}}(f) = \Sigma' + m_i(t_i - t_{i-1})$, while $d_{\mathcal{P}'}(f) = \Sigma' + \Sigma''$, where Σ'' denotes the contribution coming from those subintervals of $\mathcal{P}'$ obtained by subdividing $[t_{i-1}, t_i]$. But now $\Sigma'' \leq M_0(t_i - t_{i-1})$, and therefore

$$d_{\mathcal{P}'}(f) - d_{\mathcal{P}}(f) = \Sigma'' - m_i(t_i - t_{i-l}) \leq \Omega_0(\text{mesh}\mathcal{P}),$$

where $\Omega_0 = M_0 - m_0$ is the oscillation of f over I. Thus we obtain

$$d_{\mathcal{P}'}(f) \leq d_{\mathcal{P}}(f) + \Omega_0(\text{mesh}\mathcal{P}).$$

By mathematical induction, we find that

$$d_{\mathcal{P}'}(f) \leq d_{\mathcal{P}}(f) + r\Omega_0(\text{mesh}\mathcal{P})$$

if $\mathcal{P}'$ is a refinement of $\mathcal{P}$ obtained by further subdividing at most r of the subintervals of $\mathcal{P}$.

Now fix $\varepsilon > 0$, select a partition $\mathcal{P}_0$ of I such that $d_{\mathcal{P}_0}(f) > \underline{J}(f) - \varepsilon/2$, and suppose $\mathcal{P}_0 = \{t_0 < \ldots < t_N\}$ has exactly N subintervals. Choose $\delta > 0$ so small that $N\Omega_0\delta < \varepsilon/2$, and let $\mathcal{P}$ be a partition of I with mesh$\mathcal{P} < \delta$. If $\mathcal{P}_+$ denotes the common refinement of $\mathcal{P}$ and $\mathcal{P}_0$ obtained by adjoining the partition points $t_1, \ldots t_N$ to $\mathcal{P}$, then, as we have just seen,

$$d_{\mathcal{P}_+}(f) \leq d_{\mathcal{P}}(f) + N\Omega_0\delta < d_{\mathcal{P}}(f) + \varepsilon/2.$$

But then, of course,

$$d_{\mathcal{P}}(f) > d_{\mathcal{P}_+}(f) - \varepsilon/2 > \underline{J}(f) - \varepsilon.$$

(Note that in this calculation $\mathcal{P}$ is any partition of I having mesh less than δ.)

Fix now one nested sequence $\{\mathcal{P}_n\}_{n=1}^{\infty}$ of partitions of I with the property that $\lim_n \operatorname{mesh}\mathcal{P}_n = 0$. Consider the corresponding bounded sequence of simple functions $\{g_n = g_{\mathcal{P}_n}\}$. These functions are all defined almost everywhere $[\mu]$, and the sequence is monotone increasing almost everywhere $[\mu]$ to a limit, which we denote by g. (The exceptional set is the set Z of all the partition points of the various partitions $\mathcal{P}_n$, and Z is countable.) The bounded convergence theorem yields

$$\int_a^b g \, d\lambda_1 = \lim_n \int_a^b g_n \, d\lambda_1 = \lim_n d_{\mathcal{P}_n}(f),$$

and this latter limit, as we have just seen, is precisely the lower Darboux integral of f.

On the other hand, if t' is a point of $I \setminus Z$, and if for each index n we denote by I_n the subinterval of $\mathcal{P}_n$ that contains t', then $\{I_n\}$ is a nested sequence of neighborhoods of t' with $\operatorname{diam}I_n \to 0$, and it is easily seen that

$$g(t') = \liminf_n \{f(t) : t \in I_n\}$$

coincides with $\liminf_{t \to t'} f(t)$. Thus g is equal almost everywhere $[\mu]$ to the *lower envelope* of f (see [**I**, Proposition 7.27]).

These facts regarding lower Darboux integrals have valid duals, of course. The dual arguments show that $\overline{J}(f)$ coincides with the Lebesgue integral over I of the *upper envelope* G of f, and that this equality is realized along any sequence $\{\mathcal{P}_n\}$ of partitions of I with $\operatorname{mesh}\mathcal{P}_n \to 0$. (Alternatively, one may employ the fact that

$$-D_{\mathcal{P}}(f) = d_{\mathcal{P}}(-f)$$

for any f and $\mathcal{P}$, and hence that

$$-\overline{J}(f) = \underline{J}(-f).)$$

We note for future reference that the above arguments show that for any $\varepsilon > 0$ we have both

$$d_{\mathcal{P}}(f) > \int_a^b g \, d\lambda_1 - \varepsilon \quad \text{and} \quad D_p(f) < \int_a^b G \, d\lambda_1 + \varepsilon$$

for any partition $\mathcal{P}$ of I with sufficiently small mesh. It should also be recalled that the upper and lower envelopes g and G are semicontinuous ([**I**, Proposition 7.27]) and therefore Borel measurable (Example 2.16).

Finally we observe that

$$\overline{J}(f) - \underline{J}(f) = \int_a^b (G - g) \, d\lambda_1.$$

Thus we have proved: *A bounded real-valued function f is Darboux integrable over I if and only if G = g almost everywhere* [λ_1], *i.e., if and only if f is continuous almost everywhere* [λ_1] *on I, and that, when this is the case, the Darboux integral J(f) coincides with the Lebesgue integral of f.*

Example 4.39. (The Riemann Integral) Just as in the preceding example, let f be a real-valued function defined and bounded on an interval $I = [a, b]$, and let $\mathcal{P} = \{t_0 < \ldots < t_n\}$ be a partition of I. If for each index $i = 1, \ldots, n$ the real number τ_i belongs to the ith subinterval $[t_{i-1}, t_i]$ of $\mathcal{P}$, then

$$R = \sum_{i=1}^{n} f(\tau_i)(t_i - t_{i-1})$$

is a *Riemann sum* for f based on $\mathcal{P}$ and the sequence $\{\tau_1, \ldots, \tau_n\}$. These sums provide the basis for still another notion of integral, namely the *Riemann integral*. We say that the function f is Riemann integrable over I if there exists a number $R(f)$ such that for any given $\varepsilon > 0$ there exists $\delta > 0$ with the property that
$$|R(f) - R| < \varepsilon$$

whenever R is a Riemann sum for f based on a partition $\mathcal{P}$ of I with mesh$\mathcal{P} < \delta$ and a sequence $\tau_i \in [t_{i-1}, t_i]$, $i = 1, \ldots, n$. This number $R(f)$, which is clearly unique if it exists, is called the *Riemann integral* of f *over I* and is written

$$R(f) = \int_a^b f(t)\, dt.$$

Since f is not assumed to be continuous, and need not assume either a greatest or a least value on any subinterval of I, the Darboux sums of f based on a partition $\mathcal{P}$ need not be among the Riemann sums for f based on $\mathcal{P}$. On the other hand, it is clear that $D_{\mathcal{P}}(f)$ and $d_{\mathcal{P}}(f)$ are, respectively, the supremum and infimum of the various Riemann sums for f based on $\mathcal{P}$. Hence if f is Riemann integrable over I, then

$$\lim_{\mathcal{P}}(D_{\mathcal{P}}(f) - d_{\mathcal{P}}(f)) = 0,$$

and f is therefore Darboux integrable over I as well.

Suppose, conversely, that f is Darboux integrable over I. Then, as was seen in Example 4.37, for given $\varepsilon > 0$ there exists $\delta > 0$ such that if $\mathcal{P}$ is a partition of I with mesh $\mathcal{P} < \delta$, then

$$J(f) - \varepsilon < d_{\mathcal{P}}(f) \leq D_{\mathcal{P}}(f) < J(f) + \varepsilon.$$

But then $|J(f) - R| < \varepsilon$ for any Riemann sum R for f based on $\mathcal{P}$. Thus f is also Riemann integrable over I, and $J(f)$ is its Riemann integral. We have proved the following result: *A bounded real-valued function f on an interval I = [a, b] is Riemann integrable over I if and only if f is continuous almost everywhere* $[\lambda_1]$ *on* I, *and when this is the case, the Riemann and Lebesgue integrals of* f *over* I *coincide.*

Remark 4.40. The definition of the Riemann integral makes sense for complex-valued functions, and a moment's reflection discloses that the results of Example 4.39 hold for complex-valued functions f. According to this example, the Riemann and Darboux integrals are exactly the same (for real-valued integrands). The true advantage of the Riemann integral lies in the greater flexibility in selecting approximating finite sums that is provided by its definition. Incidentally, the requirement in the definition of the Riemann integral that the function f be bounded results in no loss of generality. It is easily seen that the set of all Riemann sums for an unbounded function based on any one partition is also unbounded, no matter how that partition is chosen.

Remark 4.41. Examples 4.37 and 4.39 show that Riemann integration (at least for bounded functions on bounded real intervals) is merely the restriction of Lebesgue integration with respect to λ_1 to certain special functions. But that does not mean that the Riemann integral is therefore rendered superfluous. For it comes equipped with a powerful evaluation mechanism— the *fundamental theorem of calculus*—and as a matter of fact, for the most part, the Lebesgue integrals (with respect to λ_1) that can actually be evaluated in the customary sense are those that are Riemann integrals, and their evaluation is effected via the fundamental theorem, just as in ordinary calculus.

Example 4.42. Let $f : [a, b] \to \mathbb{R}$ be a continuously differentiable function on some real interval $[a, b]$ ([**I**, Example 3M]). Then f satisfies the hypotheses of the mean value theorem on every subinterval of $[a, b]$. Hence for any partition $\mathcal{P} = \{a = t_0 < \ldots < t_n = b\}$ there is a point τ_i in each subinterval $[t_{i-1}, t_i]$ such that

$$f(t_i) - f(t_{i-1}) = f'(\tau_i)(t_i - t_{i-1}),$$

and the corresponding Riemann sum $R = \sum_{i=1}^n f'(\tau_i)(t_i - t_{i-1})$ telescopes into $f(b) - f(a)$. Now f' is Riemann integrable, and R is accordingly arbitrarily close to the integral of f' over $[a, b]$ provided mesh $\mathcal{P}$ is small enough. Thus we have the well-known formula

$$\int_a^b f'(t)\, dt = f(b) - f(a).$$

Example 4.43. The function $\widetilde{g}$ of [**I**, Example 7X] has Riemann integral zero over any subinterval $[a, b]$ of $(0, +\infty)$, even though it is discontinuous at every rational number. The Riemann integral over $[0, 1]$ of the characteristic function of the Cantor set C exists (and equals zero). On the other hand, if $\widetilde{C}$ is a Cantor set in $[0, 1]$ with $\lambda_1(\widetilde{C}) > 0$ (recall Problem 3Y), then $\chi_{\widetilde{C}}$

is not Riemann integrable over $[0,1]$, while, of course, $\int_0^1 \chi_{\widetilde{C}} \, d\lambda_1 = \mu(\widetilde{C})$. Thus integration with respect to Lebesgue-Borel measure on a real interval constitutes a proper extension of Riemann integration on that interval.

Example 4.44. The function f defined by $f(0) = 1$ and

$$f(t) = \frac{\sin t}{t}, \quad 0 < t < +\infty,$$

is continuous and bounded on $[0, +\infty)$. Thus the Riemann and Lebesgue-Borel integrals of f over any interval $[0, T]$ exist and coincide. Define

$$F(T) = \int_0^T f(t) \, dt = \int_0^T f \, d\lambda_1, \quad 0 < T < +\infty.$$

Then $\lim_{T \to +\infty} F(T)$ exists. (If for any positive number T a nonnegative integer n is chosen so that $n\pi \le T < (n+1)\pi$, then $|F(T) - F(n\pi)| = \int_{n\pi}^T |f(t)| \, dt$. Since $\lim_{t \to +\infty} f(t) = 0$, this integral can be made as small as desired by taking T, and therefore n, large enough. Thus $\lim_{T \to +\infty} F(T)$ coincides with the sum of the (convergent) alternating series $\sum_{n=0}^{\infty} a_n$, where $a_n = \int_{n\pi}^{(n+1)\pi} f(t) \, dt, n \in \mathbb{N}_0$.) This limit, which is by definition the value of the convergent *improper* integral $\int_0^{+\infty} f(t) \, dt$, is *not* the Lebesgue-Borel integral of f over $[0, +\infty)$. Indeed, $\int_{[0,+\infty)} |f| \, d\lambda_1$ dominates the partial sums of the series $\sum_{n=0}^{\infty} |a_n|$, where the numbers a_n are as defined above, and since $|a_n| \ge 2/\pi(n+1) > 1/2(n+1)$, this latter series is divergent. Thus $\int_{[0,+\infty)} |f| \, d\lambda_1 = +\infty$. Moreover, f^+ and f^- both have integrals with respect to μ equal to $+\infty$ over $[0, +\infty)$, so $\int_{[0,+\infty)} f \, d\lambda_1$ is undefined. Thus Lebesgue-Borel integration over a ray is *not* an extension of (improper) Riemann integration over that ray.

Remark 4.45. In order to obtain the greatest possible flexibility in their application, the main convergence theorems have been so formulated as to apply to functions that are undefined on various sets of measure zero. To maintain this degree of generality in the statement of every theorem would be both pedantic and pointless. In the following problems, and throughout most of the balance of the book, we routinely assume that we are dealing with functions that are everywhere defined, even when (indeed, especially when) it is clear that they may be allowed to be undefined on null sets.

Problems

4A. Use Proposition 4.8 to give a new proof that if a function f is integrable with respect to a measure μ, then the support of f is σ-finite with respect to μ (see Problem 3R).

4B. Let X be measure space with measure μ, and let p be a positive number. The collection of all those measurable scalar-valued functions f on X with the property that $|f|^p$ is integrable $[\mu]$ is called the *Lebesgue space* of order p on X and is denoted by $\mathcal{L}^p = \mathcal{L}^p(X, \mu)$. (Note that the space of all integrable functions with respect to μ, which we have regularly denoted by $\mathcal{L}$, coincides in this definition with the space $\mathcal{L}^1(X, \mu)$.) Show that $\mathcal{L}^1$ is a vector space (see Problem 3B). Show also that, if $\mu(X) < +\infty$, and if $p \leq q$, then $\mathcal{L}_q \subset \mathcal{L}_p$. (The assumption that $\mu(X)$ is finite cannot be dropped. Indeed, if γ denotes the counting measure on the space $\mathbb{N}$, the inclusions are all reversed; if $p \leq q$, then $\mathcal{L}_p(\mathbb{N}, \gamma) \subset \mathcal{L}_q(\mathbb{N}, \gamma)$.)

Remark 4.46. Note that the Lebesgue space $\mathcal{L}^p(\mathbb{N}, \gamma)$ consists simply of the collection of all infinite sequences $\{\xi_n\}_{n=1}^{\infty}$ with the property that $\sum |\xi_n|^p < +\infty$. This space is usually denoted by ℓ^p. For the case $p = 1$, see Problem 1I. We undertake a more thorough study of the spaces $\mathcal{L}^p$ in Chapter 9.

4C. Let X be a measure space with measure μ, and suppose that f belongs to $\mathcal{L}^p(X, \mu)$ for some $p > 0$. Show that for every positive number α the set $E_\alpha = E(|f| \geq \alpha)$ satisfies the inequality

$$\mu(E_\alpha) \leq \frac{1}{\alpha^p} \int |f|^p \, d\mu.$$

(This generalization of Problem 3N(i) is sometimes called *Markov's inequality*. The case $p = 2$ is also known as *Tchebysheff's inequality*.) Show too that if $\int |f|^p \, d\mu = 0$ for some $p > 0$, then $f = 0$ almost everywhere $[\mu]$. (In the case $p = 1$ this last result provides a converse to Proposition 4.17.)

4D. Let $(X, \mathbf{S}, \mu)$ be a measure space with $\mu(X) > 0$, and let f be a scalar-valued function that is measurable $[\mu]$ on X. Then f is said to be *essentially bounded* with respect to μ, or *essentially bounded* $[\mu]$, if there exists a real number M such that $|f| \leq M$ almost everywhere $[\mu]$. Show that if f is essentially bounded, then there is a *smallest* real number M such that $|f| \leq M$ almost everywhere $[\mu]$. This number is the *essential supremum* of $|f|$ (notation: $M = \operatorname{ess\,sup}_X |f|$). More generally, if E is a measurable set, then f is *essentially bounded* on E if $f\chi_E$ is essentially bounded on X, and the essential supremum of $|f|$ on E (notation: $\operatorname{ess\,sup}_E |f|$) is, by definition, the essential supremum of $f\chi_E$. Show that if E is a set of finite measure and if f is measurable $[\mu]$ and essentially bounded on E, then f is integrable $[\mu]$ over E, and

$$\int_E |f| \, d\mu \leq (\operatorname{ess\,sup}_E |f|)\mu(E).$$

4E. Let X be a measure space with measure μ, let E be any measurable subset of X, and let f be a real-valued function that is integrable over E. Suppose that $\alpha \leq f \leq \beta$ almost everywhere on E, where α and β denote extended real numbers. Show that

$$\alpha\mu(E) \leq \int_E f \, d\mu \leq \beta\mu(E).$$

(If α, β, and $\mu(E)$ are all finite, the assumption that f is integrable over E may be replaced by the assumption that f is measurable over E, and the assertion is trivial. It is precisely when one or more of the numbers α, β, and $\mu(E)$ is infinite that care must be exercised.)

4F. Let $(X, \mathbf{S}, \mu)$ be a measure space, let f be a nonnegative measurable function on X, and let E be a measurable subset of X. Let us define $\nu_f(E) = \int_E f \, d\mu$ if f is integrable $[\mu]$ over E, and $\nu_f(E) = +\infty$ otherwise. Show that the set function thus defined on $\mathbf{S}$ is a measure on $(X, \mathbf{S})$. This measure is still called the *indefinite integral* of f, even if f is not integrable over X. It is easy to see that if f is not integrable $[\mu]$, then ν_f need not be concentrated on sets of finite measure with respect to μ. Show by example that it also need not satisfy the conclusion of Proposition 4.3. (It is clear, however, that ν_f is *always* absolutely continuous with respect to μ; see the remark following Problem 2V. If the notation $\int f \, d\mu = +\infty$ is permitted for nonnegative measurable functions f that are not integrable, then the set function ν_f has the simpler definition $\nu_f(E) = \int_E f \, d\mu$ for all $E \in \mathbf{S}$.)

4G. If f and g are two scalar-valued functions, each defined on some subset of X of full measure, and if α and β are scalars, let us agree to write $\alpha f + \beta g$ for the function $h(x) = \alpha f(x) + \beta g(x)$ defined on the intersection of the domains of definition of f and g. Suppose that $(X, \mathbf{S}, \mu)$ is a measure space and that f and g are both measurable $[\mu]$ on X. Show that $h = \alpha f + \beta g$ as just defined is again measurable $[\mu]$. This definition does *not* turn the collection $\mathcal{M}_\mu$ of all scalar-valued functions measurable $[\mu]$ into a vector space (why not?). However, if $\dot{\mathcal{M}}_\mu$ denotes the set of equivalence classes $\dot{f}$ of functions f in $\mathcal{M}_\mu$ with respect to the equivalence relation $f = g$ almost everywhere $[\mu]$, and if we write $\alpha\dot{f} + \beta\dot{g} = \dot{h}$ where $h = \alpha f + \beta g$, then $\dot{\mathcal{M}}_\mu$ *is a* vector space. Note the relation between this space and the quotient space $\mathcal{M}/\mathcal{M}_0$ of the space of all measurable functions on X taken modulo the submanifold $\mathcal{M}_0$ of functions vanishing almost everywhere $[\mu]$. Each element $\dot{f}$ of $\dot{\mathcal{M}}_\mu$ contains exactly one element of $\mathcal{M}/\mathcal{M}_0$, this element consisting of course of the functions in $\dot{f}$ that are everywhere defined.

4H. Let $(X, \mathbf{S}, \mu)$ be a measure space. Then subsets E and F of X are *almost equal* $[\mu]$ if and only if the symmetric difference $E \triangle F$ is a subset of a set of measure zero. Hence, if E and F are measurable and almost equal $[\mu]$, then $\mu(E) = \mu(F)$. Clearly the relation of being almost equal $[\mu]$ is an equivalence relation on $\mathbf{S}$. Let $\dot{E}$ denote the equivalence class of E and let $\dot{\mathbf{S}}$ denote the set of all such equivalence classes. The pair $(\dot{\mathbf{S}}, \dot{\mu})$, where $\dot{\mu}$ is the function on $\dot{\mathbf{S}}$ defined by $\dot{\mu}(\dot{E}) = \mu(E)$, is known as the *measure algebra* of $(X, \mathbf{S}, \mu)$. Show that when $\mu(X) < \infty$, $\rho(\dot{E}, \dot{F}) = \mu(E \triangle F)$ defines a metric on $\dot{\mathbf{S}}$. Show also that $\dot{\mathbf{S}}$ is complete as a metric space with respect to the metric ρ. (Hint: Start, as one always may in a completeness argument, with a sequence of sets $\{E_n\}$ having the property that $\sum_n \rho(\dot{E}_{n+1}, \dot{E}_n) < +\infty$ ([**I**, Proposition 8.6]). Show that there exists a null set Z with the property that the sequence $\{E_n \setminus Z\}$ is convergent to a limit E_0 (see [**I**, Example 1U]). Show likewise that the sequence $\{E_m \setminus Z) \triangle (E_n \setminus Z)\}_{n=1}^\infty$ converges to $(E_m \setminus Z) \triangle E_0$.)

4I. A measure space $(X, \mathbf{S}, \mu)$ is said to be *complete* if every subset of a set of measure zero with respect to μ is measurable and is, therefore, also a set of measure zero. (Another way to say this is that the collection $\mathbf{Z}$ of null sets is a σ-ideal not only in $\mathbf{S}$, but in the power class on X.) Show that if $(X, \mathbf{S}, \mu)$ is a complete measure space and if f is a function that is measurable $[\mu]$, then every function g such that $g = f$ almost everywhere $[\mu]$ is also measurable $[\mu]$. Similarly, if $(X, \mathbf{S}, \mu)$ is complete and if E belongs to $\mathbf{S}$, then every set F that is almost equal to $E[\mu]$ also belongs to $\mathbf{S}$ and we have $\mu(E) = \mu(F)$. Again, if Z is a null set in a complete measure space $(X, \mathbf{S}, \mu)$, then an arbitrary scalar-valued function defined on an arbitrary subset of X is measurable $[\mu]$ on Z.

4J. Let $(X, \mathbf{S}, \mu)$ be a measure space. Consider the collection $\overline{\mathbf{S}}$ of those subsets of X that are almost equal $[\mu]$ to some set E in $\mathbf{S}$. If F belongs to $\overline{\mathbf{S}}$ and if E_1 and E_2 are sets in $\mathbf{S}$ such that F and $E_i, i = 1, 2$, are almost equal $[\mu]$, then E_1 and E_2 are almost equal $[\mu]$ and therefore $\mu(E_1) = \mu(E_2)$ (see Problem 4H). Thus we may and do define $\overline{\mu}(F) = \mu(E_1) = \mu(E_2)$. Show that $\overline{\mathbf{S}}$ is a σ-algebra of subsets of X and that $\overline{\mu}$ is a measure on the measurable space $(X, \overline{\mathbf{S}})$. Show also that $(X, \overline{\mathbf{S}}, \overline{\mu})$ is complete, and that if $(X, \mathbf{T}, \nu)$ is any complete measure space such that $\mathbf{S} \subset \mathbf{T}$ and $\mu = \nu|\mathbf{S}$, then $\overline{\mathbf{S}} \subset \mathbf{T}$ and $\overline{\mu} = \nu|\overline{\mathbf{S}}$. The space $(X, \overline{\mathbf{S}}, \overline{\mu})$ is the *completion* of $(X, \mathbf{S}, \mu)$. Show that a property $P(x)$ holds almost everywhere $[\mu]$ on the measure space $(X, \mathbf{S}, \mu)$ if and only if it holds almost everywhere $[\overline{\mu}]$ on the completion $(X, \overline{\mathbf{S}}, \overline{\mu})$.

4K. Let $(X, \mathbf{S}, \mu)$ be a measure space and let $(X, \overline{\mathbf{S}}, \overline{\mu})$ be its completion. Let f be a function on X that is measurable [$\mathbf{S}$] and let g be a function on X such that $g = f$ almost everywhere $[\mu]$. Show that g is measurable [$\overline{\mathbf{S}}$]. Show, conversely, that if g is a function on X that is measurable [$\overline{\mathbf{S}}$], then there exists a function f that is measurable [$\mathbf{S}$] such that $g = f$ almost everywhere $[\mu]$. (Hint: Reduce the problem to the case of simple functions by writing $g = \sum_n g_n$, where the functions g_n are simple and measurable [$\overline{\mathbf{S}}$].)

Remark 4.47. This last result may also be phrased as follows. Consider the vector space $\mathcal{M}_{\widehat{\mu}}$ of equivalence classes of functions measurable [$\widehat{\mathbf{S}}$], where the equivalence relation is that of being equal almost everywhere $[\mu]$ (see Problem 4G). Then each element $\dot{g}$ of $\dot{\mathcal{M}}_{\widehat{\mu}}$ contains exactly one element $\dot{f}$ of the like space $\mathcal{M}_\mu$ formed using μ instead of $\widehat{\mu}$. Moreover, the functions belonging to $\dot{g}$ are precisely the various possible extensions of elements of $\dot{f}$ (to arbitrary subsets of X).

4L. Show that in the monotone convergence theorem the assumption that the functions f_n are nonnegative almost everywhere can be weakened. (The phrasing of the theorem in the text was chosen simply because it is more or less traditional.) Show likewise that if $\{f_n\}$ is a sequence of real functions on a measure space $(X, \mathbf{S}, \mu)$ such that each f_n is integrable $[\mu]$, and if $\{f_n\}$ is monotone *decreasing* almost everywhere, then $\lim_n f_n$ is integrable $[\mu]$ if and only if the numerical sequence $\{\int f_n \, d\mu\}$ is bounded (below), and that in this case we have $\int (\lim_n f_n) \, d\mu = \lim_n \int f_n \, d\mu$.

4M. The assumption that the functions f_n in Fatou's lemma are nonnegative almost everywhere *cannot* be dropped (example ?), but it can be relaxed, and the conclusion can also be strengthened slightly. Here is an alternative version of Fatou's lemma frequently found in the literature. Let X be a measure space with measure μ, and let $\{f_n\}$ be a sequence of real functions on X such that each f_n is integrable $[\mu]$. Suppose that (i) there exists a function φ, integrable $[\mu]$ over X, such that $\varphi \leq f_n$ almost everywhere for all n, and (ii) $\liminf_n \int f_n \, d\mu < +\infty$. Then the pointwise inferior limit of $\{f_n\}$ is integrable $[\mu]$ and we have

$$\int \left(\liminf_n f_n \right) d\mu \leq \liminf_n \int f_n \, d\mu.$$

(Hint: In the proof of Fatou's lemma as given in the text we actually have $\int g_n \, d\mu \leq \inf_{k \geq n} \int f_k \, d\mu$.)

4N. Use the preceding problem to obtain the following theorem. Let X be a measure space with measure μ, and let $\{f_n\}$ be a sequence of real functions measurable $[\mu]$ on X. Suppose that there exist functions φ and Φ, both integrable $[\mu]$ over X, such that

$\varphi \leq f_n \leq \Phi$ almost everywhere $[\mu]$ for all n. Then both $\liminf_n f_n$ and $\limsup_n f_n$ are integrable $[\mu]$ and we have

$$\int (\liminf_n f_n)\, d\mu \leq \liminf_n \int f_n\, d\mu$$

$$\leq \limsup_n \int f_n\, d\mu \leq \int (\limsup_n f_n)\, d\mu.$$

Use this fact to give a new proof of the dominated convergence theorem.

4O. Let $(X, \mathbf{S}, \mu)$ be a measure space, and let $\{f_n\}$ be a sequence of integrable functions on X that is Cauchy in the mean and converges almost everywhere $[\mu]$ to a limit f. Use Fatou's lemma to show that $\{f_n\}$ also converges to f in the mean.

4P. (i) Give an example of a measure space $(X, \mathbf{S}, \mu)$ and a sequence of functions $\{f_n\}$ on X converging pointwise to zero such that $\int f_n\, d\mu = 0$ and $\int |f_n|\, d\mu = 1$ for every positive integer n. (Thus $f_n \to f$ pointwise, together with the condition that $\int f_n\, d\mu \to \int f\, d\mu$, does *not* imply in general that $\{f_n\}$ converges to f in the mean.)

(ii) On the other hand, if $(X, \mathbf{S}, \mu)$ is a measure space, and if $\{f_n\}$ is a sequence of real-valued integrable functions on X that converges pointwise to an integrable limit f in such a way that the numerical sequence $\{\int f_n\, d\mu\}$ converges to $\int f\, d\mu$, and *if there exists a single integrable function g such that $f_n \geq g$ almost everywhere for every n* (or such that $f_n \leq g$ almost everywhere for every n, or even such that one or the other holds for every n) then $\{f_n\}$ *does* converge to f in the mean. (Hint: Suppose $f_n \geq g$ for all n. Show that $(f_n - f)^- \leq f_n$, and hence that $\int (f_n - f)^-\, d\mu \to 0$ by the dominated convergence theorem. Then verify that the sequence $\{\int (f_n - f)^+\, d\mu\}$ also tends to zero.)

4Q. The hypotheses of Theorem 4.31 are necessary as well as sufficient. Indeed, if $\{f_n\}$ is a sequence of integrable functions that is Cauchy in the mean on a measure space $(X, \mathbf{S}, \mu)$, then the indefinite integrals of the functions $|f_n|$ are uniformly absolutely continuous and uniformly concentrated on sets of finite measure with respect to μ.

4R. Let γ denote the counting measure on the set $\mathbb{N}$ of positive integers, and for each $m \in \mathbb{N}$ let $e_m = \{e_{m,n}\}$ denote the sequence defined by

$$e_{m,n} = \begin{cases} 0, & m \neq n, \\ 1, & m = n. \end{cases}$$

The indefinite integral ν_m of e_m with respect to γ is just the unit point mass at m, and the sequence $\{\nu_m\}$ is certainly uniformly absolutely continuous with respect to γ, but $\{e_m\}_{m=1}^{\infty}$ converges pointwise to the sequence zero, while $\int e_m d\gamma \equiv 1$. Show that the requirement that the sequence $\{\nu_n\}$ be uniformly concentrated on sets of finite measure cannot be dropped in general, even in Corollary 4.32.

4S. Let δ denote the discrete measure on $\mathbb{N}$ obtained by assigning to each positive integer n the weight $w_n = 1/2^n$, and let $f_n = 2^n e_n$, with e_n as in the preceding problem. Then f_n is integrable $[\delta]$, when regarded as a function on $\mathbb{N}$, and the sequence $\{f_n\}$ tends pointwise to zero. Moreover, $\delta(\mathbb{N}) = 1$. Nevertheless, we have $\int f_n d\delta = 1$ for every n. Thus the requirement that the sequence $\{\nu_n\}$ be uniformly absolutely continuous cannot be dropped either, even in Corollary 4.32.

4T. Let $(X, \mathbf{S}, \mu)$ be a measure space such that $\mu(X) < +\infty$, and let $\{f_n\}$ be a sequence of integrable functions on X converging uniformly to a scalar-valued function f. Show that f is integrable and that $\{f_n\}$ converges to f in the mean. Show by example that if the requirement that μ be finite is dropped, then it is possible for the limit f not to be integrable. Show also that even if f is integrable, the sequence $\{\int f_n \, d\mu\}$ may fail to converge, or may converge to some limit other than $\int f \, d\mu$.

4U. Let $(X, \mathbf{S}, \mu)$ be a measure space, and let $\{f_\lambda\}_{\lambda \in \Lambda}$ be a net of real-valued functions defined everywhere on X and integrable $[\mu]$. Suppose that the net $\{f_\lambda\}$ is monotone increasing in the sense that $f_\lambda \leq f_{\lambda'}$ almost everywhere whenever $\lambda \leq \lambda'$, and suppose finally that the index family Λ is countably determined ([**I**, Problem 4K]). Show that the pointwise limit $\lim_\lambda f_\lambda$ is integrable if and only if the numerical net $\{\int f_\lambda \, d\mu\}_{\lambda \in \Lambda}$ is bounded, and in this case we have $\int (\lim_\lambda f_\lambda) \, d\mu = \lim_\lambda \int f_\lambda \, d\mu$. (Hint: See [**I**, Problems 1I and 4K].) Show by example that the assumption that Λ is countably determined cannot be dropped.

Remark 4.48. The reason for assuming the functions f_λ to be everywhere defined is to ensure that the numerical nets $\{f_\lambda(x)\}_{\lambda \in \Lambda}$ are defined on a substantial set of points. Trivial examples show that if each f_λ is permitted to be undefined on its own null set, then every point of X may fail to belong to the domain of definition of f_λ for all λ in a cofinal subset of Λ (even when Λ is countably determined). In these circumstances it is somewhat questionable what "$\lim_\lambda f_\lambda(x)$" is to mean. What can be obtained is the following highly technical result. Suppose all of the hypotheses of the theorem in Problem 4U are satisfied, except that each f_λ is only assumed to be defined almost everywhere $[\mu]$. For each point $x \in X$ let Λ_x denote the set of those λ such that f_λ is defined at x. Suppose also that the net $\{\int f_\lambda \, d\mu\}$ is bounded. Then there exists a measurable set E such that $\mu(X \setminus E) = 0$ and such that for every $x \in E$ the set Λ_x is a directed cofinal subset of Λ. Furthermore, if we define a function f by setting $f(x) = \lim_{\lambda \in \Lambda_x} f_\lambda(x)$ at each point of E, then f is integrable $[\mu]$ and we have $\int f \, d\mu = \lim_{\lambda \in \Lambda} \int f_\lambda \, d\mu$.

4V. Show that the theorem of Beppo-Levi (Corollary 4.25) holds for indexed sums of integrable functions provided that the index family is countable. (Simple examples show that it does not hold in general for uncountable indexed sums.)

4W. Let $(X, \mathbf{S}, \mu)$ be a measure space, and let Λ be a countably determined directed set of indices. Suppose given a net of functions $\{f_\lambda\}_{\lambda \in \Lambda}$ converging pointwise to a limit f, where each f_λ is everywhere defined on X and integrable $[\mu]$. For each index λ let ν_λ denote the indefinite integral of f_λ, and suppose the family $\{\nu_\lambda\}_{\lambda \in \Lambda}$ is uniformly absolutely continuous and uniformly concentrated on sets of finite measure with respect to μ. Show that f is integrable $[\mu]$ and that the net $\{f_\lambda\}$ converges to f in the mean. State and prove a dominated convergence theorem for nets in place of sequences.

4X. Let $(X, \mathbf{S}, \mu)$ be a measure space, let f be a scalar-valued function on X that is measurable $[\mu]$, and let $\{f_n\}$ be a sequence of functions each of which is measurable $[\mu]$. Write $E_{\varepsilon,n} = E(|f - f_n| \geq \varepsilon)$ for $\varepsilon > 0$ and every $n \in \mathbb{N}$. Then $\{f_n\}$ is said to converge to f in *measure* if $\lim_n \mu(E_{\varepsilon,n}) = 0$ for every $\varepsilon > 0$. If the functions f and f_n are all everywhere defined, and if the sequence $\{f_n\}$ converges uniformly to f, then it also converges to f in measure. Show that if $\mu(X) < +\infty$ and if $\{f_n\}$ merely converges to f almost everywhere, then $\{f_n\}$ also converges to f in measure. (Convergence in measure is studied in more detail in Chapter 9.)

4Y. Let $(X, \mathbf{S}, \mu)$ be a measure space. Show that if the functions f and f_n of the preceding problem are all integrable $[\mu]$, and if the sequence $\{f_n\}$ converges to f in the mean, then it also converges to f in measure. Show, in addition, that Theorems 4.29, 4.31, and 4.35 are all valid with convergence almost everywhere replaced by convergence in measure. Show, finally, that if $\{f_n\}$ converges to f in measure, then a subsequence $\{f_{n_k}\}$ converges to f almost everywhere $[\mu]$. (Hint: Choose f_{n_k} so $\mu(E(|f - f_{n_k}| \geq 1/k)) < 1/2^k$.)

4Z. Consider a function $f : I \to \mathbb{R}$, where $I = [a, b]$ is a finite interval. Show that the following conditions are equivalent.

(i) f is Riemann integrable.
(ii) Given $\varepsilon > 0$, there exist continuous functions $g, h : I \to \mathbb{R}$ such that $g \leq f \leq h$ and $\int_I (h - g) \, d\lambda_1 < \varepsilon$.
(iii) Given $\varepsilon > 0$, there exist polynomials g, h with real coefficients such that $g \leq f \leq h$ on I and $\int_I (h - g) \, d\lambda_1 < \varepsilon$.

4AA. This problem provides an interesting use of the Riemann integral in proving a theorem in analysis. Consider a power series $f(x) = \sum_{n=0}^{\infty} a_n x^n$ with $a_n \geq 0$ for $n \in \mathbb{N}_0$ and with the property that the series converges in $\mathbb{R}$ for $x \in (0, 1)$ and satisfies

$$\lim_{x \uparrow 1} (1 - x) f(x) = 1.$$

(i) Show that $\lim_{x \uparrow 1} (1 - x) \sum_{n=0}^{\infty} a_n x^n p(x^n) = \int_{[0,1]} p \, d\lambda_1$ for every polynomial p.
(ii) Show that the result in (i) also holds for any Riemann integrable function p defined on $[0, 1]$. (Hint: Use Problem 4Z(iii).)
(iii) Apply (ii), with p the characteristic function of an interval, to conclude that $\lim_N N^{-1} \sum_{n=0}^{N} a_n = 1$.

The proof suggested above is due to Karamata. Part (iii) is an example of a Tauberian theorem, which is a partial converse to the following result. If $\{a_n\} \subset \mathbb{C}$ satisfies $\lim_N N^{-1} \sum_{n=0}^{N} a_n = 1$, then $\lim_{x \uparrow 1} (1 - x) \sum_{n=0}^{\infty} a_n x^n = 1$.

4BB. Consider a compact Hausdroff topological space X, and a subalgebra $\mathcal{A}$ of the algebra $\mathcal{C}(X)$ of all continuous complex-valued functions on X. Assume that $\mathcal{A}$ contains the constant functions, and for every two distinct points $x, y \in X$ there exists a *real-valued* function $u \in \mathcal{A}$ such that $u(x) \neq u(y)$. Stone's extension of the Weierstrass approximation theorem states that, under these assumptions, every function $f \in \mathcal{C}(X)$ can be approximated uniformly on X by functions in $\mathcal{A}$. This problem sketches a proof discovered by V. Machado. Assume, to get a contradiction, that there exists a function $f \in \mathcal{C}(X)$ such that the number

$$d = \inf_{g \in \mathcal{A}} \sup_{x \in X} |f(x) - g(x)|$$

is not zero. Denote by $\mathcal{F}$ the collection of those closed nonempty subsets $C \subset X$ with the property that

$$\inf_{g \in \mathcal{A}} \sup_{x \in C} |f(x) - g(x)| = d.$$

We have $\mathcal{F} \neq \varnothing$ since $X \in \mathcal{F}$.

(i) Let $\mathcal{C} \subset \mathcal{F}$ be a collection which is totally ordered by inclusion. Show that $\bigcap_{C \in \mathcal{C}} C \in \mathcal{F}$.

(ii) Consider a minimal element $C_0 \in \mathcal{F}$, that is, C_0 has no proper closed subset which belongs to $\mathcal{F}$. Assume that C_0 contains at least two elements, and select a real-valued function $u \in \mathcal{A}$ such that $\sup_{x \in C_0} u(x) = 1$ and $\inf_{x \in C_0} u(x) = 0$. Define closed sets $C_1 = \{x \in C_0 : u(x) \leq 2/3\}$ and $C_2 = \{x \in C_0 : u(x) \geq 1/3\}$ so that C_1 and C_2 are proper closed subsets of C_0 such that $C_1 \cup C_2 = C_0$. Choose functions $g_j \in \mathcal{A}$ with the property that $\sup_{x \in C_j} |f(x) - g_j(x)| < d$ for $j = 1, 2$, and define

$$g = [1 - (1 - u^n)^{2^n}]g_1 + (1 - u^n)^{2^n} g_2 \in \mathcal{A}.$$

Show that $\sup_{x \in C_0} |f(x) - g(x)| < d$ if n is sufficiently large.

(iii) Derive a contradiction to the assumption that $d > 0$.

4CC. A trigonometric polynomial (with period 1) is a function of the form

$$q(x) = \sum_{n=-N}^{N} c_n e^{2\pi i n x}, \quad x \in \mathbb{R},$$

where $N \in \mathbb{N}$ and $c_n \in \mathbb{C}$ for $|n| \leq N$. Let $f : [0, 1] \to \mathbb{C}$ be a continuous function such that $f(0) = f(1)$. Prove the Weierstrass approximation theorem for trigonometric polynomials: *for every $\varepsilon > 0$ there exists a trigonometric polynomial q such that $|f(x) - q(x)| < \varepsilon$ for all $x \in [0, 1]$.* If f is real-valued then q can be chosen to be real-valued as well.

4DD. In Problem 4Z, assume that $I = [0, 1]$ and show that conditions (i–iii) are also equivalent to the following condition.

(iv) Given $\varepsilon > 0$, there exist real-valued trigonometric polynomials g, h such that $g \leq f \leq h$ on I and $\int_I (h - g) \, d\lambda_1 < \varepsilon$.

4EE. Consider a number $\xi \in \mathbb{R} \setminus \mathbb{Q}$, and denote by $\xi_n = n\xi - [n\xi] \in [0, 1]$ the fractional part of $n\xi$. Use Problem 4DD to show that

$$\lim_{n \to \infty} \frac{f(\xi_1) + f(\xi_2) + \cdots + f(\xi_n)}{n} = \int_0^1 f(t) \, dt$$

for every Riemann integrable function $f : [0, 1] \to \mathbb{R}$. In particular,

$$\lim_{n \to \infty} \frac{\mathrm{card}\{j : \xi_j \in [a, b], j = 1, 2, \ldots, n\}}{n} = b - a$$

for every interval $[a, b] \subset [0, 1]$. (Sequences $\{\xi_n\}$ which satisfy this last condition are said to be *equidistributed* in $[0, 1]$. The result in this problem and its proof are due to H. Weyl.)

4FF. Consider real numbers $\xi_1, \xi_2, \ldots, \xi_d$ which are linearly independent over the rational field $\mathbb{Q}$, and let $A = [a_1, b_1] \times \cdots \times [a_d, b_d]$ be a cell contained in $[0, 1]^d$. Show that the limit

$$\lim_{n \to \infty} \frac{\mathrm{card}\{j \in \{1, \ldots, n\} : j\xi_i - [j\xi_i] \in [a_i, b_i] \text{ for all } i = 1, \ldots d\}}{n}$$

exists and equals $\lambda_d(A)$.

4GG. Let $f : \mathbb{R} \to \mathbb{R}$ be an absolutely continuous function. Show that f maps every λ_1-null set to a λ_1-null set.

4HH. Let $(X, \mathbf{S}, \mu)$ be a measure space, let $\mathcal{F}$ be a family of complex-valued functions integrable $[\mu]$ over X. Show that $\mathcal{F}$ is uniformly integrable $[\mu]$ if and only if the following three conditions are satisfied.

  (i) $\sup_{f \in \mathcal{F}} \int_X |f| \, d\mu < +\infty$.
  (ii) $\lim_{R \to +\infty} \sup_{f \in \mathcal{F}} \int_{|f| \geq R} |f| \, d\mu = 0$.
  (iii) $\lim_{\varepsilon \to 0} \sup_{f \in \mathcal{F}} \int_{|f| < \varepsilon} |f| \, d\mu = 0$.

Find examples of uniformly integrable sequences $\{f_n\}$ which satisfy precisely two of the above conditions and which converge to zero almost everywhere $[\mu]$.

4II. Let $(X, \mathbf{S}, \mu)$ be a probability space. A family $\{E_n\}_{n \in \mathbb{N}}$ of measurable sets in X is said to be *independent* if the identity

$$\mu(A_1 \cap A_2 \cap \cdots \cap A_n) = \mu(A_1)\mu(A_2) \cdots \mu(A_n)$$

holds for every $n \in \mathbb{N}$ and for every choice $A_i \in \{E_i, X \setminus E_i\}$, $i = 1, \dots, n$. Suppose that the function N defined in Example 4.28 is finite on some set of positive measure. Show that $\sum_{n=1}^{\infty} \mu(E_n) < +\infty$.

Chapter 5
Existence and uniqueness of measures

As was shown in Chapter 3, for any measurable space $(X, \mathbf{S})$, the correspondence between the set of measures on $(X, \mathbf{S})$ and the set of Lebesgue integrals on $(X, \mathbf{S})$ is a bijection (Theorems 3.29 and 3.37). This knowledge is of small value, however, unless one has in hand a good supply of measures to be integrated with respect to. In this chapter we discuss some of the more important ways in which measures arise.

While such uncomplicated measures as the discrete ones introduced in Chapter 3 can be described directly, most of the measures of interest in mathematical analysis are obtained through some indirect procedure. Generally speaking, this procedure consists of a process of extension or approximation, sometimes encompassing several stages, starting from some more primitive set function, not itself a measure. (For example, a discrete measure may be thought of as obtained by extending the set function consisting of the weights on the collection of singletons.) Historically, many such processes have been used in many different constructions, and a general description covering all of them is not practical. Nearly all methods of generating measures, however, share a common penultimate stage, consisting of a set function known as an *outer measure*. Accordingly, we begin with a brief treatment of this notion.

Definition 5.1. A nonnegative, extended real-valued set function μ^* defined on the collection of all subsets of a set X is an *outer measure* on X if it is countably subadditive and satisfies the condition $\mu^*(\varnothing) = 0$. If μ^* is an outer measure on X and $A \subset X$, then the value of $\mu^*(A)$ is called the *outer measure of A*.

Observe that an outer measure μ^* on a set X is automatically finitely subadditive, and therefore monotone, on 2^X: If $A \subset B \subset X$, then $\mu^*(A) \leq \mu^*(B)$.

Example 5.2. Let $(X, \mathbf{S}, \mu)$ be a measure space, and for each $A \subset X$ set

$$\mu^*(A) = \inf\{\mu(E) \colon A \subset E \text{ and } E \in \mathbf{S}\}.$$

© Springer International Publishing Switzerland 2016
H. Bercovici et al., *Measure and Integration*,
DOI 10.1007/978-3-319-29046-1_5

If $\{A_n\}_{n=1}^{\infty}$ is an arbitrary sequence of subsets of X and ε is a positive number, then for each n there is a set E_n in $\mathbf{S}$ such that $A_n \subset E_n$ and $\mu(E_n) \leq \mu^*(A_n) + \varepsilon/2^n$. Thus we have

$$\mu\left(\bigcup_{n=1}^{\infty} E_n\right) \leq \sum_{n=1}^{\infty} \mu^*(A_n) + \varepsilon.$$

Since $\bigcup A_n \subset \bigcup E_n$, this shows that

$$\mu^*\left(\bigcup_{n=1}^{\infty} A_n\right) \leq \sum_{n=1}^{\infty} \mu^*(A_n) + \varepsilon,$$

and since ε is arbitrary, it follows that μ^* is countably subadditive. Thus μ^* is an outer measure on X, and it is clear that μ^* extends the given measure μ. Furthermore, it follows easily from this argument that for every subset $A \subset X$ there is a set $E \in \mathbf{S}$ such that $A \subset E$ and $\mu^*(E) = \mu(E) = \mu^*(A)$ (see Problem 3I).

Example 5.3. Let μ^* be an outer measure on a set X, and let $A \subset X$. Then the restriction $\mu^*|2^A$ of μ^* to the power class on A is clearly an outer measure on A. We denote this outer measure by $\mu^*|A$.

Example 5.4. Let μ^* be an outer measure on a set X and let $A \subset X$. Then setting

$$\mu_A^*(M) = \mu^*(M \cap A)$$

for each $M \subset X$ defines a new outer measure μ_A^*. It is important to note that μ_A^*—the *concentration* of μ^* on A, as we call it—is an outer measure on X, and must be distinguished, in general, from the restriction $\mu^*|A$ introduced in Example 5.3.

Definition 5.5. Let μ^* be an outer measure on a set X, and let $E \subset X$. Then E is *measurable with respect to* μ^*, or, briefly, *measurable* $[\mu^*]$, if E splits every set additively with respect to μ^*, that is, if for every $A \subset X$ we have

$$\mu^*(A) = \mu^*(A \cap E) + \mu^*(A \setminus E).$$

Note that it follows from the subadditivity of μ^* that to show that a set E is measurable $[\mu^*]$, it suffices to verify the inequality

$$\mu^*(A \cap E) + \mu^*(A \setminus E) \leq \mu^*(A)$$

for every $A \subset X$.

Example 5.6. Let μ^* be an outer measure on a set X and suppose Z is a subset of X such that $\mu^*(Z) = 0$. Then for any $A \subset X$ we have $\mu^*(A \cap Z) \leq \mu^*(Z) = 0$, and therefore

$$\mu^*(A \cap Z) + \mu^*(A \setminus Z) = \mu^*(A \setminus Z) \le \mu^*(A),$$

by monotonicity. Thus every set Z of outer measure zero is measurable $[\mu^*]$.

Example 5.7. Let μ^* be an outer measure on a set X, let $A \subset X$, and let μ_A^* be the concentration of μ^* on A (Example 5.4). If $E \subset X$ is measurable $[\mu^*]$, then for any subset $M \subset X$, the intersection $M \cap A$ is split additively by E, so

$$\mu^*((M \cap A) \cap E) + \mu^*((M \cap A) \setminus E) = \mu^*(M \cap A).$$

Equivalently,

$$\mu_A^*(M \cap E) + \mu_A^*(M \setminus E) = \mu_A^*(M),$$

so that E is also measurable $[\mu_A^*]$.

Proposition 5.8. *Let μ^* be an outer measure on a set X and let* **M** *denote the collection of all sets measurable with respect to μ^*. Then* **M** *is a σ-algebra of subsets of X containing all sets of outer measure zero with respect to μ^*, so $(X, \mathbf{M})$ is a measurable space. Moreover, $\mu^*|\mathbf{M}$ is a complete measure on $(X, \mathbf{M})$ (Problem 4I). (The measure $\mu^*|\mathbf{M}$ is called the measure defined by μ^*.)*

Proof. It is clear from the definition that the collection **M** is complemented, and it was shown in Example 5.6 that **M** contains all sets of outer measure zero with respect to μ^*. In particular, both $\varnothing$ and X belong to **M**. It remains to show that **M** is a σ-ring and that μ^* is countably additive on **M** (the completeness of $\mu^*|\mathbf{M}$ is immediate).

Suppose, to begin with, that $E, F \in \mathbf{M}$. Then E and F partition X into four subsets as follows: $E_{11} = E \cap F$, $E_{12} = E \setminus F$, $E_{21} = F \setminus E$, and $E_{22} = X \setminus (E \cup F)$. For an arbitrary subset A of X let $A_{ij} = A \cap E_{ij}$, $i, j = 1, 2$. Then, since A_{11} and A_{12} are the sets into which F splits $A \cap E$, and since F is measurable $[\mu^*]$, we have

$$\mu^*(A \cap E) = \mu^*(A_{11}) + \mu^*(A_{12}).$$

Similarly, $\mu^*(A \setminus E) = \mu^*(A_{21}) + \mu^*(A_{22})$, and thus, since E is also measurable $[\mu^*]$,

$$\mu^*(A) = \mu^*(A_{11}) + \mu^*(A_{12}) + \mu^*(A_{21}) + \mu^*(A_{22}). \tag{5.1}$$

From this equation it follows readily that all four of the sets E_{ij}, as well as their various unions, are measurable $[\mu^*]$, and hence that **M** is a ring. (For example, we have $\mu^*(A \setminus E_{11}) \le \mu^*(A_{12}) + \mu^*(A_{21}) + \mu^*(A_{22})$ and therefore $\mu^*(A \cap E_{11}) + \mu^*(A \setminus E_{11}) \le \mu^*(A)$, and since this is true for an arbitrary subset $A \subset X$, it follows that E_{11} is measurable $[\mu^*]$.) There is yet another consequence of (5.1). If E and F are disjoint and if A is a subset of $E \cup F$, then $A_{11} = A_{22} = \varnothing$, and we have

$$\mu^*(A) = \mu^*(A_{12}) + \mu^*(A_{21}).$$

More generally, by mathematical induction we obtain

$$\mu^*(A) = \mu^*(A \cap E_1) + \cdots + \mu^*(A \cap E_n), \tag{5.2}$$

where $\{E_1, \ldots, E_n\}$ is any finite pairwise disjoint collection of sets belonging to $\mathbf{M}$ and A is an arbitrary subset of $E_1 \cup \cdots \cup E_n$. In particular, μ^* is finitely additive on $\mathbf{M}$.

Thus far it has been shown that $\mathbf{M}$ is a ring of sets and that $\mu^*|\mathbf{M}$ is a finitely additive set function on $\mathbf{M}$. By a standard argument (Problem 1D), the proof will be complete if we show that for an arbitrary pairwise disjoint infinite sequence $\{E_n\}_{n=1}^\infty$ of sets in $\mathbf{M}$, the union $E = \bigcup_n E_n$ belongs to $\mathbf{M}$ and $\mu^*(E) = \sum_n \mu^*(E_n)$. To see that this is indeed so, set

$$F_n = E_1 \cup \cdots \cup E_n, \quad n \in \mathbb{N},$$

and let $A \subset X$. Then $\mu^*(A) = \mu^*(A \cap F_n) + \mu^*(A \setminus F_n)$ since $F_n \in \mathbf{M}$. Hence, writing $A_n = A \cap E_n$, $n \in \mathbb{N}$, we obtain

$$\mu^*(A \setminus E) + \mu^*(A_1) + \cdots + \mu^*(A_n) \leq \mu^*(A)$$

by virtue of (5.2). Since this holds for every n, we conclude that

$$\mu^*(A \setminus E) + \mu^*(A \cap E) \leq \mu^*(A \setminus E) + \sum_{n=1}^\infty \mu^*(A_n) \leq \mu^*(A), \tag{5.3}$$

whence it follows that equality holds throughout (5.3). Hence E is measurable $[\mu^*]$. Finally, for $A = E$, (5.3) reduces to $\mu^*(E) = \sum_{n=1}^\infty \mu^*(E_n)$. $\square$

Example 5.9. Let $(X, \mathbf{S}, \mu)$ be a measure space, let μ^* be the outer measure on X introduced in Example 5.2, and let $E \in \mathbf{S}$. For an arbitrary subset A of X there exists a set $F \in \mathbf{S}$ such that $A \subset F$ and $\mu(F) = \mu^*(A)$. But then $A \cap E \subset F \cap E$ and $A \setminus E \subset F \setminus E$, and we have

$$\mu^*(A \cap E) + \mu^*(A \setminus E) \leq \mu(F \cap E) + \mu(F \setminus E) = \mu(F) = \mu^*(A),$$

which shows that E is measurable $[\mu^*]$. Thus the given σ-algebra $\mathbf{S}$ is contained in the σ-algebra $\mathbf{M}$ of sets measurable $[\mu^*]$, and the measure $\widetilde{\mu} = \mu^*|\mathbf{M}$ defined by μ^* is an extension of the given measure μ. Moreover, since $\widetilde{\mu}$ is automatically complete, it is an extension of the completion of μ as well (Problem 4J).

Example 5.10. It can happen that an outer measure μ^* on a set X is actually a measure (on $(X, 2^X)$). When this is the case, μ^* is itself the measure defined by μ^*.

According to Proposition 5.8, an outer measure always gives rise in a natural way to a measure. We next focus attention on the problem of generating outer measures. Our main tool in this investigation will be a very primitive type of set function known as a *gauge*.

Definition 5.11. Let $\mathcal{C}$ be a collection of subsets of a set X such that $\varnothing \in \mathcal{C}$. A *gauge* on X is a function $\gamma : \mathcal{C} \to [0, +\infty]$ such that $\gamma(\varnothing) = 0$.

Example 5.12. Let f be a monotone increasing, real-valued function defined on $\mathbb{R}$ and, for each pair of real numbers a, b such that $a \le b$, set $\gamma_f((a, b]) = f(b) - f(a)$. Then ρ_f is a nonnegative real-valued set function defined on the collection $\mathcal{H}$ of all half-open intervals $(a, b]$, and it is obvious that ρ_f is a gauge on $\mathbb{R}$. This gauge is easily seen to extend to a finitely additive set function ν_f on the ring $\mathbf{H}_r$ generated by $\mathcal{H}$.

Example 5.13. Let f be as in the preceding example and, for each pair of rational numbers a, b such that $a \le b$, set $\widetilde{\gamma}_f((a, b] \cap \mathbb{Q}) = f(b) - f(a)$. Then $\widetilde{\gamma}_f$ is a gauge on $\mathbb{Q}$, and this gauge also extends at once to a finitely additive set function $\widetilde{\nu}_f$ on the ring $\mathbf{H}_r^{\sim}$ generated by the collection $\widetilde{\mathcal{H}}$ consisting of sets of the form $(a, b]^{\sim} = (a, b] \cap \mathbb{Q}$ with $a, b \in \mathbb{Q}$, $a < b$.

Example 5.14. Let $\mathcal{B}$ denote the collection of all bounded sets in a metric space X and set $\gamma_0(\varnothing) = 0$ and $\gamma_0(B) = 1$ for all $B \ne \varnothing$ in $\mathcal{B}$. Then γ_0 is a gauge on X. So is the set function $\gamma_1(B) = \mathrm{diam}(B)$, $B \in \mathcal{B}$.

There is a standard procedure for generating an outer measure from a given gauge.

Proposition 5.15. *Let $\mathcal{C}$ be a collection of subsets of a set X, and let γ be a gauge defined on $\mathcal{C}$. We define $\mu^* : 2^X \to [0, +\infty]$ as follows:*

1. $\mu^*(\varnothing) = 0$,
2. $\mu^*(A) = +\infty$ *if $A \ne \varnothing$ has no countable covering by sets in $\mathcal{C}$, and*
3. *if $A \ne \varnothing$ has such a covering,*

$$\mu^*(A) = \inf \left\{ \sum_{C \in \mathcal{D}} \gamma(C) : \mathcal{D} \subset \mathcal{C}, A \subset \bigcup_{C \in \mathcal{D}} C \right\}, \tag{5.4}$$

where the infimum is taken over all the countable coverings $\mathcal{D} \subset \mathcal{C}$ of A.

Then μ^* is an outer measure on X. (The outer measure μ^* defined by (5.4) is called the outer measure *generated by γ.)*

Proof. Clearly $\mu^* : 2^X \to [0, +\infty]$, and $\mu^*(\varnothing) = 0$. Moreover, if C is any set in $\mathcal{C}$, then $\mu^*(C) \le \gamma(C)$ since the singleton $\{C\}$ is one of the countable coverings employed in (5.4). It remains to verify that μ^* is countably subadditive. To this end let $\{A_m\}_{m=1}^{\infty}$ be an arbitrary infinite sequence of subsets of X, set $A = \bigcup_m A_m$ and fix $\varepsilon > 0$. It suffices to consider the case $\mu^*(A) < \infty$, in which case $\mu^*(A_m) < \infty$ for all $m \in \mathbb{N}$. Choose for each m

a countable covering $\{C_n^m\}_{n=1}^\infty$ of A_m consisting of sets belonging to $\mathcal{C}$ and satisfying the inequality

$$\sum_{n=1}^\infty \gamma(C_n^m) \le \mu^*(A_m) + \varepsilon/2^m.$$

But then the doubly indexed family $\{C_n^m\}$ is a countable covering of A, and we have

$$\sum_{m,n} \gamma(C_n^m) = \sum_m \sum_n \gamma(C_n^m) \le \sum_m \mu^*(A_m) + \varepsilon.$$

Hence $\mu^*(A) \le \sum_m \mu^*(A_m) + \varepsilon$ and the proposition follows since ε is arbitrary.   □

Example 5.16. If $(X, \mathbf{S}, \mu)$ is a measure space, then μ is a gauge on X, and it is clear that the outer measure μ^* generated by μ coincides with the one introduced in Example 5.2.

Propositions 5.8 and 5.15 open a royal road to the construction of measures. Start with any gauge γ—such set functions are easily constructed—allow it to generate an outer measure μ^*, and then extract from μ^* the measure μ that it defines. It may happen however that the measure μ we obtain is worthless, either because it measures too few sets, or because it bears no useful relation to the originally given gauge.

Example 5.17. Let X be a metric space and let γ_0 be the gauge defined on X in Example 5.14. The outer measure μ^* generated by γ_0 can be described as follows: If $A \subset X$, then

$$\mu^*(A) = \begin{cases} 0 & \text{if } A = \varnothing, \\ 1 & \text{if } \varnothing \ne A \in \mathcal{B}, \\ +\infty & \text{if } A \notin \mathcal{B}. \end{cases}$$

Now let $E \subset X$ be such that $\varnothing \ne E \ne X$, let $x \in E$ and let $y \in X \setminus E$. Then $A = \{x, y\}$ is not split additively by E with respect to μ^* ($\mu^*(A) = \mu^*(A \cap E) = \mu^*(A \setminus E) = 1$). It follows that $\varnothing$ and X are the only two subsets of X that are measurable $[\mu^*]$.

Example 5.18. Let $X = \mathbb{Q}$, let f be a monotone increasing real-valued function on $\mathbb{R}$, and let $\tilde{\gamma}_f$ be the gauge on X introduced in Example 5.13. We compute $\mu^*(\{r\})$, where μ^* denotes the outer measure generated by $\tilde{\gamma}_f$ and $r \in \mathbb{Q}$. Clearly $\mu^*(\{r\})$ is the infimum of the differences $f(b) - f(a)$ where a and b are rational numbers such that $a < r \le b$. But then

$$\mu^*(\{r\}) = \inf\{f(r) - f(a) \colon a \in \mathbb{Q}, a < r\},$$

so that $\mu^*(\{r\}) = f(r) - \lim_{t \uparrow r} f(t)$, the so-called *left jump* of f at r. The left jump of f at r is denoted $d_-(r)$ for the remainder of this example.

Next let $F = \{r_1, \ldots, r_N\}$ be a finite subset of X, and suppose, as we may, that $r_1 < \cdots < r_N$. If F is covered by a sequence $\{\tilde{H}_n\}$ of half-open intervals in $\tilde{\mathcal{H}}$, then there is at least one $\tilde{H}_{n_0}$ containing an interval $(a_1, r_1]^\sim$ where $a_1 < r_1$, and replacing $\tilde{H}_{n_0}$ by that interval and all of the other intervals $\tilde{H}_n$ by $\tilde{H}_n \setminus (-\infty, r_1]$ gives rise to a new covering $\{\tilde{H}_n'\}$ of F by intervals in $\tilde{\mathcal{H}}$ such that

$$\sum_n \tilde{\gamma}_f(H_n') \leq \sum_n \tilde{\gamma}_f(H_n).$$

By an obvious mathematical induction we conclude that there exist disjoint intervals $(a_i, r_i]^\sim$ such that $a_i < r_i, i = 1, \ldots, N$, and such that

$$\sum_{i=1}^N \tilde{\gamma}_f((a_i, r_i]^\sim) \leq \sum_n \tilde{\gamma}_f(\tilde{H}_n),$$

whence it follows that $\mu^*(F) \geq \sum_{i=1}^N d_-(r_i)$. We conclude that

$$\mu^*(A) = \sum_{r \in A} d_-(r) \quad A \subset X.$$

Thus the outer measure generated by $\tilde{\gamma}_f$ is not without interest, but it reflects only discontinuous behavior—left discontinuous behavior, at that— of the given function f. In particular, if f is left continuous at every rational number r, then the outer measure generated by $\tilde{\gamma}_f$ is identically zero.

It should be added that the real trouble in this example lies not in the choice of the gauge $\tilde{\gamma}_f$, but rather in the countability of the space X. Nevertheless, the example does show rather dramatically how, in the passage from gauge to outer measure to measure, vital information can be lost. The gauge $\tilde{\gamma}_f$ itself permits the reconstruction of the function $f|\mathbb{Q}$ up to an additive constant, no matter what f one starts with.

There are two problems to be solved in connection with the utility of our scheme for defining measures: Starting with a gauge γ, when does the outer measure μ^* generated by γ *extend* γ? And when does the measure μ defined by μ^* do so? The former question is easy to answer. The latter is not, and we content ourselves with giving some reasonably useful sufficient conditions.

Proposition 5.19. *Let γ be a gauge defined on $\mathcal{C} \subset 2^X$, and let μ^* be the outer measure generated by γ. A necessary and sufficient condition that μ^* agree with γ on $\mathcal{C}$ is that γ be countably subadditive.*

Proof. As has already been noted (see the proof of Proposition 5.15), we always have $\mu^* \leq \gamma$ setwise on $\mathcal{C}$. If γ is assumed to be countably subadditive, then according to (5.4) we also have $\gamma \leq \mu^*$. This establishes the sufficiency of the condition; its necessity is obvious. $\square$

Example 5.20. Let $\mathcal{Z}$ denote the collection consisting of all cells

$$Z = (a_1, b_1] \times \cdots \times (a_d, b_d], \quad a_i < b_i, \quad i = 1, \ldots, d, \tag{5.5}$$

in $\mathbb{R}^d$. The formula $|Z| = (b_1 - a_1) \cdots (b_d - a_d)$ for each cell Z given by (5.5) defines a gauge $|\cdot|$ on $\mathbb{R}^d$. The outer measure generated by this gauge is *Lebesgue outer measure* on $\mathbb{R}^d$, or *d-dimensional Lebesgue outer measure*, and is henceforth denoted λ_d^*. This is the most important outer measure of all, both historically and mathematically, and it is appropriate to give some attention to the properties of the gauge $|\cdot|$ that generates it.

Suppose first that each of the edges $(a_i, b_i]$ of a cell Z given by (5.5) is partitioned in some way—say by a sequence $a_i = t_0^{(i)} < \cdots < t_{N_i}^{(i)} = b_i$—and we then form all of the cells

$$(t_{j_1-1}^{(1)}, t_{j_1}^{(1)}] \times \cdots \times (t_{j_d-1}^{(d)}, t_{j_d}^{(d)}], \quad j_i = 1, \ldots, N_i, \quad i = 1, \ldots, d,$$

having for their edges the various subintervals of these partitions. The resulting set of closed subcells of Z is called a *cellular partition* of Z. Clearly if $\mathcal{P}$ is such a cellular partition of Z, then $|Z| = \sum_{W \in \mathcal{P}} |W|$.

Suppose now that we are given some subcells $Z_1, \ldots, Z_p \subset Z$. It is a simple matter to define a cellular partition $\mathcal{P}$ of Z such that each $Z_j, j = 1, \ldots, p$, is the union of the cells in $\mathcal{P}$ that are contained in it. (The partition of the ith edge of Z used to construct $\mathcal{P}$ must contain all of the endpoints of the ith edges of the various cells $Z_1, \ldots, Z_p$.) But then, using this partition $\mathcal{P}$, we immediately learn two important facts:

(A) If the subcells $Z_1, \ldots, Z_p$ *cover* Z, then $|Z| \le |Z_1| + \cdots + |Z_p|$. (For in this case every summand $|W| \ne 0$ in $\sum_{W \in \mathcal{P}} |W|$ appears in at least one of the subsums $\sum\limits_{W \in \mathcal{P}, W \subset Z_i} |W|$.)

(B) If the subcells $Z_1, \ldots, Z_p$ are *nonoverlapping*—that is, if $Z_i \cap Z_j$ is contained in a face of both Z_i and Z_j whenever $i \ne j, i, j = 1, \ldots, p$—then $|Z_1| + \cdots + |Z_p| \le |Z|$. (For in this case each summand $|W| \ne 0, W \in \mathcal{P}$, appears in *at most one* of the subsums $\sum_{W \in \mathcal{P}, W \subset Z_i} |W|, i = 1, \ldots, p$.)

From these two observations we derive two further facts.

(C) If a cell Z in $\mathbb{R}^d$ is covered by a finite collection $Z_1, \ldots, Z_p$ of nonoverlapping subcells of Z, then $|Z| = |Z_1| + \cdots + |Z_p|$.

(D) If a cell Z in $\mathbb{R}^d$ is covered in any fashion by a finite collection $Z_1, \ldots, Z_p$ of cells in $\mathbb{R}^d$, then $|Z| \le |Z_1| + \cdots + |Z_p|$. (For Z is then also covered by the subcells $Z \cap Z_1, \ldots, Z \cap Z_p$.)

On the basis of this final observation it is easy to show that the gauge $|\cdot|$ is countably subadditive on $\mathcal{Z}$. Indeed, let

$$Z = (a_1, b_1] \times (a_2, b_2] \times \cdots \times (a_d, b_d]$$

be a cell in $\mathbb{R}^d$, let $\{Z_n\}_{n=1}^\infty$ be a covering of Z by cells, and fix $\varepsilon > 0$. Then for each n it is a simple matter to construct a cell $\widehat{Z}_n$ containing Z_n in its interior and such that $|\widehat{Z}_n| < |Z_n| + \varepsilon/2^n$. (Replace each edge of Z_d by an interval having the same center as that edge but t times longer for some $t > 1$. The cell $\widehat{Z}_n$ thus obtained satisfies $|\widehat{Z}_n| = t^d |Z_n|$. Then take t to be sufficiently close to 1.) The set Z is covered by the open interiors $(\widehat{Z}_n)^\circ$, and consequently

$$[a_1 + \varepsilon, b_1] \times \cdots \times [a_d + \varepsilon, b_d] \subset (\widehat{Z}_1)^\circ \cup \cdots \cup (\widehat{Z}_N)^\circ$$

for some suitably large N, by the Heine-Borel theorem ([**I**, Corollary 8.33]). Thus, by (D),

$$\prod_{j=1}^d (b_j - a_j - \varepsilon) \le |Z_1| + \cdots + |Z_N| + \varepsilon \le \sum_{n=1}^\infty |Z_n| + \varepsilon.$$

The countable subadditivity of $|\cdot|$ on $\mathcal{Z}$ follows by letting ε tend to zero. Hence for every cell Z in $\mathbb{R}^d$ we have $\lambda_d^*(Z) = |Z|$. Lebesgue outer λ_d^* measure is *translation invariant* on $\mathbb{R}^d$, that is

$$\lambda_d^*(E) = \lambda_d^*(E + x), \quad E \subset \mathbb{R}^d, x \in \mathbb{R}^d,$$

where $E + x = \{y + x : y \in E\}$. Indeed, this equality is easily verified when E is a cell, and its extension to arbitrary sets is obtained directly from the definition of λ_d^*.

Example 5.21. Let f be a monotone increasing, real-valued function defined on $\mathbb{R}$, and suppose f is *right continuous* on $\mathbb{R}$ (that is, $f(t) = \lim_{s \downarrow t} f(s)$ for every $t \in \mathbb{R}$). Let γ_f be the gauge associated with f as in Example 5.12. The proof of the countable subadditivity of γ_f is quite like that for the gauge $|\cdot|$ of Example 5.20. Let $\{(a_n, b_n]\}_{n=1}^\infty$ be a sequence of intervals in $\mathcal{H}$ covering an interval $(a, b]$, where we assume $a < b$ and $a_n < b_n$, $n \in \mathbb{N}$, and fix $\varepsilon > 0$. For each n, choose $b_n' > b_n$ such that $\gamma_f((a_n, b_n']) < \gamma_f((a_n, b_n]) + \varepsilon/2^{n+1}$ (using the right continuity of f at b_n). Similarly, choose $a' \in (a, b)$ such that $\gamma_f((a', b]) > \gamma_f((a, b]) - \varepsilon/2$ (using right continuity of f at a). The sequence $\{(a_n, b_n')\}$ of open intervals covers the closed interval $[a', b]$, and it follows by the Heine-Borel theorem ([**I**, Corollary 8.33]) that some finite number $(a_1, b_1'), \ldots, (a_N, b_N')$ of these intervals covers $[a', b]$. But then the half-open intervals $(a_1, b_1'], \ldots, (a_N, b_N']$ also cover the half-open interval $(a', b]$. As noted in Example 5.12, the gauge γ_f extends to a finitely additive set function ν_f on the ring $\mathbf{H}_r$, and therefore

$$\gamma_f((a', b]) \le \gamma_f((a_1, b_1']) + \cdots + \gamma_f((a_N, b_N']).$$

From this it follows that

$$\gamma_f((a,b]) \le \sum_{n=1}^{\infty} \gamma_f((a_n, b_n]) + \varepsilon,$$

and letting $\varepsilon \to 0$ we conclude that γ_f is countably subadditive. Hence the outer measure μ_f^* generated by γ_f extends γ_f. By Proposition

Example 5.22. Let f be the monotone increasing function on $\mathbb{R}$ obtained by setting

$$f(t) = \begin{cases} 0 & \text{if } t \le 0, \\ 1 & \text{if } t > 0, \end{cases}$$

and let γ_f be the gauge associated with f as in Example 5.12. The sequence of intervals $\{(-n, 0]\}_{n=1}^{\infty}$ covers the ray $(-\infty, 0]$, and $\gamma_f((-n, 0]) = 0$ for every index n. Similarly, the sequence of intervals $\{(1/n, n]\}_{n=1}$ provides a covering of the ray $(0, +\infty)$ with $\gamma_f((1/n, n]) = 0$ for every n. Thus, applying (5.4) one sees that the outer measure generated by γ_f is identically zero. It follows from Proposition 5.19 that γ_f is not countably subadditive. (Note that f is *left continuous* rather than right continuous.)

If it is assumed that the outer measure μ^* generated by a gauge γ extends γ, then it is obvious that the measure μ defined by μ^* also extends γ if and only if the sets belonging to the domain $\mathcal{C}$ of γ are all measurable $[\mu^*]$. (Note that in this case it follows that the entire σ-algebra generated by $\mathcal{C}$ is contained in the σ-algebra $\mathbf{M}$ of sets measurable $[\mu^*]$.) In this connection the following result provides an important beginning.

Proposition 5.23. *Let γ be a gauge defined on $\mathcal{C} \subset 2^X$, and let μ^* be the outer measure on X generated by γ. Then a set $A \subset X$ is measurable $[\mu^*]$ if and only if it splits all of the sets in $\mathcal{C}$ additively, that is, if and only if*

$$\mu^*(C) = \mu^*(C \cap A) + \mu^*(C \setminus A), \quad C \in \mathcal{C}.$$

Proof. The necessity of the condition is obvious. To prove its sufficiency, let A have the stated property, let B be an arbitrary subset of X, and fix $\varepsilon > 0$. It is immediate that A splits B additively if $\mu^*(B) = +\infty$, so we assume that $\mu^*(B)$ is finite. Then there exists a covering $\{C_n\}$ of B consisting of sets in $\mathcal{C}$ such that $\sum_n \gamma(C_n) \le \mu^*(B) + \varepsilon$. For each n set $C_n' = C_n \cap A$ and $C_n'' = C_n \setminus A$. Then

$$B \cap A \subset \bigcup_n C_n' \quad \text{and} \quad B \setminus A \subset \bigcup_n C_n'',$$

so

$$\mu^*(B \cap A) \le \sum_n \mu^*(C_n') \quad \text{and} \quad \mu^*(B \setminus A) \le \sum_n \mu^*(C_n'').$$

Hence

$$\mu^*(B \cap A) + \mu^*(B \setminus A) \leq \sum_n [\mu^*(C_n') + \mu^*(C_n'')] = \sum_n \mu^*(C_n).$$

But $\mu^*(C_n) \leq \gamma(C_n)$ for each n as was noted above. Thus

$$\mu^*(B \cap A) + \mu^*(B \setminus A) \leq \sum_n \gamma(C_n) \leq \mu^*(B) + \varepsilon,$$

and since ε is arbitrary, this shows that A also splits B additively, and hence that A is measurable $[\mu^*]$. $\qquad\qquad\qquad\qquad\qquad\qquad\qquad\qquad\qquad\qquad\square$

Example 5.24. The gauge in Example 5.20 satisfies the hypothesis of Proposition 5.23, and the σ-ring generated by $\mathcal{Z}$ is $\mathbf{B}_{\mathbb{R}^d}$. The measure λ_d generated by this gauge is Lebesgue measure on $\mathbb{R}^d$, or d-*dimensional Lebesgue measure*. The restriction of λ_d to $\mathbf{B}_{\mathbb{R}^d}$ is d-*dimensional Lebesgue-Borel measure*. It is convenient to use the notation λ_d for both of these measures. It will be clear from the context whether we work with Lebesgue measure or with its restriction to $\mathbf{B}_{\mathbb{R}^d}$. It was noted in Example 5.20 that Lebesgue outer measure is translation invariant. It follows that a set $E \subset \mathbb{R}^d$ is Lebesgue measurable if and only if $E + x$ is Lebesgue measurable for every $x \in \mathbb{R}^d$, in which case

$$\lambda_d(E) = \lambda_d(E + x), \quad x \in \mathbb{R}^d.$$

Example 5.25. The measure μ_f that arises from the gauge γ_f in Example 5.22 is called the *Lebesgue-Stieltjes measure* determined by the function f. The integral of a function h relative to μ_f is usually written as $\int_{\mathbb{R}} h(t) \, df(t)$. More generally, suppose that $I \subset \mathbb{R}$ is an interval and $f : I \to \mathbb{R}$ is a monotone increasing, right-continuous function. Then one can use the procedure of Example 5.22 to construct the Lebesgue-Stieltjes measure μ_f on I.

Example 5.26. Let $I = [a, b]$, $a < b$, be a compact interval and let $f : I \to \mathbb{R}$ be a monotone increasing, right-continuous function. For every point $t \in I$, $\mu_f(\{t\})$ equals the jump of f at the point t, that is, the difference between $f(t)$ and the left-hand limit of f at t. It follows that f is continuous on I if and only if μ_f does not assign positive measure to any singleton. We can similarly characterize those functions f with the property that $\mu_f|\mathbf{B}_I$ is absolutely continuous relative to the Lebesgue-Borel measure $\lambda_1|\mathbf{B}_I$. This happens precisely when f is *absolutely continuous* in the following sense: for every $\varepsilon > 0$ there exists $\delta > 0$ such that for every finite collection $\{(a_i, b_i)\}_{i=1}^n$ of pairwise disjoint subintervals of I satisfying $\sum_{i=1}^n (b_i - a_i) < \delta$ we have $\sum_{i=1}^n (f(b_i) - f(a_i)) < \varepsilon$. The equivalence of these notions of absolute continuity, as defined above, follows easily from Proposition 4.3.

The notion of an absolutely continuous function can be extended to arbitrary functions $f : I \to \mathbb{C}$, where $I \subset \mathbb{R}$ is an arbitrary interval. We say that such an f is absolutely continuous on I if for every $\varepsilon > 0$ there exists $\delta > 0$

such that for every finite collection $\{(a_i, b_i]\}_{i=1}^n$ of pairwise disjoint subintervals of I satisfying $\sum_{i=1}^n (b_i - a_i) < \delta$ we have $\sum_{i=1}^n |f(b_i) - f(a_i)| < \varepsilon$. It is obvious that f is absolutely continuous on I if and only if its real and imaginary parts are.

Example 5.27. Lebesgue measure on $\mathbb{R}^d$ is a proper extension of Lebesgue-Borel measure. Indeed, the σ-algebra $\mathbf{B}_{\mathbb{R}^d}$ is generated by countably many open sets, and it follows easily that the cardinality of $\mathbf{B}_{\mathbb{R}^d}$ equals $2^{\aleph_0}$. On the other hand, Problem 3Y shows that the ternary Cantor set C satisfies $\lambda_1(C) = 0$. Since Lebesgue measure is complete, it follows that every subset of C is measurable $[\lambda_1]$. Thus the cardinality of the σ-algebra of Lebesgue measurable sets is at least $2^{\operatorname{card} C} = 2^{2^{\aleph_0}}$. Of course, $2^{2^{\aleph_0}}$ is the cardinality of $2^{\mathbb{R}}$. The Cantor-Bernstein theorem [**I**, Theorem 4.1] implies that there are exactly $2^{2^{\aleph_0}}$ Lebesgue measurable sets in $\mathbb{R}$. One may wonder whether *all* subsets of $\mathbb{R}$ are Lebesgue measurable. Using the axiom of choice, one can construct sets which are not Lebesgue measurable. This is done as follows. Given two real numbers $x, y \in \mathbb{R}$, write $x \equiv_{\mathbb{Q}} y$ if $y - x \in \mathbb{Q}$. Then $\equiv_{\mathbb{Q}}$ is an equivalence relation, and the equivalence classes are of the form $\mathbb{Q} + x$ with $x \in \mathbb{R}$. Since each such set is dense in $\mathbb{R}$, the axiom of choice implies the existence of a set $E \subset (0, 1)$ which contains exactly one element in each equivalence class. Observe now that

$$\mathbb{R} = \bigcup_{x \in \mathbb{Q}} (E + x),$$

so that

$$+\infty = \lambda_1(\mathbb{R}) \le \sum_{x \in \mathbb{Q}} \lambda_1^*(E + x).$$

All the terms in the sum above are equal to $\lambda_1^*(E)$, and thus $\lambda_1^*(E) > 0$. On the other hand, the sets $E + x$, $x \in \mathbb{Q}$, are pairwise disjoint, and $E + x \subset (0, 2)$ if $x \in (0, 1)$. If E were measurable $[\lambda_1]$ we could conclude that

$$\sum_{x \in \mathbb{Q} \cap (0,1)} \lambda_1^*(E + x) \le \lambda_1((0, 2)) = 2,$$

and therefore $\lambda_1^*(E) = 0$. This contradiction shows that E is not measurable $[\lambda_1]$.

The set E constructed above shows, more generally, that there does not exist a translation-invariant measure μ on $2^{\mathbb{R}}$ such that $\mu((0, 1)) = 1$. Banach proved that there do exist translation invariant, finitely additive set functions $\mu : 2^{\mathbb{R}} \to [0, +\infty]$ such that $\mu((0, 1)) = 1$.

One easy consequence of Proposition 5.23 is the following somewhat special result.

Proposition 5.28. *Let* $\mathbf{R}$ *be a ring of subsets of a set* X *and let* γ *be a finitely additive gauge defined on* $\mathbf{R}$. *Then the sets in* $\mathbf{S}(\mathbf{R})$ *are all measurable with respect to the outer measure* μ^* *generated by* γ.

Proof. Let P and Q be sets in $\mathbf{R}$. According to Proposition 5.23 it suffices to show that P splits Q additively with respect to μ^*, and it suffices to consider the case $\mu^*(Q) < +\infty$. To this end fix $\varepsilon > 0$ and let $\{R_n\}$ be a countable covering of Q by sets in $\mathbf{R}$ such that $\sum_n \gamma(R_n) \leq \mu^*(Q) + \varepsilon$. The sets $R_n \cap P$ belong to $\mathbf{R}$ and cover $Q \cap P$, so $\mu^*(Q \cap P) \leq \sum_n \gamma(R_n \cap P)$. Similarly, $\mu^*(Q \setminus P) \leq \sum_n \gamma(R_n \setminus P)$, and therefore

$$\mu^*(Q \cap P) + \mu^*(Q \setminus P) \leq \sum_n [\gamma(R_n \cap P) + \gamma(R_n \setminus P)] = \sum_n \gamma(R_n) \leq \mu^*(Q) + \varepsilon$$

since $\gamma(R_n \cap P) + \gamma(R_n \setminus P) = \gamma(R_n)$ for each n by the finite additivity of γ. The result follows. □

Proposition 5.28 answers a question that may well not be worth asking—for instance, when the outer measure μ^* is itself without interest (recall Example 5.22). When Proposition 5.28 is combined with Proposition 5.19, however, a useful result, which needs no further proof, is obtained.

Corollary 5.29. *Let* $\mathbf{R}$ *be a ring of subsets of a set* X *and let* γ *be a countably subadditive and finitely additive gauge defined on* $\mathbf{R}$. *Then the outer measure* μ^* *generated by* γ *and the measure* μ *defined by* μ^* *both extend* γ. *In particular, every set in* $\mathbf{S}(\mathbf{R})$ *is measurable* $[\mu]$.

Remark 5.30. The hypotheses of Corollary 5.29 were selected so as to make as clear as possible how this result follows from those that precede it. It should be observed that a gauge γ (defined on a ring of sets $\mathbf{R}$) that is both finitely additive and countably subadditive on $\mathbf{R}$ is, in fact, *countably additive* on $\mathbf{R}$, and that the converse is also true. That is, a countably additive gauge defined on a ring of sets is also finitely additive and countably subadditive. Thus Corollary 5.29 might equally well have been stated for a countably additive gauge on $\mathbf{R}$. Such a set function is usually called a *ring measure* on $\mathbf{R}$.

Definition 5.31. A gauge γ defined on a collection $\mathcal{C} \subset 2^X$ is said to be *extendible* if there exists a measure μ on a σ-algebra $\mathbf{S}$ in X such that $\mathcal{C} \subset \mathbf{S}$ and $\mu|\mathcal{C} = \gamma$. Equivalently, γ is extendible if there exists an outer measure μ^* on X such that $\mu^*|\mathcal{C} = \gamma$ and every set $E \in \mathcal{C}$ is measurable $[\mu^*]$.

Proposition 5.32. *Assume that* $\mathcal{C} \subset 2^X$ *and* $\gamma : \mathcal{C} \to [0, +\infty]$ *is an extendible gauge. Denote by* μ^* *the outer measure defined by* γ. *Then:*

(1) *For every set* $A \subset X$ *there exists a set* $B \subset X$ *which is measurable* $[\mu^*]$, $B \supset A$, *and* $\mu^*(B) = \mu^*(A)$.

(2) *For every increasing sequence* $\{A_n\} \subset 2^X$ *with union* A, *we have* $\mu^*(A) = \lim_n \mu^*(A_n)$.

Proof. To prove (1), assume first that $\mu^*(A) < +\infty$. It follows from the definition of μ^* that, given $m \in \mathbb{N}$, there exists a sequence $\{C_{m,n}\}_{n=1}^{\infty} \subset \mathcal{C}$ such that $A \subset \bigcup_n C_{m,n}$ for all $n \in \mathbb{N}$ and $\sum_n \gamma(C_{m,n}) \leq \mu^*(A) + 1/m$. We then set $B = \bigcap_m \bigcup_n C_{m,n}$. This set is measurable $[\mu^*]$ because γ was assumed extendible. If $\mu^*(A) = +\infty$ we can choose $B = X$.

Consider next an increasing sequence $\{A_n\} \subset 2^X$, and use (1) to find sets $B_n \supset A_n$ which are measurable $[\mu^*]$ and $\mu^*(B_n) = \mu^*(A_n)$. Replacing B_n by $\bigcap_{k \geq n} B_k$, we may assume that the sequence $\{B_n\}$ is increasing as well. We have then

$$\mu^*(A) \leq \mu^* \left(\bigcup_n B_n \right) = \lim_n \mu^*(B_n) = \lim_n \mu^*(A_n),$$

where we used Proposition 3.28 in the first equality. The inequality $\mu^*(A) \geq \lim_n \mu^*(A_n)$ is true because $A \supset A_n$ for all n. $\qquad\qquad\square$

The reader will recall from Definition 1.24 in Chapter 1 the notion of an analytic set in a metric space. Analytic sets play an important role in certain areas of mathematics like descriptive set theory, fractals, etc., and therefore at times it is useful to know that such analytic sets are measurable. For measures on metric spaces defined on the Borel subsets of the space, one can prove in some generality that analytic subsets are measurable as well (relative to the corresponding outer measure).

Proposition 5.33. *Consider a separable, complete metric space X, an extendible gauge γ defined on the collection $\mathcal{G}$ of open subsets of X, and the outer measure μ^* determined by γ. Then every analytic set $A \subset X$ with finite outer measure is measurable $[\mu^*]$.*

Proof. Assume that the set A is the result of the Suslin operation (Definition 1.24) applied to the system $\{F_{n_1,\dots,n_k}\}$ of closed subsets of X. We may assume in addition that each set $F_{n_1,\dots,n_k}$ is nonempty, has diameter less than $1/k$, and contains $F_{n_1,\dots,n_k,n_{k+1}}$ for all $n_{k+1} \in \mathbb{N}$ (Problem 1Z(i)). Fix now $\varepsilon > 0$, and use the notation $F_{\mathbf{n}} = \bigcap_{k=1}^{\infty} F_{n_1,\dots,n_k}$ for $\mathbf{n} = (n_1, n_2, \dots) \in \mathbb{N}^{\mathbb{N}}$. Proposition 5.32 implies that

$$\lim_{m \to \infty} \mu^* \left[\bigcup_{n_1 \leq m} F_{\mathbf{n}} \right] = \mu^*(A),$$

and thus there exists $m_1 \in \mathbb{N}$ such that

$$\mu^* \left[\bigcup_{n_1 \leq m_1} F_{\mathbf{n}} \right] > \mu^*(A) - \varepsilon.$$

Proceeding inductively, we find a sequence $\mathbf{m} \in \mathbb{N}^{\mathbb{N}}$ such that

$$\mu^* \left[\bigcup_{n_1 \leq m_1, \ldots, n_k \leq m_k} F_{\mathbf{n}} \right] > \mu^*(A) - \varepsilon - \varepsilon^2 \cdots - \varepsilon^k > \mu^*(A) - \frac{\varepsilon}{1 - \varepsilon}. \quad (5.6)$$

Define now a subset of A by setting $K_\varepsilon = \bigcup_{n_j \leq m_j, j \in \mathbb{N}} F_{\mathbf{n}}$. We claim that

$$K_\varepsilon = \bigcap_{k=1}^{\infty} \bigcup_{n_1 \leq m_1, \ldots, n_k \leq m_k} F_{n_1, \ldots, n_k}. \quad (5.7)$$

The inclusion $F_{\mathbf{n}} \subset K_\varepsilon$ is obvious since $n_j \leq m_j$ for all j. Conversely, fix a point x in the set on the right-hand side of (5.7), so that for each $k \in \mathbb{N}$ we have $x \in F_{n_1(k), \ldots, n_k(k)}$ for some numbers $n_1(k) \leq m_1, \ldots, n_k(k) \leq m_k$. A simple diagonal argument shows that there exists $\mathbf{n} \in \mathbb{N}^{\mathbb{N}}$ such that, for each j, the equality $n_j(k) = n_j$ is true for infinitely many values of $k \geq j$. Then clearly $x \in F_{\mathbf{n}}$.

The equality (5.7) shows that K_ε is closed and totally bounded, and therefore compact because X is complete. Let G be an arbitrary open set containing K_ε. The compactness of K_ε implies that $\{x \in X : \operatorname{dist}(x, K_\varepsilon) \leq 1/k\} \subset G$ for some integer k, and therefore

$$\bigcup_{n_1 \leq m_1, \ldots, n_k \leq m_k} F_{\mathbf{n}} \subset \bigcup_{n_1 \leq m_1, \ldots, n_k \leq m_k} F_{n_1, \ldots, n_k} \subset G.$$

By (5.6), $\mu^*(G) \geq \mu^*(A) - \varepsilon/(1 - \varepsilon)$, and thus $\mu^*(K_\varepsilon) \geq \mu^*(A) - \varepsilon/(1 - \varepsilon)$. Repeating this construction for a sequence $\varepsilon_n \to 0$, we obtain a measurable subset $C = \bigcup_n K_{\varepsilon_n} \subset A$ such that $\mu^*(C) = \mu^*(A)$. We know from Proposition 5.32 that $\mu^*(A) = \mu^*(B)$ for some measurable $B \supset A$, and we conclude that $\mu^*(B \setminus A) \leq \mu^*(B \setminus C) = \mu^*(B) - \mu^*(C) = 0$. The conclusion follows because sets of outer measure zero are measurable $[\mu^*]$ by Proposition 5.8. $\square$

One way to prove that a gauge is extendible is to prove that it is countably additive on $\mathcal{C}$, but the direct verification of that fact can be tedious. In this connection the following somewhat technical result is sometimes useful. In it we assume that the collection of sets on which the given gauge is defined is a *lattice* of sets (Problem 1K).

Proposition 5.34. *Let* $\mathbf{L}$ *be a lattice of subsets of a set* X *and let* γ *be a countably subadditive gauge on* $\mathbf{L}$ *satisfying the condition*

(S) $\gamma(L) + \gamma(M) \leq \gamma(L \cup M) + \gamma(L \cap M)$

for every pair of sets L, M *in* $\mathbf{L}$. *Suppose in addition that the following condition is satisfied for some set* L_0 *belonging to* $\mathbf{L}$:

(H) *For every* $\varepsilon > 0$ *there exists a subset* $Q \subset L_0$ *such that* $L \setminus Q \in \mathbf{L}$ *for every* $L \in \mathbf{L}$ *and* $\gamma(L_0 \setminus Q) < \varepsilon$.

Then L_0 is measurable $[\mu^*]$, where μ^* is the outer measure on X generated by γ.

Proof. Proposition 5.19 implies that μ^* extends γ. By Proposition 5.23 it suffices to show that L_0 splits the sets of $\mathbf{L}$ additively. Fix $L \in \mathbf{L}$ such that $\mu^*(L) < +\infty$, and $\varepsilon > 0$. Select $Q \subset L_0$ satisfying the conditions in (H), and set $M = L \setminus Q, L' = L \cap L_0, L'' = L \setminus L_0$. Then $L'' \subset M$ and $L' \cup M = L$. Moreover, $L' \cap M \subset L_0 \setminus Q$ so $\mu^*(L' \cap M) < \varepsilon$. An application of (S) yields $\mu^*(L') + \mu^*(M) \le \mu^*(L) + \varepsilon$ and therefore

$$\mu^*(L') + \mu^*(L'') \le \mu^*(L) + \varepsilon.$$

Since ε is arbitrary, this implies $\mu^*(L') + \mu^*(L'') \le \mu^*(L)$, and the result follows. $\qquad\square$

Corollary 5.35. *Let $\mathbf{L}, \gamma$ and μ^* be as in the foregoing proposition, and let $\mathcal{L}_0$ denote the collection of all those sets L_0 in $\mathbf{L}$ that satisfy (H). Then:*

(1) All of the sets in the σ-ring $\mathbf{S}(\mathcal{L}_0)$ are measurable $[\mu^*]$.
(2) If $\mathcal{L}_0$ contains all sets $L_0 \in \mathcal{L}$ satisfying $\gamma(L_0) < +\infty$, all the sets in the σ-ring $\mathbf{S}(\mathcal{L})$ are measurable $[\mu^*]$.

Proof. Part (1) follows immediately from Proposition 5.19. To prove part (2) we need a slight modification of the preceding proof. Fix $L \in \mathbf{L}$ such that $\mu^*(L) < \infty$, and $\varepsilon > 0$. Under the hypothesis of (2), $L' = L \cap L_0$ belongs to $\mathcal{L}_0$, and therefore we can select $Q \subset L'$ such that $\mu^*(L' \setminus Q) < \varepsilon$. Apply now (S) to L' and $L \setminus Q$ to obtain

$$\mu^*(L') + \mu^*(L \setminus Q) \le \mu^*(L) + \mu^*((L \cap L_0) \setminus Q) < \mu^*(L) + \varepsilon,$$

where we used the obvious identity $L' \cap (L \setminus Q) = (L \cap L_0) \setminus Q$. The desired conclusion follows again by using $\mu^*(L'') \le \mu^*(L \setminus Q)$ and letting $\varepsilon \to 0$. $\quad\square$

Definition 5.36. A complex- or extended real-valued set function λ defined on a lattice $\mathbf{L}$ of subsets of a set X is said to be *modular*, or to satisfy the *modular law* if

(M) $\lambda(L) + \lambda(M) = \lambda(L \cup M) + \lambda(L \cap M), \quad L, M \in \mathbf{L}.$

Any modular set function on $\mathbf{L}$ satisfies (S) (for "semimodularity") of Proposition 5.34 (Clearly too, any finitely additive set function ν on a ring $\mathbf{R}$ of subsets of X must be modular, and, in fact, Corollary 5.35 can be used to give a new proof of Proposition 5.34 Indeed, to see that every set R_0 in $\mathbf{R}$ also satisfies (H) simply set $Q = R_0$.) We observe that if ρ is a gauge on a lattice $\mathbf{L}$ and if $L, M \in \mathbf{L}$ are disjoint, then (S) reduces to

$$\gamma(L) + \gamma(M) \le \gamma(L \cup M).$$

Thus condition (S) together with countable subadditivity implies countable additivity. (Of course, if γ is to be extendible, it must be countably additive.) Condition (S) was chosen for the formulation of Proposition 5.34 principally because even for certain modular gauges it is easier to verify than (M).

Example 5.37. Let X be a locally compact topological space, and denote by $\mathcal{C}_c(X,\mathbb{R})$ the vector space of continuous functions $f : X \to \mathbb{R}$ with the property that the support $N_f = E(f \neq 0)$ is relatively compact in X. Consider a positive linear functional $\varphi : \mathcal{C}_c(X,\mathbb{R}) \to \mathbb{R}$. For each open set U in X we define

$$\gamma_\varphi(U) = \sup\{\varphi(f) \colon 0 \leq f \leq \chi_U, f \in \mathcal{C}_c(X,\mathbb{R})\}.$$

Then γ_φ is a gauge defined on the collection $\mathcal{G}$ of all open subsets of X, and $\gamma_\varphi(U) < +\infty$ if U is relatively compact. Indeed, if U is relatively compact, by Urysohn's lemma (see [**I**, Chapter 9]), there exists $f \in \mathcal{C}_c(X,\mathbb{R})$ such that $\chi_U \leq f \leq 1$, and the positivity assumption implies that $\gamma_\varphi(U) \leq \varphi(f)$. We show that γ_φ is extendible.

In order to verify the countable subadditivity of γ_φ, let V be open in X, let $\{U_n\}_{n \in \mathbb{N}}$ be an arbitrary open covering of V, let $f \in \mathcal{C}_c(X,\mathbb{R})$ satisfy $0 \leq f \leq \chi_V$, and let $\varepsilon > 0$. We set $K = \{x \in X \colon f(x) \geq \varepsilon\}$. If $f_\varepsilon = (f - \varepsilon) \vee 0$, then $0 \leq f_\varepsilon \leq \chi_K$ and $f \leq f_\varepsilon + \varepsilon$. Since K is compact there exists a positive integer N such that

$$K \subset U_1 \cup \cdots \cup U_N,$$

and there exist functions $g_1, \ldots, g_N \in \mathcal{C}_c(X,\mathbb{R})$ such that $0 \leq g_i \leq \chi_{U_i}, i = 1, \ldots, N$, and such that if $g = g_1 + \cdots + g_N$, then $0 \leq g \leq 1$ on X, while $g(x) = 1$ for all $x \in K$. (This is just a partition of unity subordinate to $\{U_1, \ldots, U_N\}$; see [**I**, Problem 9I].) Thus $f_\varepsilon = f_\varepsilon g = f_\varepsilon g_1 + \cdots + f_\varepsilon g_N$, whence it follows that $\varphi(f_\varepsilon) \leq \gamma_\varphi(U_1) + \cdots + \gamma_\varphi(U_N)$. Since $f \leq f_\varepsilon + \varepsilon$, we conclude that

$$\varphi(f) \leq \gamma_\varphi(U_1) + \cdots + \gamma_\varphi(U_N) + \varepsilon \leq \sum_{n=1}^{\infty} \gamma_\varphi(U_n) + \varepsilon.$$

By the definition of γ_φ this shows that $\gamma_\varphi(V) \leq \sum_{n=1}^{\infty} \gamma_\varphi(U_n)$ because ε is arbitrary.

Next, if $U \subset X$ satisfies $\gamma_\varphi(U) < +\infty$ and $\varepsilon > 0$, there exists a function $f \in \mathcal{C}_c(X,\mathbb{R})$ such that $0 \leq f \leq \chi_U$ and such that $\varphi(f) > \gamma_\varphi(U) - \varepsilon$. Fix $\varepsilon > 0$, and set $K = \{x \in X \colon f(x) \geq \varepsilon/2\}$. Then K is closed, so $U \setminus K$ is open along with U. If $g \in \mathcal{C}_c(X,\mathbb{R})$ and $0 \leq g \leq \chi_{U \setminus K}$, then it is readily verified that $f + g \leq \chi_U + \varepsilon/2$. Hence, if we set $h = (f + g - \varepsilon/2) \vee 0$, so $0 \leq h \leq \chi_U$ and $f + g \leq h + \varepsilon/2$, then $\varphi(f) + \varphi(g) \leq \varphi(h) + \varepsilon/2 \leq \gamma_\varphi(U) + \varepsilon/2$. But then

$$\varphi(g) \leq \gamma_\varphi(U) - \varphi(f) + \varepsilon/2 < \varepsilon,$$

whence it follows that $\gamma_\varphi(U \setminus K) \leq \varepsilon$, and thus U satisfies condition (H) of Proposition 5.34.

Finally, to show that γ_φ satisfies condition (S) of Proposition 5.34, let U and V be two open subsets of X, and let f and g be any two continuous functions on X such that $0 \le f \le \chi_U$ and $0 \le g \le \chi_V$. If we set

$$h = (f + g) \wedge 1 \quad \text{and} \quad k = f + g - h,$$

then it is easy to see that $0 \le h \le \chi_{U \cup V}$ and $0 \le k \le \chi_{U \cap V}$. Since $f + g = h + k$, we have

$$\varphi(f) + \varphi(g) = \varphi(h) + \varphi(k) \le \gamma_\varphi(U \cup V) + \gamma_\varphi(U \cap V),$$

and (S) follows. Thus ρ_φ is indeed extendible by Corollary 5.35(2).

So far in this chapter we have investigated at some length the question of the existence of various measures. It is desirable to complement these results with a uniqueness theorem. The following result settles the matter quite effectively.

Theorem 5.38. *Let μ and ν be two measures on the same space X (each measure defined, in general, on its own σ-algebra of measurable sets). Suppose that μ and ν agree and are finite-valued on a collection $\mathcal{D}$ of subsets of X that is closed with respect to the formation of (finite) intersections. Then μ and ν agree on the σ-algebra $\mathbf{S}(\mathcal{D})$.*

Proof. Let D_0 be a fixed set belonging to $\mathcal{D}$, and let $\mathcal{D}_0$ denote the trace of $\mathcal{D}$ on D_0. Since $\mathcal{D}$ is closed with respect to the formation of intersections, $\mathcal{D}_0$ has the same property, consisting as it does of all those sets in $\mathcal{D}$ that are subsets of D_0. Next let $\mathcal{Q}_0$ denote the collection of all those subsets of D_0 on which μ and ν agree. Since μ and ν are measures, and since $\mu(D_0) = \nu(D_0) < +\infty$, it is clear that $\mathcal{Q}_0$ is a σ-quasiring (see Chapter 1 for definitions). Since $\mathcal{D}_0 \subset \mathcal{Q}_0$ we have $\mathbf{S}_0 \subset \mathcal{Q}_0$, where $\mathbf{S}_0 = \mathbf{S}(\mathcal{D}_0)$, by Theorem 1.23. But $\mathbf{S}_0$ is just the trace of $\mathbf{S}(\mathcal{D})$ on D_0. Thus μ and ν agree on every set in $\mathbf{S}(\mathcal{D})$ that is a subset of D_0. Since D_0 was an arbitrary set in $\mathcal{D}$, this shows that $\mu(E) = \nu(E)$ for every E in $\mathbf{S}(\mathcal{D})$ for which there exists $D \in \mathcal{D}$ such that $E \subset D$.

To complete the proof, let E be an arbitrary set in $\mathbf{S}(\mathcal{D})$. There exists a sequence $\{D_n\}$ in $\mathcal{D}$ that covers E (Problem 1C(iii)). Let $\{E_n\}$ be the disjointification of the sequence $\{E \cap D_n\}$. Then $E_n \subset D_n$, and therefore $\mu(E_n) = \nu(E_n)$ for every index n. But $\{E_n\}$ is a pairwise disjoint sequence of sets with union E, and it follows that $\mu(E) = \sum_n \mu(E_n) = \sum_n \nu(E_n) = \nu(E)$. □

Example 5.39. The collection of closed cells in $\mathbb{R}^d$ is closed with respect to the formation of intersections, so Theorem 5.38 applies and says that Lebesgue-Borel measure λ_d is uniquely determined by the fact that $\lambda_d(Z) = |Z|$ for all such cells.

Example 5.40. Let $X = \mathbb{R}$ and consider the counting measure γ on X along with the measure 2γ. These two measures agree at every set of the ring $\mathbf{H}_\ell$ of finite unions of half-open intervals $[a, b), a < b$, since both are identically equal to $+\infty$ on $\mathbf{H}_\ell$ except at the empty set $\varnothing$. But γ and 2γ do not agree, of course, on the σ-ring generated by $\mathbf{H}_\ell$, since this ring is precisely $\mathbf{B}_\mathbb{R}$ and contains all finite sets. Thus the requirement that μ and ν be finite-valued on the collection $\mathcal{D}$ in Theorem 5.38 is indispensable for the validity of the theorem. (The requirement that $\mathcal{D}$ be closed with respect to the formation of intersections also cannot be dropped from the hypotheses of Theorem 5.38, as may be learned by experimenting with a three point space.)

To this point we have only used outer measures as a source of abstract measures via Proposition 5.23. As we have seen before, however, some things are much more interesting in the context of a metric space. We conclude this chapter with a brief sketch of Carathéodory's theory of outer measures on metric spaces.

Definition 5.41. Let X be a metric space and let μ^* be an outer measure on X. Then μ^* is a *Carathéodory outer measure* if every Borel set in X is measurable $[\mu^*]$.

The following simple proposition is useful.

Proposition 5.42. *An outer measure μ^* on a metric space X is a Carathéodory outer measure if all the open sets in X are measurable $[\mu^*]$, or if all the closed sets are. If X is separable, and if all of the sets in some topological base for X are measurable $[\mu^*]$, then μ^* is a Carathéodory outer measure.*

Proof. Each of the stated collections of sets generates the σ-algebra $\mathbf{B}_X$. $\square$

Example 5.43. Lebesgue outer measure on $\mathbb{R}^d$ is a Carathéodory outer measure. Similarly, the outer measures ν_f^* associated with the various monotone increasing real-valued functions f on $\mathbb{R}$ are all Carathéodory outer measures.

If μ^* is a Carathéodory outer measure on a metric space (X, ρ), and if E and F are disjoint closed sets in X, then

$$\mu^*(E \cup F) = \mu^*(E) + \mu^*(F),$$

since $E \cup F$ is split additively by the open set $X \setminus E$. Similarly, if A and B are arbitrary subsets of X such that $d(A, B) = \inf\{\rho(x, y) : x \in A, y \in B\} > 0$, then

$$\mu^*(A \cup B) = \mu^*(A) + \mu^*(B),$$

since $A \cup B$ is split additively by $X \setminus A^-$. This last named fact has a useful converse. To prove it we need a lemma.

Lemma 5.44. *Let μ^* be an outer measure on a metric space X satisfying the condition*

$$\mu^*(A \cup B) = \mu^*(A) + \mu^*(B) \ \text{whenever} \ d(A, B) > 0, \tag{5.8}$$

and let $\{A_n\}_{n=1}^{\infty}$ be a sequence of subsets of X such that $d(A_m, A_n) > 0$ whenever $m \neq n$ and neither A_m nor A_n is empty. Then

$$\mu^*\left(\bigcup_{n=1}^{\infty} A_n\right) = \sum_{n=1}^{\infty} \mu^*(A_n).$$

Proof. Since $d(A_1 \cup \cdots \cup A_n, A_{n+1}) > 0$ whenever $A_1 \cup \cdots \cup A_n$ and A_{n+1} are both nonempty, it is clear (mathematical induction) that

$$\mu^*(A_1 \cup \cdots \cup A_n) = \sum_{k=1}^{n} \mu^*(A_k)$$

for each index n. But then $\sum_{k=1}^{n} \mu^*(A_k) \leq \mu^*(\bigcup_{k=1}^{\infty} A_k)$ for every n, so

$$\sum_{k=1}^{\infty} \mu^*(A_k) \leq \mu^*\left(\bigcup_{k=1}^{\infty} A_k\right),$$

and the lemma follows.

Proposition 5.45. *If μ^* is an outer measure on a metric space (X, ρ) and if condition (5.8) holds, then μ^* is a Carathéodory outer measure.*

Proof. We show that if A is a subset of X and U is an open subset of X, then U splits A additively with respect to μ^*, whence the statement follows by Proposition 5.42. In this endeavor we may and do assume without loss of generality that $\varnothing \neq U \neq X$ (for otherwise U doesn't split A at all) and that $\mu^*(A) < +\infty$ (since any splitting of a set of infinite outer measure is additive). Let $F_n = \{x \in X : d(x, X \setminus U) \geq 1/n\}$ for each $n \in \mathbb{N}$, so the sequence $\{F_n\}$ increases and has union U, and define $D_1 = F_1$ and

$$D_{n+1} = F_{n+1} \setminus F_n, \quad n \in \mathbb{N}.$$

The sequence $\{D_n\}_{n=1}^{\infty}$ need not satisfy the hypothesis of Lemma 5.44, but if m and n are positive integers such that $n \geq m + 2$, and if neither D_m nor D_n is empty, then $d(D_m, D_n) > 0$. (If $x \in F_m$ and $y \notin F_{m+1}$ then $\rho(x, y) \geq (1/m) - (1/(m+1)) = 1/m(m+1)$.) Hence the lemma does apply to the sequence $\{A \cap D_{2n}\}_{n=1}^{\infty}$ as well as the sequence $\{A \cap D_{2n-1}\}_{n=1}^{\infty}$, and we conclude that

$$\sum_{n=1}^{\infty} \mu^*(A \cap D_{2n}) = \mu^*\left(A \cap \bigcup_{n=1}^{\infty} D_{2n}\right) \leq \mu^*(A) < +\infty$$

and also that

$$\sum_{n=1}^{\infty} \mu^*(A \cap D_{2n-1}) = \mu^*\left(A \cap \bigcup_{n=1}^{\infty} D_{2n-1}\right) \leq \mu^*(A) < +\infty.$$

But then $\sum_{n=1}^{\infty} \mu^*(A \cap D_n) < +\infty$ too, and it follows (Problem 5N) that the sum of the series $\sum_{n=1}^{\infty} \mu^*(A \cap D_n)$, that is, the limit of the sequence

$$\{\mu^*(A \cap (D_1 \cup \cdots \cup D_n)) = \mu^*(A \cap F_n)\},$$

coincides with $\mu^*\big(A \cap \bigcup_{n=1}^{\infty} F_n\big) = \mu^*(A \cap U)$. From this it follows that the sequence $\{\mu^*(A \cap F_n) + \mu^*(A \setminus U)\}$ tends upward to $\mu^*(A \cap U) + \mu^*(A \setminus U)$. But $d(F_n, X \setminus U) > 0$ for every n (for which $F_n \neq \varnothing$), so

$$\mu^*(A \cap F_n) + \mu^*(A \setminus U) \leq \mu^*(A)$$

for every n, and we finally obtain the desired result

$$\mu^*(A \cap U) + \mu^*(A \setminus U) = \mu^*(A). \qquad \square$$

Thus far we have dealt only with the generation of outer measures from gauges on abstract sets. On a metric space there is a different way of generating measures called the *Carathéodory procedure*. In outlining this construction it will be convenient to employ the following notation. If $\mathcal{C}$ is a collection of subsets of a metric space X, and if r is an arbitrary positive number, then $\mathcal{C}_r$ will denote the subcollection of consisting of those sets C in $\mathcal{C}$ such that $\mathrm{diam}(C) \leq r$.

Definition 5.46. Let X be a metric space and let γ be a gauge defined on a collection $\mathcal{C} \subset 2^X$. Then for each positive number r the restriction of γ to $\mathcal{C}_r$ is a gauge on X and generates an outer measure μ_r^*. Moreover, if $0 < r < s$, then $\mathcal{C}_r \subset \mathcal{C}_s$, so $\mu_r^*(A) \geq \mu_s^*(A)$ for every $A \subset X$. Thus the net $\{\mu_r^*\}_{r>0}$ is monotone increasing, and therefore converges to an outer measure μ_0^*. (See Problem 5E; the directed set employed here is the ray $(0, +\infty)$ directed downward.) The outer measure μ_0^* will be said to be generated by γ via the *Carathéodory procedure*, and the measure μ_0 defined by μ_0^* will be referred to as the measure generated by γ via the *Carathéodory procedure*.

Example 5.47. Consider the collection $\mathcal{Z}$ of closed cells in $\mathbb{R}^d$ and the gauge $|\cdot|$. In any covering $A \subset \bigcup_n Z_n$ of a set A by closed cells, we can always replace each cell Z_n in the covering by any cellular partition of itself without changing the value of the sum of the volumes of the cells. Thus each of the approximating outer measures $\mu_r^*, r > 0$, in this case coincides with Lebesgue outer measure, and so therefore does the outer measure obtained by the Carathéodory procedure.

Example 5.48. In a metric space X define a gauge $\gamma : 2^X \to [0, +\infty]$ by $\gamma(A) = \mathrm{diam}\, A$, $A \in 2^X$. The outer measure ξ_1^* generated by this gauge by the Carathéodory procedure is known as (*one-dimensional*) *Hausdorff outer measure* on X.

In the special case $X = \mathbb{R}^d$, consider the restriction of the gauge γ to the collection of closed convex subsets of $\mathbb{R}^d$. If C is a subset of $\mathbb{R}^d$, and if $\widehat{C}$ denotes the closed convex hull of C, then $\operatorname{diam} \widehat{C} = \operatorname{diam} C$ (since every closed ball that contains C also contains $\widehat{C}$), whence it follows that the Hausdorff outer measure ξ_1^* is also the outer measure generated by this restricted gauge. But if σ is a line segment in $\mathbb{R}^d$, and $\{C_n\}$ is a countable covering of σ by closed convex sets, then $\{C_n \cap \sigma\}$ is a countable covering of σ by subsegments of σ, and it follows that $\xi_1^*(\sigma)$ is just the length of σ. In particular, ξ_1^* coincides with Lebesgue outer measure on $\mathbb{R}$. In this connection see Problems 5S and 5T.

The Carathéodory procedure is more complicated than the method described in Proposition 5.15, but it frequently yields a more useful end product. In particular, outer measures generated by the Carathéodory procedure always define a measure on Borel sets.

Proposition 5.49. *If γ is a gauge defined on a collection $\mathcal{C}$ of subsets of a metric space X, and if μ_0^* is the outer measure on X generated by γ by the Carathéodory procedure, then μ_0^* is a Carathéodory outer measure.*

Proof. By Proposition 5.45 it suffices to verify condition (5.8) for μ_0^*. Let $A, B \subset X$ satisfy $d(A, B) = d_0 > 0$ and let $r \in (0, d_0)$. It suffices to prove (5.8) when $\mu_0^*(A \cup B) < +\infty$. Fix $\varepsilon > 0$, and let $\{C_n\}$ be a countable covering of $A \cup B$ by sets belonging to $\mathcal{C}_r$ such that $\sum_n \gamma(C_n)$ is no greater than $\mu_r^*(A \cup B) + \varepsilon$. Denote by $\{C_n'\}$ the collection of those sets in the covering $\{C_n\}$ that meet A, and by $\{C_n''\}$ the collection of those that meet B. Then $\{C_n'\}$ and $\{C_n''\}$ have no sets in common, since a set of diameter less than d_0 cannot meet both A and B. Thus $\{C_n'\}$ is a countable covering of A, $\{C_n''\}$ is a countable covering of B, and they have no sets in common. Moreover,

$$\mu_r^*(A) \le \sum_n \gamma(C_n') \quad \text{and} \quad \mu_r^*(B) \le \sum_n \gamma(C_n'').$$

But then

$$\mu_r^*(A) + \mu_r^*(B) \le \sum_n \gamma(C_n) \le \mu_r^*(A \cup B) + \varepsilon \le \mu_0^*(A \cup B) + \varepsilon.$$

Letting first r and then ε tend to zero, the result follows. $\square$

As has already been suggested, a central difficulty in dealing with an outer measure generated by the Carathéodory procedure lies in relating the values of that outer measure to those of the gauge that generates it. In this connection we offer the following elementary result.

Proposition 5.50. *Let γ be a gauge defined on a collection $\mathcal{C}$ of subsets of a metric space X, and let μ_0^* denote the outer measure generated by γ via the Carathéodory procedure. If μ^* is the outer measure generated by γ in the elementary manner of Proposition 5.15, then $\mu^* \le \mu_0^*$. Hence if γ is countably subadditive, then $\gamma \le \mu_0^*$ setwise on $\mathcal{C}$.*

Proof. The outer measure μ^* is dominated setwise by each of the preliminary outer measures μ_r^* used in the construction of μ_0^*. If γ is countably subadditive, then $\mu^*|\mathcal{C} = \gamma$ (Proposition 5.19). □

So far we have focused on constructing measures from 'raw data' such as gauges. There is a basic construction which yields a new measure from a given measure and a measurable map. Assume that $(X, \mathbf{S}, \mu)$ is a measure space, $(Y, \mathbf{T})$ is a measurable space, and $\varphi : X \to Y$ is a measurable map. Define $\nu : \mathbf{T} \to [0, +\infty]$ by setting

$$\nu(E) = \mu(\varphi^{-1}(E)) = \mu(\{x \in X : \varphi(x) \in E\}), \quad E \in \mathbf{T}.$$

Then ν is a measure. Indeed, if $\{E_n\}_{n=1}^\infty \subset \mathbf{T}$ is a sequence of pairwise disjoint sets with union E, then the sets $\{\varphi^{-1}(E_n)\}_{n=1}^\infty \subset \mathbf{S}$ are pairwise disjoint as well with union $\varphi^{-1}(E)$. Countable additivity follows immediately from this observation.

Definition 5.51. The measure ν constructed above is called the *image* or *push-forward* of μ via φ, and it is denoted $\mu \circ \varphi^{-1}$ or μ_φ. If $(Y, \mathbf{T}) = (\mathbb{R}, \mathbf{B}_\mathbb{R})$, the measure μ_φ is called the *distribution* of φ. More generally, if $\varphi = (\varphi_1, \varphi_2, \ldots, \varphi_d) : X \to \mathbb{R}^d$ is measurable $[\mathbf{S}, \mathbf{B}_{\mathbb{R}^d}]$, the measure μ_φ is called the *joint distribution* of $\varphi_1, \varphi_2, \ldots, \varphi_d$.

Distributions of random variables are frequently used in probability. A particularly important case is the distribution $\mu_{|f|}$, where $f : X \to \mathbb{C}$ is a measurable function. The numbers

$$\mu_{|f|}((t, +\infty]) = \mu(\{x \in X : |f(x)| > t\})$$

can be used to determine whether the function f is integrable $[\mu]$; see Example 8.13.

Example 5.52. Assume that $(X, \mathbf{S}, \mu)$ is a measure space and $f : X \to \mathbb{R}$ is a constant function—say $f(x) = \alpha$. Then μ_f is a point mass $\mu(X)\delta_\alpha$ at α:

$$\mu_f(E) = \begin{cases} \mu(X), & \alpha \in E, \\ 0, & \alpha \notin E. \end{cases}$$

More generally, if $f = \sum_{j=1}^n \alpha_j \chi_{\sigma_j}$, where the sets $\{\sigma_j\}_{j=1}^n \subset \mathbf{S}$ form a partition of X, we have

$$\mu_f = \sum_{j=1}^n \mu(\sigma_j)\delta_{\alpha_j}.$$

Here, of course, δ_{α_j} is the unit point mass at α_j, also known as the *Dirac measure* at α_j.

Proposition 5.53. *Let* $(X, \mathbf{S}, \mu)$ *be a measure space, let* $(Y, \mathbf{T})$ *be a measurable space, and let* $\varphi : X \to Y$ *be a measurable map. A measurable function* $f : Y \to \mathbb{C}$ *is integrable* $[\mu \circ \varphi^{-1}]$ *if and only if* $f \circ \varphi$ *is integrable* $[\mu]$, *and*

$$\int_Y f \, d(\mu \circ \varphi^{-1}) = \int_X f \circ \varphi \, d\mu$$

when either of these functions is integrable.

Proof. It suffices to prove the result for positive functions f. Then the monotone convergence theorem shows that we can restrict ourselves to simple functions, and linearity allows us to further restrict ourselves to characteristic functions. When $f = \chi_E$ with $E \in \mathbf{T}$, we have $f \circ \varphi = \chi_{\varphi^{-1}(E)}$, and the result amounts to the definition of $\mu \circ \varphi^{-1}$ in this special case.

Proposition 5.54. *Let $(X, \mathbf{S}, \mu)$ be a probability space, and let $f : X \to \mathbb{R}$ be a measurable function. Endow the segment $(0, 1)$ with the restriction of Lebesgue measure λ_1. There exists a unique monotone decreasing, right continuous function $g : (0, 1) \to \mathbb{R}$ such that $\mu_g = \mu_f$.*

Proof. Simply define

$$g(t) = \inf\{\beta \in \mathbb{R} : \mu_f((\beta, +\infty)) > t\}, \quad t \in (0, 1).$$

The assertions of the proposition are now routine verifications.   □

The function g constructed above is called the *decreasing rearrangement* of f. It can be defined when $\mu(X) = +\infty$ provided that $\mu(\{x \in X : |f(x)| > t\}) < +\infty$ for every $t > 0$, in which case g is monotone decreasing and right-continuous on $(0, +\infty)$.

Problems

5A. Let μ^* be an outer measure on a space X. Show that if $A, B \subset X$, and if either A or B is measurable, then $\mu^*(A \cup B) + \mu^*(A \cap B) = \mu^*(A) + \mu^*(B)$.

5B. Let μ^* be an outer measure on a space X, and let A be a subset of X. Then the restriction of μ^* to the power class on A is an outer measure on A. (We denote this outer measure by $\mu^*|A$.) Show that if E is measurable $[\mu^*]$, then $E \cap A$ is measurable $[\mu^*|A]$. In the event that A is itself a measurable set, the measure defined by $\mu^*|A$ coincides with the contraction of μ to A, where μ denotes the measure defined by μ^*.

5C. Let μ^* be an outer measure on a space X, and let A be a subset of X. Define $\nu^*(B) = \mu^*(A \cap B)$ for every subset B of X. Show that ν^* is an outer measure on X. (This outer measure is called the *concentration* of μ^* on A.) Show that if E is measurable $[\mu^*]$, then $E, E \cap A$, and $E \setminus A$ are all measurable $[\nu^*]$. How is ν^* related to the contraction $\mu^*|A$ of μ^* to A as defined in the preceding problem?

5D. Let μ^* be an outer measure on a space X, and let $\{E_n\}$ be a disjoint sequence of $[\mu^*]$-measurable subsets of X with union E. Then for any subset $A \subset X$, measurable or not, we have $\mu^*(E \cap A) = \sum_n \mu^*(E_n \cap A)$ and $\mu^*(E \setminus A) = \sum_n \mu^*(E_n \setminus A)$. Similarly, if $\{F_n\}$ is an increasing sequence of measurable sets, then $\lim_n \mu^*(F_n \cap A) = \mu^*(\lim_n F_n \cap A)$ for arbitrary A. (The same goes for decreasing sequences provided some $\mu^*(F_n \cap A)$ is finite.)

5E. If μ_1^* and μ_2^* are two outer measures on a space X, then $\mu_1^* + \mu_2^*$ is also an outer measure on X. Show likewise that if $\{\mu_\lambda^*\}_{\lambda \in \Lambda}$ is a monotone increasing net of outer measures on X, then the limit $\lim_\lambda \mu_\lambda^*$ is an outer measure. In particular, if $\{\mu_\alpha^*\}_{\alpha \in A}$ is an arbitrary indexed family of outer measures on X, then the setwise indexed sum $\sum_{\alpha \in A} \mu_\alpha^*$ is an outer measure on X. (Hint: The finite subsets F of A form a directed set under the inclusion ordering.)

5F. Let μ^* be an outer measure on a space X, and let $\{A_n\}_{n=1}^\infty$ be an infinite sequence of subsets of X. Define $B_n = \bigcup_{i=1}^n A_i$, $n \in \mathbb{N}$, and set $B = \lim_n B_n = \bigcup_{n=1}^\infty A_n$. Prove that if $\sum_{n=1}^\infty \mu^*(A_n) < +\infty$, then $\lim_n \mu^*(B_n) = \mu^*(B)$. (Hint: Show that $\mu^*(B \setminus B_n) \to 0$.)

5G. The following are all examples of outer measures. Show in each case that the only measurable sets are the sets of outer measure zero and their complements.

(i) Let X be an arbitrary set. Set $\mu^*(\varnothing) = 0$ and let $\mu^*(A) = 1$ for every nonempty subset $A \subset X$.

(ii) Let γ denote the counting measure on $\mathbb{N}$, and for each set $A \subset \mathbb{N}$, let $\delta^*(A) = \limsup_n \gamma(A \cap \{1, \ldots, n\})/n$.

(iii) Let X be a set whose cardinal number is $\aleph_\alpha$ where $\alpha > 0$ (see the paragraph following [I, Theorem 5.8]). If $A \subset X$ is infinite, then $\operatorname{card} A = \aleph_\eta$ for some $0 \le \eta \le \alpha$. For such a set, define

$$\mu^*(A) = \begin{cases} n, & \eta = n < \omega \\ +\infty, & \eta \ge \omega, \end{cases}$$

and let $\mu^*(A) = 0$ if $\operatorname{card} A < \aleph_0$.

5H. The following are examples of gauges defined on collections of subsets of various spaces. In each case describe the outer measure generated by the gauge and find the σ-algebra of measurable sets.

(i) Let X denote an arbitrary set containing at least two distinct points, and let $\gamma(\varnothing) = 0$ and $\gamma(A) = 1$ for $\varnothing \ne A \subset X$.

(ii) Let $\mathcal{C}$ denote the collection consisting of the empty set $\varnothing$ and all closed intervals $[a, b] \subset \mathbb{R}$. Define γ on $\mathcal{C}$ by setting $\gamma(\varnothing) = 0$ and $\gamma([a, b]) = 1$ for $a \le b$.

(iii) Let X denote the open ray of all positive real numbers. If $0 < a \le b$, let $\gamma([a, b]) = 1/a$ and set $\gamma(\varnothing) = 0$.

5I. Define $\gamma([a, b]) = (b - a)^2$ for every closed interval $[a, b] \subset \mathbb{R}$ (and set $\gamma(\varnothing) = 0$, of course). Determine the outer measure generated by this gauge. Which sets are measurable with respect to this outer measure?

5J. Let γ be a gauge defined on a collection $\mathcal{C}$ of subsets of a space X, and let μ^* be the outer measure generated by γ. Show that for every subset $A \subset X$ there exists a set E belonging to the σ-algebra $\mathbf{S}$ generated by $\mathcal{C}$ such that $A \subset E$ and $\mu^*(A) = \mu^*(E)$. Show, similarly, that if X is a metric space, then the same result holds for the outer measure μ_0^* generated by γ by means of the Carathéodory procedure.

5K. If μ^* is an outer measure on a space X, if $\mathbf{S}$ denotes the σ-algebra of measurable sets with respect to μ^*, and if $\mu = \mu^*|\mathbf{S}$ is the measure defined by μ^*, then the measure space $(X, \mathbf{S}, \mu)$ is always complete. Let $(X, \mathbf{S}, \mu)$ be an arbitrary measure space, let μ^* denote the outer measure generated by μ (a measure is clearly an extendible gauge), and let $(X, \widetilde{\mathbf{S}}, \widetilde{\mu})$ be the measure space defined by μ^*. Show that if $(X, \widehat{\mathbf{S}}, \widehat{\mu})$ denotes the completion of $(X, \mathbf{S}, \mu)$ (Problem 4J), then $\widehat{\mathbf{S}}$ is contained in $\widetilde{\mathbf{S}}$ and $\widetilde{\mu}$ is an extension of $\widehat{\mu}$. Show, furthermore, that if X is σ-finite with respect to μ, then $\widetilde{\mu} = \widehat{\mu}$. (That this result is not true in general may be seen by considering the measure space $(X, \mathbf{S}_0, \gamma_0)$, where X is an arbitrary noncountable set, $\mathbf{S}_0$ denotes the σ-algebra of all the countable subsets of X and their complements (Examples 1.6 and 1.13), and γ_0 denotes the restriction of the counting measure on X to $\mathbf{S}_0$.)

5L. Modify the proof of Proposition 5.23 given in the text so as to make the set to which condition (ii) is applied the intersection L' rather than L_0. Using the fact that a set of infinite outer measure is split additively by every set, show that Proposition 5.23 remains valid if (ii) holds for all sets of finite outer measure.

5M. Let μ^* be an outer measure on a space X, and let $\mathbf{F}$ denote the ideal of those subsets $A \subset X$ for which there exists a measurable set E such that $A \subset E$ and $\mu^*(E) < +\infty$. Show that if A belongs to $\mathbf{F}$ and if E and F are measurable sets of finite measure such that $A \subset E$ and $A \subset F$, then

$$\mu^*(E) - \mu^*(E \setminus A) = \mu^*(F) - \mu^*(F \setminus A).$$

Thus the number $\mu^*(E) - \mu^*(E \setminus A)$ is determined by A alone and is independent of E. For a set A belonging to $\mathbf{F}$ it is customary to define the *inner measure* $\mu_*(A)$ of A by writing $\mu_*(A) = \mu^*(E) - \mu^*(E \setminus A)$. Show that $0 \leq \mu_*(A) \leq \mu^*(A)$ for every A in $\mathbf{F}$, and that $\mu_*(E) = \mu^*(E)$ whenever E is a measurable set of finite measure. Show also that μ_* is a monotone set function on $\mathbf{F}$.

5N. Let X be a metric space and let $\{A_n\}_{n=1}^\infty$ be a monotone increasing sequence of subsets of X. We say that $\{A_n\}$ is *metrically nested* if $d(A_n, X \setminus A_{n+1}) > 0$ for every positive integer n such that $A_n \neq \varnothing$ and $A_{n+1} \neq X$. Let μ^* be a Carathéodory outer measure on X, let $\{A_n\}$ be a metrically nested sequence of subsets of X, and let $A = \bigcup_n A_n$. Prove that $\mu^*(A) = \lim_n \mu^*(A_n)$.

5O. What are the outer measures generated by the gauges in Problems 5H(ii) and 5H(iii) by means of the Carathéodory procedure?

5P. Let X be a separable metric space and let $\mathcal{C}$ denote the collection of all bounded subsets of X. For each positive real number p, we define $\gamma_p(\varnothing) = 0$ and $\gamma_p(C) = (\operatorname{diam} C)^p$ for all $C \in \mathcal{C}, C \neq \varnothing$. The outer measure ξ_p^* generated by γ_p via the Carathéodory procedure is known as the *Hausdorff outer measure of dimension p* on X, and the measure ξ_p defined by ξ_p^* is the *Hausdorff measure of dimension p* on X. Show that the same outer measure is obtained if the gauge is restricted to the closed [open] sets in $\mathcal{C}$.

Remark 5.55. If λ_d^* denotes Lebesgue outer measure on $\mathbb{R}^d$ then λ_d^* and the Hausdorff outer measure ξ_d^* on $\mathbb{R}^d$ are not identical, except when $d = 1$. It can be shown, however, that they are constant multiples of one another. Thus it is not uncommon to find in the literature a slightly different definition of the Hausdorff measure of dimension greater than one based on a gauge $\widetilde{\gamma}_d = v(d) \cdot \gamma_d$, where $v(d)$ is a proportionality constant chosen so as to make Hausdorff measure of dimension d coincide with Lebesgue measure in Euclidean space of dimension d. Incidentally, a 0-dimensional version of Hausdorff measure on an arbitrary separable metric space X is obtained by defining $\gamma_0(\varnothing) = 0$ and $\gamma_0(C) = 1$ for all nonempty sets $C \subset X$. It is easily seen, however, that the outer measure generated by this gauge by means of the Carathéodory procedure is simply the counting measure on X.

5Q. Suppose that X is a separable metric space. Let $\mathbf{B}$ denote the σ-algebra of Borel sets in X, let γ be a countably subadditive gauge defined on $\mathbf{B}$, and let μ_0^* denote the outer measure generated by γ via the Carathéodory procedure. Let E be a Borel set in X, and let $\{\mathcal{P}_n\}_{n \in \mathbb{N}}$ be a sequence of partitions of E into countably many Borel subsets such that

  (i) all the elements of $\mathcal{P}_n$ have diameter less than $1/n$, and
  (ii) $\mathcal{P}_{n+1}$ is a *refinement* of $\mathcal{P}_n$ in the sense that every element of $\mathcal{P}_{n+1}$ is a subset of an element of $\mathcal{P}_n$

  for all $n \in \mathbb{N}$. Show that

$$\mu_0^*(E) = \lim_n \sum_{F \in \mathcal{P}_n} \gamma(F).$$

5R. (i) Let X be a metric space and let $f : X \to Y$ be a mapping of X into an arbitrary set Y. If E is a subset of X and if $\mathcal{P}$ is a partition of $E \subset X$, then $g = \sum_{F \in \mathcal{P}} \chi_{f(F)}$ has for its value at each point y of Y the number (either a positive integer or $+\infty$) of those sets F belonging to $\mathcal{P}$ such that $F \cap f^{-1}(\{y\}) \neq \varnothing$. Let $\{\mathcal{P}_n\}$ be a sequence of partitions of E that satisfies properties (i) and (ii) in the preceding problem. Show that the functions $g_n = \sum_{F \in \mathcal{P}_n} \chi_{f(F)}$ form a monotone increasing sequence which converges pointwise on Y to $N(y) = \gamma(E \cap f^{-1}(\{y\}))$, where γ denotes the counting measure on X.

  (ii) Let X be a separable metric space, let $(Y, \mathbf{S}, \mu)$ be a measure space, and let $f : X \to Y$ have the property that $f(E)$ belongs to $\mathbf{S}$ for every Borel set $E \subset X$. Define $\gamma(E) = \mu(f(E))$ for every Borel set $E \subset X$, and let μ_0^* denote the outer measure generated by the gauge γ via the Carathéodory procedure. Show that if $E \subset X$ is a Borel set, then $\mu_0^*(E) < +\infty$ if and only if the function N (of part (i)) is integrable $[\mu]$, in which case we have

$$\mu_0^*(E) = \int_Y N \, d\mu. \tag{5.9}$$

  (If we employ the notation $\int_Y g \, d\mu = +\infty$ when g is a nonnegative function on Y that is measurable $[\mathbf{S}]$ but not integrable $[\mu]$, then (5.9) holds without exception. It should be noted that the requirement that f map Borel sets onto measurable sets is not as restrictive as it might seem. Indeed, if both X and Y are separable, complete metric spaces, μ is σ-finite, and $\mathbf{S} \supset \mathbf{B}_Y$, then every continuous mapping f meets the requirement. See Problems 1Z and 1AA and Proposition 5.33.)

5S. Assume that μ is a measure on $(\mathbb{R}^d, \mathbf{B}_{\mathbb{R}^d})$ such that $\mu([0,1]^d) = 1$ and μ is *translation invariant*, that is $\mu(A + x) = \mu(A)$ for all $A \in \mathbf{B}_{\mathbb{R}^d}$ and $x \in \mathbb{R}^d$. Prove that $\mu(A) = \lambda_d(A)$ for every $A \in \mathbf{B}_{\mathbb{R}^d}$.

5T. Denote by **L** the lattice of elementary figures (that is, finite unions of cells) in $\mathbb{R}^d$. We are interested in functions $\gamma : \mathbf{L} \to \mathbb{C}$ which are modular (see Definition 5.36), translation invariant (as in Problem 5S), invariant under permutations of the coordinates, and such that $\gamma([a_1, b_1] \times \cdots \times [a_d, b_d])$ is a continuous function of the variables $a_1, \ldots, a_d$ and $b_1, \ldots, b_d$. Denote by V the vector space consisting of all functions γ satisfying these properties. For instance, the restriction $\lambda_d|\mathbf{L}$ belongs to V.

(i) Assume that $\gamma \in V$ and show that there exist scalars $\lambda_j \in \mathbb{C}$, $j = 0, 1, \ldots, d$, such that
(†) $\gamma([0, x_1] \times \cdots \times [0, x_d]) = \lambda_0 + \sum_{k=1}^d \lambda_k s_k$, $x_1, \ldots, x_d \in \mathbb{R}$,
where $s_k = \sum_{i_1 < i_2 < \cdots < i_k} x_{i_1} x_{i_2} \cdots x_{i_k}$ is the elementary symmetric function of degree k.

(ii) Conversely, given $\lambda_0, \lambda_1, \ldots, \lambda_d \in \mathbb{C}$, show that there exists a *unique* function $\gamma \in V$ satisfying (†). (Hint: The functions $\gamma \in V$ satisfy the inclusion–exclusion formula $\gamma(A_1 \cup \cdots \cup A_n) = \sum_{j=1}^n (-1)^{j-1} \sum_{i_1 < i_2 < \cdots < i_j} \gamma(A_{i_1} \cap \cdots \cap A_{i_j})$.)

Remark 5.56. In addition to λ_d, another very interesting element of V is the *Euler characteristic* corresponding to $\lambda_0 = 1$ and $\lambda_k = 0$ for $k = 1, \ldots, d$. The functions $\gamma \in V$ have modular extensions to the larger lattice consisting of finite unions of compact convex sets in $\mathbb{R}^d$. These extensions are invariant under all rigid motions, not just translations and permutations of the coordinates.

5U. Consider a separable metric space X and for each $p > 0$ let $\xi_p : \mathbf{B}_X \to [0, +\infty]$ denote the corresponding Hausdorff measure. Given a set $A \in \mathbf{B}_X$, show that there exists $p_A \in [0, \infty]$ such that $\xi_p(A) = 0$ for $p < p_A$ and $\xi_p(A) = +\infty$ for $p > p_A$. The number p_A is called the *Hausdorff dimension* of A. When $p_A \in (0, +\infty)$, the Hausdorff measure $\xi_{p_A}(A)$ may take any value in $[0, +\infty]$.

5V. Determine the Hausdorff dimension of the usual Cantor ternary set C in $\mathbb{R}$. Show that there exist sets $A \in \mathbf{B}_\mathbb{R}$ of any given Hausdorff dimension $p \in [0, 1]$.

5W. Let (X, d) and (X', d') be metric spaces and let $f : X \to X'$ be a Lipschitz map. Show that the Hausdorff dimension of $f(E)$ is at most equal to that of E for every $E \subset X$. Prove a similar result under the assumption that f is α-Hölder for some $\alpha \in (0, 1)$.

5X. Fix an integer $d \in \mathbb{N}$ and a function $f : \mathbb{R}^d \to \mathbb{R}^d$ which satisfies a Lipschitz condition, that is, there exists a constant $c > 0$ such that $\|f(x) - f(y)\|_2 \leq c\|x - y\|_2$ for every $x, y \in \mathbb{R}^n$. Show that, for any Lebesgue measurable set E with $\lambda_d(E) = 0$, we also have $\lambda_d(f(E)) = 0$.

5Y. Let $f : \mathbb{R} \to \mathbb{R}$ be a monotone increasing function that is absolutely continuous on every finite interval. Show that f maps every λ_1-null set to a λ_1-null set.

Chapter 6
Signed measures, complex measures, and absolute continuity

Given a measure, a signed measure, or a complex measure μ on a measurable space $(X, \mathbf{S})$, there are frequently other measures nearby that are closely related to μ in one way or another. These measures can often be put to good use in explicating various properties of μ. In the same vein, given two measures μ and ν (on the same or different measurable spaces), there are often useful ways in which they can be related and other ways of constructing new measures from the pair. In this chapter we explore some of the most important of these constructions and relations.

Signed measures and complex measures on a measurable space were defined in Chapter 4 (see Definition 4.1). The study of these objects can be reduced to that of positive measures using the powerful Radon-Nikodým Theorem 6.10, a particular case of which is the Hahn Decomposition Theorem 6.3. We describe these results along with some of the special forms they take for Borel measures defined on an interval. The Radon-Nikodým theorem is intimately related to the differentiation of functions on $\mathbb{R}^d$, and this allows us to treat the fundamental theorem of calculus in considerable generality. We also prove two closely related covering lemmas which are useful in the study of maximal functions (see Theorem 9.40).

Proposition 6.1. *Consider a measurable space $(X, \mathbf{S})$ and a signed measure $\mu : \mathbf{S} \to \mathbb{R}^{\natural}$ such that $\mu(X) < +\infty$ $[\mu(X) > -\infty]$. Then*

$$0 \le \sup\{\mu(E) : E \in \mathbf{S}\} < +\infty \quad [0 \ge \inf\{\mu(E) : E \in \mathbf{S}\} > -\infty].$$

Proof. Replacing μ by $-\mu$ we see that it suffices to consider the case $\mu(X) < +\infty$. Assume to the contrary that $\sup\{\mu(E) : E \in \mathbf{S}\} = +\infty$, and denote by $\mathcal{C}$ the collection of sets $F \in \mathbf{S}$ with the property that

$$\sup\{\mu(E) : E \in \mathbf{S}, E \subset F\} = +\infty,$$

© Springer International Publishing Switzerland 2016
H. Bercovici et al., *Measure and Integration*,
DOI 10.1007/978-3-319-29046-1_6

so $X \in \mathcal{C}$. If $X = F_1 \cup F_2 \cup \cdots \cup F_m$ is any finite partition of X by sets $F_j \in \mathbf{S}$, then at least one of the sets F_j must also belong to $\mathcal{C}$ because

$$\mu(E) = \sum_{j=1}^{m} \mu(E \cap F_j) \leq \sum_{j=1}^{m} \sup\{\mu(E_j) : E_j \in \mathbf{S}, E_j \subset F_j\}, \quad E \in \mathbf{S},$$

and the desired conclusion is reached by taking the supremum over $E \in \mathbf{S}$.

There exist sets $G_n \in \mathbf{S}$, $n \in \mathbb{N}$, such that $\mu(G_1) \geq 1$, $\mu(G_n) \geq 1 + \mu(G_{n-1})$ for $n > 1$, and for each $m < n$ we have either $G_n \subset G_m$ or $G_n \cap G_m = \varnothing$. These sets can be constructed by induction. The existence of G_1 satisfying $\mu(G_1) \geq 1$ follows from the fact that $X \in \mathcal{C}$. Assume that the sets $G_1, G_2, \ldots, G_{n-1}$ have been constructed, and partition the set X into 2^{n-1} subsets (some of them empty), each of the form

$$F = A_1 \cap A_2 \cap \cdots \cap A_{n-1},$$

where A_j is either G_j or $X \setminus G_j$, $j = 1, 2, \ldots, n-1$. As noted above, there is one such set F which belongs to $\mathcal{C}$, and therefore we can choose $G_n \in \mathbf{S}$ such that $G_n \subset F$ and $\mu(G_n) \geq 1 + \mu(G_{n-1})$.

Observe that there cannot exist an infinite sequence $n_1 < n_2 < \cdots$ such that the sets $\{G_{n_j}\}_j$ are pairwise disjoint because in that case

$$+\infty > \mu(X) = \mu\left(X \setminus \bigcup_{j=1}^{\infty} G_{n_j}\right) + \sum_{j=1}^{\infty} \mu(G_{n_j})$$

$$= \mu\left(X \setminus \bigcup_{j=1}^{\infty} G_{n_j}\right) + \infty = +\infty.$$

We conclude that there exists an infinite sequence $n_1 < n_2 < \cdots$ such that $G_{n_j} \supset G_{n_{j+1}}$ for all $j \in \mathbb{N}$. We have then $\mu(G_{n_j} \setminus G_{n_{j+1}}) \leq -1$ and considering the partition

$$G_{n_1} = \left(\bigcap_{j=1}^{\infty} G_{n_j}\right) \cup \left(\bigcup_{j=1}^{\infty}(G_{n_j} \setminus G_{n_{j+1}})\right)$$

we obtain

$$1 \leq \mu(G_{n_1}) = \mu\left(\bigcap_{j=1}^{\infty} G_{n_j}\right) + \sum_{j=1}^{\infty} \mu(G_{n_j} \setminus G_{n_{j+1}})$$

$$= \mu\left(\bigcap_{j=1}^{\infty} G_{n_j}\right) - \infty = -\infty,$$

a contradiction. $\qquad\qquad\qquad\qquad\qquad\qquad\qquad\qquad\qquad\qquad\qquad\quad$ $\square$

If μ is any complex measure on a measurable space $(X, \mathbf{S})$, the set functions $E \mapsto \Re\mu(E)$ and $E \mapsto \Im\mu(E)$, $E \in \mathbf{S}$, are clearly finite signed measures.

Corollary 6.2. *If μ is a complex measure on the measurable space $(X, \mathbf{S})$ then there exists a constant $c \geq 0$ such that*

$$|\mu(E_1)| + |\mu(E_2)| + \cdots + |\mu(E_n)| \leq c$$

for every finite collection $\{E_j\}_{j=1}^n \subset \mathbf{S}$ of pairwise disjoint sets. The smallest constant c with this property is called the total variation *of μ.*

Proof. It suffices to prove the corollary when μ is real-valued because $|\mu(E)| \leq |\Re\mu(E)| + |\Im\mu(E)|$ for $E \in \mathbf{S}$. Let $\{E_j\}_{j=1}^n \subset \mathbf{S}$ be pairwise disjoint, and set $F_1 = \bigcup\{E_j : \mu(E_j) \geq 0\}$, $F_2 = \bigcup\{E_j : \mu(E_j) < 0\}$. We have then

$$\sum_{j=1}^n |\mu(E_j)| = \mu(F_1) - \mu(F_2) \leq \sup\{\mu(E) : E \in \mathbf{S}\} - \inf\{\mu(E) : E \in \mathbf{S}\} < +\infty$$

by Proposition 6.1. $\qquad\square$

One way to build a signed measure on a measurable space $(X, \mathbf{S})$ is to start with two measures μ and ν on $(X, \mathbf{S})$, one of which is finite, and form the difference $\mu - \nu$. It is of interest to know that this simple construction produces the most general signed measure.

Theorem 6.3. (Hahn Decomposition) *Given a signed measure μ on a measurable space $(X, \mathbf{S})$, there exists a set $A \in \mathbf{S}$ such that, for every measurable set E, we have $\mu(E) \geq 0$ when $E \subset A$ and $\mu(E) \leq 0$ when $E \subset X \setminus A$. If $A' \in \mathbf{S}$ is another set with the same property, we have $\mu(E) = 0$ for every measurable set $E \subset A \triangle A'$. The partition $X = A \cup (X \setminus A)$ is called a Hahn decomposition of X with respect to μ.*

Proof. The last assertion is immediate. Indeed, every subset $E \subset A \triangle A'$ can be written as $E = E_1 \cup E_2$, with $E_1 \subset A \setminus A'$ and $E_2 \subset A' \setminus A$. Then $\mu(E_1) \geq 0$ because $E_1 \subset A$ and $\mu(E_1) \leq 0$ because $E_1 \subset X \setminus A'$. Similarly, $\mu(E_2) = 0$.

In proving existence we assume without loss of generality that $\mu(X) < +\infty$, and thus

$$0 \leq M = \sup\{\mu(E) : E \in \mathbf{S}\} < +\infty$$

by Proposition 6.1. It suffices to produce a set $A \in \mathbf{S}$ such that $\mu(A) = M$. Indeed, if A is such a set and $E \subset A$ is measurable, we have

$$M = \mu(A) = \mu(E) + \mu(A \setminus E) \leq \mu(E) + M$$

by the definition of M, and thus $\mu(E) \geq 0$. On the other hand, if $E \subset X \setminus A$ is measurable, then

$$M \geq \mu(A \cup E) = M + \mu(E),$$

so $\mu(E) \leq 0$.

To produce a set A with the desired properties, choose sets $A_n \in \mathbf{S}$ such that $\mu(A_n) > M - 1/2^n$ for $n \in \mathbb{N}$. We conclude the proof by showing that $\mu(A) = M$, where

$$A = \limsup_n A_n = \bigcap_{n=1}^{\infty} \bigcup_{m=n}^{\infty} A_m.$$

Given a measurable set $E \subset A_{n+1}$, we have

$$M - \frac{1}{2^n} \leq \mu(A_{n+1}) = \mu(A_{n+1} \setminus E) + \mu(E) \leq M + \mu(E)$$

so $\mu(E) \geq -1/2^{n+1}$. Fix now $n \in \mathbb{N}$ and define $B_m = A_{m+1} \setminus (A_n \cup A_{n+1} \cup \cdots \cup A_m)$ for $m \geq n$, so $\mu(B_m) \geq -1/2^{m+1}$ by what has just been proved. Then

$$\mu\left(\bigcup_{m=n}^{\infty} A_m\right) = \mu\left(A_n \cup \bigcup_{m=n}^{\infty} B_m\right) = \mu(A_n) + \sum_{m=n}^{\infty} \mu(B_m),$$

and thus the sets $C_n = \bigcup_{m \geq n} A_m$ satisfy $M \geq \mu(C_n) \geq M - 1/2^{n-1}$. The sequence $\{C_n\}$ is decreasing with intersection A, and the calculation

$$\mu(C_1) = \mu(A) + \sum_{n=1}^{\infty} \mu(C_n \setminus C_{n+1}) = \mu(A) + \lim_{N \to \infty} \sum_{n=1}^{N} \mu(C_n \setminus C_{n+1})$$

$$= \mu(A) + \lim_{N \to \infty} [\mu(C_1) - \mu(C_{N+1})] = \mu(A) + \mu(C_1) - M.$$

shows that $\mu(A) = M$, as claimed. $\qquad \square$

The uniqueness assertion in the preceding result allows us to give the following definition.

Definition 6.4. Let μ be signed measure on a measurable space $(X, \mathbf{S})$, and let $X = A \cup (X \setminus A)$ be a Hahn decomposition of X with respect to μ. The measures defined (unambiguously) for $E \in \mathbf{S}$ by

$$\mu_+(E) = \mu(E \cap A), \mu_-(E) = -\mu(E \setminus A), \text{ and } |\mu|(E) = \mu_+(E) + \mu_-(E),$$

are called the *positive variation*, *negative variation*, and *variation* of μ, respectively.

The following result is essentially a summary of what has been proved about signed measures.

Theorem 6.5. *The positive and negative variations μ_+ and μ_- of a signed measure μ on a measurable space $(X, \mathbf{S})$ are measures on $(X, \mathbf{S})$, at least one of which is finite, and $\mu = \mu_+ - \mu_-$ setwise on $\mathbf{S}$. Moreover, μ_+ and μ_- are minimal in this respect in the sense that if ν_1 and ν_2 are any two measures on $(X, \mathbf{S})$ such that $\mu = \nu_1 - \nu_2$, then there exists a finite measure δ on $(X, \mathbf{S})$*

such that $\nu_1 = \mu_+ + \delta$ *and* $\nu_2 = \mu_- + \delta$. *If* μ *is finite or* σ-*finite, then so are* μ_+ *and* μ_-.

Proof. Assume that $\mu = \nu_1 - \nu_2$, where ν_1 and ν_2 are measures on $(X, \mathbf{S})$, and let $X = A \cup (X \setminus A)$ be a Hahn decomposition of X with respect to μ. Then the measure

$$\delta(E) = \nu_2(E \cap A) + \nu_1(E \setminus A) \quad E \in \mathbf{S},$$

satisfies the required conditions. The remaining assertions of the theorem are immediate.   □

The representation of a signed measure μ as the difference of its positive and negative variations is called the *Jordan decomposition* of μ.

Example 6.6. Let $f : \mathbb{R} \to \mathbb{R}$ be a right-continuous, monotone increasing function. We have seen in Example 5.21 that there exists a Borel measure μ on $\mathbb{R}$ satisfying the relation

$$\mu((a, b]) = f(b) - f(a), \quad a, b \in \mathbb{R}, a \le b. \tag{6.1}$$

Two functions define the same Borel measure if and only if they differ by a constant. Indeed, every (positive) Borel measure μ which assigns finite measure to compact subsets of $\mathbb{R}$ is obtained in this way. Thus we can recover the function f (normalized by the condition $f(0) = 0$) by the formula

$$f(t) = \begin{cases} \mu((0, t]), & t \ge 0, \\ -\mu((t, 0]), & t < 0. \end{cases} \tag{6.2}$$

Assume now that μ is a signed Borel measure on $\mathbb{R}$, which we assume finite for simplicity, and define a function $f : \mathbb{R} \to \mathbb{R}$ by (6.2). The function f is still right-continuous, but it is not increasing if μ takes some negative values. Instead, f has *bounded variation* in the sense that there exists a constant $c \ge 0$ such that

$$\sum_{j=1}^{n} |f(b_j) - f(a_j)| \le c$$

for any finite collection $\{[a_j, b_j)\}_{j=1}^{n}$ of pairwise disjoint intervals. (The smallest constant c in this inequality is called the *total variation* of the function f.) Indeed, the sum above is nothing but

$$\sum_{j=1}^{n} |\mu((a_j, b_j])| \le |\mu|(\mathbb{R}).$$

The Jordan decomposition of μ shows that there exist bounded, right continuous, monotone increasing functions $f_1, f_2 : \mathbb{R} \to \mathbb{R}$ with the property that $f(t) = f_+(t) - f_-(t)$. Indeed, this fact can be proved using solely the fact that f has bounded variation (Problem 6A). Therefore any right continuous function f of bounded variation is seen to determine a Borel signed measure μ satisfying (6.1). This discussion can be extended to signed Borel measures μ which assign finite measure to every bounded interval, in which case the corresponding function f has finite variation on every bounded interval but its total variation may be infinite.

The fact that a signed measure has a Jordan decomposition enables us to define the integral of a function with respect to it.

Definition 6.7. Let μ be a signed measure on a measurable space $(X, \mathbf{S})$ A measurable function $f : X \to \mathbb{C}$ is said to be *integrable* with respect to μ (or integrable $[\mu]$) if it is integrable $[|\mu|]$ or, equivalently, it is integrable both $[\mu_+]$ and $[\mu_-]$. If f is integrable $[\mu]$, then the integral of f with respect to μ is defined to be

$$\int_X f \, d\mu = \int_X f \, d\mu_+ - \int_X f \, d\mu_-.$$

Similarly, if $\zeta : \mathbf{S} \to \mathbb{C}$ is a complex measure on $(X, \mathbf{S})$ written as $\zeta = \mu + i\nu$ in terms of its real and imaginary parts, a measurable function $f : X \to \mathbb{C}$ is said to be *integrable* $[\zeta]$ if it is integrable with respect to both μ and ν. If f is integrable $[\zeta]$, then the integral of f with respect to ζ is defined to be

$$\int_X f \, d\zeta = \int_X f \, d\mu + i \int_X f \, d\nu.$$

Just as in the case of signed measures, integrability $[\zeta]$ is equivalent to integrability relative to a (positive) measure $|\zeta|$ called the variation of ζ. Observe that for a signed measure μ and a measurable set E we have

$$|\mu|(E) = \sup \left\{ \sum |\mu(F)| : F \in \mathcal{F} \right\},$$

where the supremum is taken over the collection of all finite families $\mathcal{F}$ of pairwise disjoint measurable subsets of E. Thus the following definition does extend Definition 6.4.

Definition 6.8. Consider a measurable space $(X, \mathbf{S})$ and a complex measure $\zeta : \mathbf{S} \to \mathbb{C}$. The *variation* of ζ is the set function $|\zeta| : \mathbf{S} \to [0, +\infty)$ defined by

$$|\zeta|(E) = \sup \left\{ \sum |\zeta(F)| : F \in \mathcal{F} \right\}, \quad E \in \mathbf{S},$$

where the supremum is taken over the collection of all finite families $\mathcal{F}$ of pairwise disjoint measurable subsets of E.

The number $|\zeta|(X)$ is precisely the total variation of ζ defined earlier.

Proposition 6.9. *Let ζ be a complex measure on the measurable space $(X, \mathbf{S})$, and set $\mu = \Re \zeta$, $\nu = \Im \zeta$. Then:*

(1) *The function $|\zeta|$ is a finite (positive) measure on $(X, \mathbf{S})$.*
(2) *We have $|\mu|(E) \vee |\nu|(E) \le |\zeta|(E) \le |\mu|(E) + |\nu|(E)$ for $E \in \mathbf{S}$.*
(3) *A function $f : X \to \mathbb{C}$ is integrable $[\zeta]$ if and only if it is integrable $[|\zeta|]$.*
(4) *The map $f \mapsto \int_X f \, d\zeta$, defined on the linear space $\mathcal{L}$ of functions that are integrable $[|\zeta|]$, is a linear functional satisfying*

$$\left| \int_X f \, d\zeta \right| \le \int_X |f| \, d|\zeta|, \quad f \in \mathcal{L}.$$

Proof. To prove (1) we verify that $|\zeta|$ is countably additive. Let $\{E_n\}_{n=1}^{\infty} \subset \mathbf{S}$ be a collection of pairwise disjoint sets, set $E = \bigcup_n E_n$, and let $\mathcal{F}_n$ be a finite collection of pairwise disjoint subsets of E_n, $n \in \mathbb{N}$. Fix $N \in \mathbb{N}$, and observe that $\mathcal{F}_1 \cup \cdots \cup \mathcal{F}_N$ is a finite collection of disjoint subsets of E, and therefore

$$|\zeta|(E) \ge \sum_{n=1}^{N} \sum_{F \in \mathcal{F}_n} |\zeta(F)|.$$

Take now the supremum over all collections $\{\mathcal{F}_1, \dots, \mathcal{F}_N\}$ to conclude that $|\zeta|(E) \ge \sum_{n=1}^{N} |\zeta|(E_n)$, and let $N \to \infty$ to obtain $|\zeta|(E) \ge \sum_n |\zeta|(E_n)$. For the opposite inequality, let $\mathcal{F}$ be a finite collection of pairwise disjoint measurable subsets of E, and denote $\mathcal{F}_n = \{F \cap E_n : F \in \mathcal{F}\}$. The countable additivity of ζ yields

$$\sum_{F \in \mathcal{F}} |\zeta(F)| \le \sum_{F \in \mathcal{F}} \sum_{n=1}^{\infty} |\zeta(F \cap E_n)|$$

$$= \sum_{n=1}^{\infty} \sum_{F \in \mathcal{F}_n} |\zeta(F)| \le \sum_{n=1}^{\infty} |\zeta|(E_n).$$

Taking the supremum over all such collections $\mathcal{F}$ we obtain the countable additivity of $|\zeta|$.

Parts (2), (3), and the linearity of the map $f \mapsto \int_X f \, d\zeta$ follow easily from the definitions. If $s = \sum_{j=1}^{n} \alpha_j \chi_{E_j}$ is such that the sets $E_j \in \mathbf{S}$ are pairwise disjoint, then

$$\left| \int_X s \, d\zeta \right| = \left| \sum_{j=1}^{n} \alpha_j \zeta(E_j) \right| \le \sum_{j=1}^{n} |\alpha_j| |\zeta(E_j)| \le \sum_{j=1}^{n} |\alpha_j| |\zeta|(E_j) = \int_X |s| \, d|\zeta|.$$

The inequality is extended to an arbitrary function $f \in \mathcal{L}$ by writing f as the limit of simple functions s_n satisfying $s_n \le |f|$ and applying the dominated convergence theorem (Theorem 4.35). $\qquad \square$

The Jordan decomposition of a measure has an analog for complex measures. In order to describe the proper setting of this analog, we recast this decomposition as follows. Let μ be a signed measure on $(X, \mathbf{S})$, and let

$X = A \cup (X \setminus A)$ be a Hahn decomposition of X with respect to μ. Then the measure μ is equal to the indefinite integral of the function $f = \chi_A - \chi_{X \setminus A}$ with respect to $|\mu|$. That is, we have

$$\mu(E) = \int_E f \, d|\mu|, \quad E \in \mathbf{S}.$$

The extension to the complex case requires the remarkable theorem of Radon and Nikodým which we prove next. We recall that the notation $\nu \ll \mu$ for two measures indicates that ν is absolutely continuous with respect to μ, that is, $\mu(E) = 0$ implies $\nu(E) = 0$ for every measurable set E (see Definition 4.2).

Theorem 6.10. (Radon-Nikodým Theorem). *Consider two measures μ, ν on a measurable space $(X, \mathbf{S})$ such that μ is σ-finite and $\nu \ll \mu$. There exists a measurable function $f : X \to [0, +\infty]$ such that*

$$\nu(E) = \int_E f \, d\mu, \quad E \in \mathbf{S}. \tag{6.3}$$

In other words, $\nu = \nu_f$ is the indefinite integral of f with respect to μ. The function f, which is uniquely determined almost everywhere $[\mu]$, is called the Radon-Nikodým derivative of ν with respect to μ, and is denoted by $d\nu/d\mu$. A function $g : X \to \mathbb{C}$ is integrable $[\nu]$ if and only if gf is integrable $[\mu]$, and

$$\int_X g \, d\nu = \int_X gf \, d\mu \left(= \int_X g \frac{d\nu}{d\mu} \, d\mu \right) \tag{6.4}$$

when either of these integrals is defined.

Proof. The uniqueness statement is easily verified. Indeed, let f and f_0 be two Radon-Nikodým derivatives of ν with respect to μ, and let $\varepsilon > 0$. Consider the measurable set
$$A = \{x : f(x) > f_0(x) + \varepsilon\},$$
and let $B \subset A$ have finite measure with respect to μ. Then we have

$$\varepsilon \mu(B) \leq \int_B f \, d\mu - \int_B f_0 \, d\mu = \mu(B) - \mu(B) = 0,$$

and we conclude that $\mu(A) = 0$ since μ is σ-finite. In other words, $f \leq f_0 + \varepsilon$ almost everywhere $[\mu]$, and therefore $f \leq f_0$ almost everywhere $[\mu]$ because ε is arbitrary. By symmetry, $f = f_0$ almost everywhere $[\mu]$.

Next, we reduce the proof to the case of a finite measure μ. Let $\{E_n\}_{n=1}^\infty$ be a measurable partition of X into sets of finite measure with respect to μ, and assume the theorem valid for each of the subspaces E_n. Denote by f_n the Radon-Nikodým derivative of $\nu|E_n$ with respect to $\mu|E_n$, and define $f : X \to [0, +\infty]$ by setting

$$f(x) = f_n(x), \quad x \in E_n, n \in \mathbb{N}.$$

Then clearly f is a Radon-Nikodým derivative of ν with respect to μ.

It remains to prove the theorem under the additional assumption that μ is a finite measure. Under this assumption, use the Hahn decomposition (Theorem 6.3) for the signed measure $r\mu - \nu$ to find, for each positive rational number r, a set $A_r \in \mathbf{S}$ such that $\nu(E) \leq r\mu(E)$ for every measurable set $E \subset A_r$, and $\nu(E) \geq r\mu(E)$ for every measurable set $E \subset X \setminus A_r$. These measurable sets are not necessarily nested. However, if $s, r \in \mathbb{Q}_+$ and $s > r$, we have

$$s\mu(A_r \setminus A_s) \leq \nu(A_r \setminus A_s) \leq r\mu(A_r \setminus A_s),$$

and thus $\mu(A_r \setminus A_s) = \nu(A_r \setminus A_s) = 0$. Using this fact it is not difficult to verify that the set

$$B_t = \bigcup_{0 < r < t} A_r, \quad t \in (0, +\infty),$$

where the union is extended over rational numbers r in the interval $(0, t)$, provides a Hahn decomposition for the signed measure $t\mu - \nu$, and it satisfies the additional requirement that

$$B_t = \bigcup_{0 < u < t} B_u, \quad t \in (0, +\infty).$$

We now define $f : X \to [0, +\infty]$ by setting

$$f(x) = \begin{cases} \inf\{t : x \in B_t\} & \text{if } x \in B_t \text{ for some } t \in (0, +\infty), \\ +\infty & \text{otherwise.} \end{cases}$$

It is easy to see that

$$\{x \in X : f(x) < t\} = B_t, \quad t \in (0, +\infty),$$

which shows that f is measurable. We complete the proof by verifying the validity of (6.3). Assume first that E is a measurable subset of B_t for some $t \in (0, +\infty)$, fix $\varepsilon > 0$, and let $0 = t_0 < t_1 < \cdots < t_n = t$ be a partition of $[0, t]$ with mesh less than ε. Write $E_i = A \cap (B_{t_i} \setminus B_{t_{i-1}})$, $i = 1, 2, \ldots, n$, so both $\nu(E_i)$ and $\nu_f(E_i)$ are straddled by the numbers $t_{i-1}\mu(E_i)$ and $t_i\mu(E_i)$. Then

$$|\nu(E_i) - \nu_f(E_i)| \leq (t_i - t_{i-1})\mu(E_i) \leq \varepsilon\mu(E_i), \quad i = 1, 2, \ldots, n,$$

and therefore

$$|\nu(E) - \nu_f(E)| \leq \varepsilon\mu(E)$$

because the sets E_i form a partition of E. Letting $\varepsilon \to 0$, we conclude that $\nu(E) = \nu_f(E)$ because $\mu(E) < +\infty$. Assume next that $E \subset \bigcup_{n=1}^{\infty} B_n$. In this case,

$$\nu(E) = \lim_n \nu(E \cap B_n) = \lim_n \nu_f(E \cap B_n) = \nu_f(E)$$

by what has just been shown and the semicontinuity of measures (Proposition 3.28). Finally, assume that $E \subset X \setminus \bigcup_{n=1}^{\infty} B_n$. If $\mu(E) = 0$ then $\nu(E) = \nu_f(E) = 0$ by absolute continuity. On the other hand, if $\mu(E) > 0$, then $\nu(E) \geq n\mu(E)$ for every $n \in \mathbb{N}$ and therefore $\nu(E) = +\infty = \nu_f(E)$. Any measurable set E can be partitioned as $E = E_1 \cup E_2$, where $E_1 = E \setminus \bigcup_{n=1}^{\infty} B_n$, and the preceding arguments show that

$$\nu(E) = \nu(E_1) + \nu(E_2) = \nu_f(E_1) + \nu_f(E_2) = \nu_f(E).$$

The last assertion of the theorem is obviously true if $g = \chi_E, E \in \mathbf{S}$, and therefore it is true for measurable simple functions. The extension to arbitrary positive functions follows immediately from the theorem of Beppo-Levi, and the general case is obtained by an application of the dominated convergence theorem. The easy details are left to the reader.   □

If μ and ν are as in the statement of the Radon-Nikodým theorem, and if g is either integrable [ν] or nonnegative and measurable [$\mathbf{S}$], then

$$\int g \, d\nu = \int g \frac{d\nu}{d\mu} \, d\mu \tag{6.5}$$

by Theorem 6.10.

Example 6.11. Consider a Borel measure ν on $\mathbb{R}$ which assigns finite measure to all bounded intervals, and let $f : \mathbb{R} \to \mathbb{R}$ be a monotone increasing function satisfying the equality $\nu((a, b]) = f(b) - f(a)$ for all $a, b \in \mathbb{R}$ such that $a \leq b$. Assume that the function f is continuously differentiable. In this case

$$\nu((a, b]) = f(b) - f(a) = \int_a^b f'(t) \, dt$$

by the fundamental theorem of calculus. We conclude that $\nu \prec \lambda_1$ and the Radon-Nikodým derivative of f is precisely the ordinary derivative of f:

$$\frac{d\nu}{d\lambda_1}(t) = f'(t), \quad t \in \mathbb{R}.$$

Corollary 6.12. *Let $(X, \mathbf{S}, \mu)$ be a σ-finite measure space, and let $\nu : \mathbf{S} \to \mathbb{C}$ be a complex measure satisfying $\nu \ll \mu$. Then there exists a $[\mu]$-integrable function $f : X \to \mathbb{C}$ such that $\nu(E) = \int_E f \, d\mu$ for every $E \in \mathbf{S}$. The function f is uniquely determined almost everywhere $[\mu]$.*

Proof. Writing ν in terms of its real and imaginary parts, it is clear that we only need to consider the case in which ν is a finite signed measure. In this case, we write $\nu = \nu_+ - \nu_-$ as the difference between its positive and negative variations. Thus, $\nu_+(E) = \nu(E \cap A), \nu_-(E) = -\nu(E \setminus A)$ for $E \in \mathbf{S}$, where $X = A \cup (X \setminus A)$ is a Hahn decomposition of X relative to ν. We claim that we also have $\nu_+ \ll \mu$ and $\nu_- \ll \mu$. Indeed, assume that $E \in \mathbf{S}$

and $\mu(E) = 0$. Then $\mu(E \cap A) = \mu(E \setminus A) = 0$ because μ is a positive measure, and thus $\nu_+(E) = \nu_-(E) = 0$. The existence of f now follows from Theorem 6.10 applied to μ_+ and μ_-. To prove uniqueness, assume f_1 and f_2 are integrable $[\mu]$ and satisfy $\int_E f_1 \, d\mu = \int_E f_2 \, d\mu = \nu(E)$ for all $E \in \mathbf{S}$. The function $g = f_1 - f_2$ satisfies $\int_E g \, d\mu = 0$ for all $E \in \mathbf{S}$. If $\{x : g(x) \neq 0\}$ has positive measure, at least one of the sets $\{x : \Re g(x) > 0\}$, $\{x : \Re g(x) < 0\}$, $\{x : \Im g(x) > 0\}$, $\{x : \Im g(x) < 0\}$ must also have positive measure $[\mu]$, and a contradiction is obtained by integrating g over such a set. $\qquad \square$

An important consequence of the Radon-Nikodým theorem is the existence of conditional expectations, which we now define.

Definition 6.13. Consider a measure space $(X, \mathbf{S}, \mu)$, a σ-algebra $\mathbf{T} \subset \mathbf{S}$, and a $[\mu]$-integrable function $f : X \to \mathbb{C}$. A $[\mu]$-integrable function $g : X \to \mathbb{C}$ is called a *conditional expectation* of f relative to $\mathbf{T}$ if

(1) g is measurable $[\mathbf{T}]$, and
(2) $\int_E f \, d\mu = \int_E g \, d\mu$ for every $E \in \mathbf{T}$.

We use the notation $g = \mathbb{E}[f | \mathbf{T}]$ when a conditional expectation exists.

One can of course modify $\mathbb{E}[f | \mathbf{T}]$ on a $[\mu]$-null set in $\mathbf{T}$ and obtain another conditional expectation. This becomes an issue when we deal with uncountably many functions. Conditional expectations are most commonly used in probability spaces where they always exist. The following statement is somewhat more general.

Corollary 6.14. *Let* $(X, \mathbf{S}, \mu)$ *be a measure space, and let* $\mathbf{T} \subset \mathbf{S}$ *be a σ-algebra such that* $\mu | \mathbf{T}$ *is σ-finite. Then every $[\mu]$-integrable function $f : X \to \mathbb{C}$ has a conditional expectation relative to $\mathbf{T}$, and the following equalities hold almost everywhere $[\mu]$*

$$\mathbb{E}[(f + g) | \mathbf{T}] = \mathbb{E}[f | \mathbf{T}] + \mathbb{E}[g | \mathbf{T}],$$
$$\mathbb{E}[\lambda f | \mathbf{T}] = \lambda \mathbb{E}[f | \mathbf{T}],$$
$$\mathbb{E}[h f | \mathbf{T}] = h \mathbb{E}[f | \mathbf{T}],$$

provided that f, g are integrable $[\mu]$, $\lambda \in \mathbb{C}$, and $h : X \to \mathbb{C}$ is bounded and measurable $[\mathbf{T}]$.

Proof. The existence and uniqueness almost everywhere $[\mu]$ follow from Corollary 6.12 applied with $\mu | \mathbf{T}$ in place of μ and $\nu(E) = \int_E f \, d\mu$, $E \in \mathbf{T}$, since ν is clearly absolutely continuous relative to $\mu | \mathbf{T}$. The first two equalities follow immediately from the uniqueness of conditional expectations. It suffices, via Proposition 2.36, to verify the third one when $h = \chi_F$ for some $F \in \mathbf{T}$, and in this case it follows again from the uniqueness of conditional expectations. Indeed,

$$\int_E \chi_F \mathbb{E}[f | \mathbf{T}] \, d\mu = \int_{E \cap F} \mathbb{E}[f | \mathbf{T}] \, d\mu = \int_{E \cap F} f \, d\mu = \int_E \chi_F f \, d\mu$$

for every $E \in \mathbf{T}$. $\qquad \square$

Definition 6.15. Two measures μ and ν on a measurable space $(X, \mathbf{S})$ are said to be *equivalent* (notation: $\mu \equiv \nu$) if $\mu \ll \nu$ and $\nu \ll \mu$.

Two measures on $(X, \mathbf{S})$ are equivalent if and only if they possess exactly the same null sets, and this is an equivalence relation on the collection of all measures on $(X, \mathbf{S})$.

Proposition 6.16. *Let λ, μ and ν be σ-finite measures on a measurable space $(X, \mathbf{S})$ and assume that $\lambda \ll \mu \ll \nu$. Then*

$$\frac{d\lambda}{d\nu} = \left(\frac{d\lambda}{d\mu}\right)\left(\frac{d\mu}{d\nu}\right)$$

almost everywhere with respect to all three measures. In particular, if μ and ν are equivalent, then

$$\frac{d\mu}{d\nu} = \frac{1}{d\nu/d\mu}$$

almost everywhere with respect to both μ and ν.

Proof. For every $E \in \mathbf{S}$,

$$\lambda(E) = \int_E \left(\frac{d\lambda}{d\mu}\right) d\mu = \int_E \left(\frac{d\lambda}{d\mu}\right)\left(\frac{d\mu}{d\nu}\right) d\nu.$$

This proves the desired identity by the uniqueness assertion in Theorem 6.10. □

The antithesis of the concept of equivalence for measures is that of singularity.

Definition 6.17. Consider two measures μ, ν on a measurable space $(X, \mathbf{S})$. We say that μ and ν are *(mutually) singular*, or that μ is *singular* with respect to ν (notation: $\mu \perp \nu$ or $\nu \perp \mu$), if there exists a set $A \in \mathbf{S}$ such that $\mu(A) = \nu(X \setminus A) = 0$.

The proof of the following result is sketched in Problem 6CC.

Proposition 6.18. (Lebesgue decomposition) *Consider two σ-finite measures μ, ν on a measurable space $(X, \mathbf{S})$. There exist unique measures ν_1, ν_2 on $(X, \mathbf{S})$ such that $\nu = \nu_1 + \nu_2$, $\nu_1 \ll \mu$, and $\nu_2 \perp \mu$. The expression $\nu = \nu_1 + \nu_2$ is called the* Lebesgue decomposition *of ν relative to μ.*

We return now to the complex analog of the Jordan decomposition.

Proposition 6.19. *Let ζ be a complex measure on a measurable space $(X, \mathbf{S})$. Then there exists a measurable function $\varphi : X \to \mathbb{C}$ such that $|\varphi| = 1$ almost everywhere $[|\zeta|]$ and $\zeta(E) = \int_E \varphi\, d|\zeta|$ for all $E \in \mathbf{S}$. The function φ is uniquely determined almost everywhere $[|\zeta|]$.*

Proof. The conclusion of the proposition, except for the statement that $|\varphi| = 1$ almost everywhere $[|\zeta|]$, follows from Corollary 6.12. To prove that $|\varphi| = 1$ almost everywhere $[|\zeta|]$, write the set $\{z \in \mathbb{C} : |z| \neq 1\}$ as a countable union of balls of the form $B = \{z : |z - re^{it}| \leq \varepsilon\}$ where $r \neq 1$, $t \in \mathbb{R}$, and $\varepsilon < |r - 1|/2$. It suffices to show that $|\zeta|(\varphi^{-1}(B)) = 0$ for any such ball B. For a point $x \in \varphi^{-1}(B)$ we have $\Re(e^{-it}\varphi(x)) \geq 1 + \varepsilon$ $[|\varphi(x)| \leq 1 - \varepsilon]$ if $r > 1$ $[r < 1]$. For every measurable set $F \subset \varphi^{-1}(B)$ we obtain

$$\Re(e^{-it}\zeta(F)) = \int_F \Re(e^{-it}\varphi)d|\zeta| \geq (1 + \varepsilon)|\zeta|(F) \quad [|\zeta(F)| \leq (1 - \varepsilon)|\zeta|(F)]$$

for $r > 1$ $[r < 1]$. Thus, if $\mathcal{F}$ is a finite collection of pairwise disjoint measurable subsets of $\varphi^{-1}(B)$,

$$\sum_{F \in \mathcal{F}} |\zeta(F)| \geq \sum_{F \in \mathcal{F}} \Re(e^{-it}\zeta(F)) \geq (1 + \varepsilon) \sum_{F \in \mathcal{F}} |\zeta|(F)$$

$$\left[\sum_{F \in \mathcal{F}} |\zeta(F)| \leq (1 - \varepsilon) \sum_{F \in \mathcal{F}} |\zeta|(F) \right]$$

for $r > 1$ $[r < 1]$. Taking the supremum over all such collections $\mathcal{F}$ yields $|\zeta|(\varphi^{-1}(B)) \geq (1 + \varepsilon)|\zeta|(\varphi^{-1}(B))$ $[|\zeta|(\varphi^{-1}(B)) \leq (1 - \varepsilon)|\zeta|(\varphi^{-1}(B))]$, and this only happens when $|\zeta|(\varphi^{-1}(B)) = 0$. $\square$

We have seen in Example 6.11 that Radon-Nikodým derivatives are related to the derivatives encountered in differential calculus. We investigate now to what extent this relation holds for Borel measures on $\mathbb{R}$. We first introduce some notation related to limits of functions $\varphi : \mathcal{I}_x \to \mathbb{R}$, where $\mathcal{I}_x$ is the collection of all finite open intervals in $\mathbb{R}$ containing a point $x \in \mathbb{R}$. Given such a function and $\alpha \in \mathbb{R}$, we write

$$\lim_{I \to x} \varphi(I) = \alpha$$

if for every $\varepsilon > 0$ there exists $\delta > 0$ such that $|\varphi(I) - \alpha| < \varepsilon$ for every interval $I \in \mathcal{I}_x$ such that $\lambda_1(I) < \delta$. The relations

$$\lim_{I \to x} \varphi(I) = \pm\infty$$

are defined analogously. Note that $\mathcal{I}_x$ is a directed set, directed downwards by inclusion, and to say of a function $\varphi : \mathcal{I}_x \to \mathbb{R}$ that $\lim_{I \to x} \varphi(I) = \alpha$ is equivalent to saying that $\varphi(I)$ converges to α along the net $\mathcal{I}_x$. The limit $\lim_{I \to x} \varphi(I)$ exists if and only if

$$\limsup_{I \to x} \varphi(I) = \inf_{\delta > 0} \sup\{\varphi(I) : I \in \mathcal{I}_x, \lambda_1(I) < \delta\}$$

and

$$\liminf_{I \to x} \varphi(I) = \sup_{\delta > 0} \inf\{\varphi(I) : I \in \mathcal{I}_x, \lambda_1(I) < \delta\}$$

are equal, in which case $\lim_{I \to x} \varphi(I)$ is the common value of these two extended real numbers. Expressions of the form $\lim_{I \to x} \varphi(I)$ can also be defined when φ is defined on sufficiently small intervals containing x. The fraction $\mu(I)/\nu(I)$ in the following statement is defined to be 0 if $\mu(I) = \nu(I) = 0$ and $+\infty$ if $\mu(I) = 0 < \nu(I)$.

Theorem 6.20. *Consider two Borel measures μ, ν on $\mathbb{R}$, both of which assign finite measure to each bounded interval, and let*

$$d\nu = h \, d\mu + d\rho$$

be the Lebesgue decomposition of ν relative to μ; that is, h is a Borel non-negative function and $\rho \perp \mu$. Then the limit

$$\lim_{I \to x} \frac{\nu(I)}{\mu(I)}$$

exists and equals $h(x)$ for $[\mu]$-almost every $x \in \mathbb{R}$. This limit exists and equals $+\infty$ for $[\rho]$-almost every $x \in \mathbb{R}$.

The proof requires a covering lemma which we discuss first. Assume that $A \subset \mathbb{R}$ and $\mathcal{C}$ is a collection of open intervals. We say that $\mathcal{C}$ is a *Vitali covering* of the set A if, for every $x \in A$ and every $\varepsilon > 0$, there exists $I \in \mathcal{C}$ such that $x \in I$ and $\lambda_1(I) < \varepsilon$.

Proposition 6.21. *Consider a Borel measure on $\mathbb{R}$ which assigns finite measure to each bounded interval, a Borel set $A \subset \mathbb{R}$ with $\mu(A) < +\infty$, and a Vitali covering $\mathcal{C}$ of A. Then for every $\varepsilon > 0$ there exists a finite collection $\{I_j\}_{j=1}^n$ of pairwise disjoint intervals in $\mathcal{C}$ such that*

$$\mu\left(A \setminus \bigcup_{j=1}^n I_j\right) < \varepsilon.$$

Proof. With $\varepsilon > 0$ given, set $\delta = \varepsilon/10$. We have

$$\mu(A) = \lim_{N \to \infty} \mu(A \cap [-N, N]),$$

and thus there exists $N \in \mathbb{N}$ such that $\mu(A \cap [-N, N]) > \mu(A) - \delta$. An application of Proposition 7.10(2) (with $X = [-N, N]$) allows us to choose a compact set $B \subset A \cap [-N, N]$ satisfying $\mu(B) > \mu(A) - \delta$. Note first that $\sum_{x \in S} \mu(\{x\}) \leq \mu(B)$ for every finite set $S \subset B$, and therefore $\sum_{x \in A} \mu(\{x\}) \leq \mu(B) < +\infty$. We can thus choose a finite set $S \subset A$ such that

$$\sum_{x \in B \setminus S} \mu(\{x\}) < \delta.$$

It is possible to find pairwise disjoint intervals $J_1, J_2, \ldots, J_M$ in $\mathcal{C}$ such that $S \subset \bigcup_{m=1}^{M} J_m$. Taking into account that some endpoints of the intervals J_m may be atoms, we see that

$$\mu\left(B \setminus \bigcup_{m=1}^{M} J_m\right) \leq \mu\left(B \setminus \bigcup_{m=1}^{M} \overline{J_m}\right) + \delta.$$

Choose a compact set $K \subset B \setminus \bigcup_{m=1}^{M} \overline{J_m}$ such that

$$\mu\left(\left[B \setminus \bigcup_{m=1}^{M} \overline{J_m}\right] \setminus K\right) < \delta,$$

and denote by $\mathcal{C}_1$ the collection of those intervals $I \in \mathcal{C}$ with the property that

$$I \cap \left(\bigcup_{m=1}^{M} \overline{J_m}\right) = \varnothing.$$

Then $\mathcal{C}_1$ is a Vitali covering of K, and we see next that it suffices to prove the proposition with $K, \mathcal{C}_1$, and $\varepsilon - 3\delta$ in place of $A, \mathcal{C}$, and ε, respectively. Indeed, if $\{I_i\}_{i=1}^{n} \subset \mathcal{C}_1$ satisfies the conclusion of the proposition after these replacements, then the collection $\{J_m\}_{m=1}^{M} \cup \{I_i\}_{i=1}^{n}$ satisfies the original conclusion of the proposition. Note for later use that $\sum_{x \in K} \mu(\{x\}) \leq \sum_{x \in B \setminus S} \mu(\{x\}) < \delta$. Set

$$\eta = \inf \mu\left(K \setminus \bigcup_{j=1}^{n} I_j\right),$$

where the infimum is taken over all finite collections $\{I_j\}_{j=1}^{n}$ of pairwise disjoint intervals from $\mathcal{C}_1$. We argue by contradiction, assuming that $\eta > \varepsilon - 3\delta$. Choose a pairwise disjoint collection $\{I_i\}_{i=1}^{n} \subset \mathcal{C}_1$ such that

$$\mu\left(K \setminus \bigcup_{j=1}^{n} I_j\right) < \eta + \delta,$$

and then choose a compact set $K_1 \subset K \setminus \bigcup_{j=1}^{n} \overline{I_j}$ such that

$$\mu\left(\left[K \setminus \bigcup_{j=1}^{n} \overline{I_j}\right] \setminus K_1\right) < \delta.$$

The collection $\mathcal{C}_2$ consisting of those intervals in $\mathcal{C}_1$ which are disjoint from $\bigcup_{j=1}^{n} I_j$ form a Vitali covering of K_1. By compactness of K_1, there exist intervals $L_1, L_2, \ldots, L_m \in \mathcal{C}_2$ such that

$$K_1 \subset \bigcup_{k=1}^{m} L_k.$$

We may assume without loss of generality that m is the smallest integer for which such a cover exists, and thus it follows that no three of the intervals $\{L_k\}_{k=1}^{m}$ have a common point. (Indeed, if three intervals (a_1, b_1), (a_2, b_2), (a_3, b_3) have a point in common, their union is

$$(\min\{a_1, a_2, a_3\}, \max\{b_1, b_2, b_3\}) = (a_p, b_q) = (a_p, b_p) \cup (a_q, b_q)$$

for some $p, q \in \{1, 2, 3\}$.) Therefore $\{1, 2, \ldots, m\} = B_1 \cup B_2$, where B_1 and B_2 are disjoint, the intervals $\{L_k\}_{k \in B_1}$ are pairwise disjoint, and so are the intervals $\{L_k\}_{k \in B_2}$. The sets $K_1 \setminus \bigcup_{k \in B_1} L_k$ and $K_1 \setminus \bigcup_{k \in B_2} L_k$ are disjoint, so one of them has measure at most $\mu(K_1)/2$. Assume for definiteness that

$$\mu\left(K_1 \setminus \bigcup_{k \in B_1} L_k \right) \leq \frac{\mu(K_1)}{2}.$$

We consider next the collection of pairwise disjoint intervals

$$\mathcal{F} = \{I_j\}_{j=1}^{n} \cup \{L_k\}_{k \in B_1}$$

for which we have

$$\mu\left(K \setminus \bigcup_{I \in \mathcal{F}} I \right) \leq \mu\left(\left[K \setminus \bigcup_{j=1}^{n} \overline{I}_j \right] \setminus \bigcup_{k \in B_1} L_k \right) + \delta$$

$$\leq \mu\left(K_1 \setminus \bigcup_{k \in B_1} L_k \right) + 2\delta$$

$$\leq \frac{\mu(K_1)}{2} + 2\delta < \frac{\eta + \delta}{2} + 2\delta < \eta.$$

The first inequality is true because the total measure of the points in K_1 which are endpoints of the intervals I_j is less than δ. The last inequality above is true if $\delta < \varepsilon/8$, and it contradicts the definition of η, thus concluding the proof. $\qquad\square$

We proceed now with the proof of Theorem 6.20.

Proof. Given a positive rational number α, we define Borel sets

$$E_\alpha = \left\{ x \in \mathbb{R} : \liminf_{I \to x} \frac{\nu(I)}{\mu(I)} < \alpha \right\}, \quad F_\alpha = \left\{ x \in \mathbb{R} : \limsup_{I \to x} \frac{\nu(I)}{\mu(I)} > \alpha \right\}.$$

We show next that every compact set $K \subset E_\alpha$ satisfies the inequality $\nu(K) \leq \alpha\mu(K)$. Indeed, fix such a set K, let $G \supset K$ be an open set, and note that the collection $\mathcal{C}$ of intervals $I \subset G$ satisfying $\nu(I) \leq \alpha\mu(I)$ forms a Vitali covering of K. Thus, given $\varepsilon > 0$, we can find pairwise disjoint intervals $\{I_j\}_{j=1}^n$ in $\mathcal{C}$ such that

$$\nu\left(K \setminus \bigcup_{j=1}^n I_j\right) < \varepsilon.$$

We deduce that

$$\nu(K) < \varepsilon + \sum_{j=1}^n \nu(I_j) \leq \varepsilon + \alpha \sum_{j=1}^n \mu(I_j) \leq \varepsilon + \alpha\mu(G).$$

We can now let $\varepsilon \to 0$ and $\mu(G) \to \mu(K)$ to conclude that $\nu(K) \leq \alpha\mu(K)$.

Switching the roles of μ and ν and replacing α by $1/\alpha$, we deduce that every compact set $K \subset F_\alpha$ satisfies the inequality $\nu(K) \geq \alpha\mu(K)$. With these observations out of the way, we are ready to show that

$$\limsup_{I \to x} \frac{\nu(I)}{\mu(I)} = \liminf_{I \to x} \frac{\nu(I)}{\mu(I)}$$

for $[\mu]$-almost and $[\nu]$-almost every $x \in \mathbb{R}$. For this purpose, it suffices to show that, given positive rational numbers $\alpha < \beta$, the set $E_\alpha \cap F_\beta$ is null $[\mu]$ and $[\nu]$. Indeed, for a compact set $K \subset E_\alpha \cap F_\beta$ we have

$$\mu(K) \leq \nu(K)/\beta \leq (\alpha/\beta)\mu(K),$$

which is only possible when $\mu(K) = \nu(K) = 0$.

Next we show that the limit in the statement of the theorem is infinite $[\rho]$-almost everywhere. It suffices to show that the set E_α is null $[\rho]$ for every rational $\alpha > 0$. The inequality $\nu(K) \leq \alpha\mu(K)$ for compact sets $K \subset E_\alpha$ shows that the concentration of ν on such a set K is absolutely continuous $[\mu]$. Thus $\rho(E_\alpha) = 0$, as was to be proved. Observe also for further use that the inequality $h(x) \leq \alpha$ must also be satisfied $[\mu]$-almost everywhere on E_α. Indeed, if $K \subset E_\alpha$ is a compact set such that $\mu(K) > 0$ and $h(x) > \alpha$ for $x \in K$, we conclude that

$$\nu(K) \geq \int_{K_1} h \, d\mu > \alpha\mu(K),$$

contrary to what we just proved. Thus we have $h \leq \alpha$ almost everywhere $[\mu]$ on E_α. Since α is arbitrary, we conclude that

$$h(x) \leq \lim_{I \to x} \frac{\nu(I)}{\mu(I)} \quad \text{almost everywhere } [\mu].$$

Finally, given a positive rational number β, we claim that $h \geq \beta$ almost everywhere $[\mu]$ on F_β. Indeed, otherwise there would exist a compact set

$K \subset F_\beta$ such that $h < \beta$ on K, $\mu(K) > 0$, and $\rho(K) = 0$. Thus

$$\nu(K) = \int_K h \, d\mu < \beta\mu(K),$$

contrary to the inequality $\nu(K) \geq \beta\mu(K)$, which concludes the proof. $\quad\square$

The special case $\mu = \lambda_1$ of Theorem 6.20 is due to Lebesgue. It can be viewed as an extension of the fundamental theorem of calculus using Lebesgue integration in place of Riemann integration.

Theorem 6.22.(1) *Let* $u : \mathbb{R} \to \mathbb{C}$ *be integrable* $[\lambda_1]$. *Then the function* $t \mapsto \int_{-\infty}^t u \, d\lambda_1$ *is differentiable almost everywhere* $[\lambda_1]$, *and*

$$\frac{d}{dt} \int_{-\infty}^t u \, d\lambda_1 = u(t) \quad \text{almost everywhere } [\lambda_1].$$

(2) *Let* $f : \mathbb{R} \to \mathbb{R}$ *be a monotone increasing function. Then* f *is differentiable almost everywhere* $[\lambda_1]$, *the derivative* f' *is integrable on every finite interval, and*

$$f(b) - f(a) \geq \int_a^b f' \, d\lambda_1$$

for all $a, b \in \mathbb{R}$ *such that* $a < b$. *Equality holds in this equation precisely when* f *is absolutely continuous on* $[a, b]$.

Proof. It suffices to prove (1) when $u \geq 0$, in which case the measure ν defined by $d\nu = u \, d\lambda_1$ satisfies

$$\int_{-\infty}^t u \, d\lambda_1 = \nu((-\infty, t]), \quad t \in \mathbb{R}.$$

Theorem 6.20 implies the following assertion for $[\lambda_1]$-almost every $t \in \mathbb{R}$: given sequences of real numbers $\{a_n\}$ and $\{b_n\}$ such that $a_n < t < b_n$ and $\lim_{n\to\infty}(b_n - a_n) = 0$, we have

$$\lim_{n\to\infty} \frac{\int_{a_n}^{b_n} u \, d\lambda_1}{b_n - a_n} = u(t).$$

Fix $t_0 \in \mathbb{R}$ for which this assertion is true, and let $\{b_n\}$ be a sequence of real numbers such that $t_0 < b_n$ and $\lim_{n\to\infty} b_n = t_0$. We choose for each $n \in \mathbb{N}$ a number $a_n < t_0$ such that $t_0 - a_n < 1/n$ and

$$\left| \frac{\int_{a_n}^{b_n} u \, d\lambda_1}{b_n - a_n} - \frac{\int_{t_0}^{b_n} u \, d\lambda_1}{b_n - t_0} \right| < \frac{1}{n}.$$

We see that

$$\lim_{n \to \infty} \frac{\int_t^{b_n} u \, d\lambda_1}{b_n - t} = u(t)$$

and this verifies that the right derivative of $t \mapsto \int_{-\infty}^t u \, d\lambda_1$ at t_0 equals $u(t_0)$. A similar argument for the left derivative concludes the proof of (1).

For (2) we observe that the function g defined by

$$g(t) = \lim_{h \downarrow 0} f(t + h), \quad t \in \mathbb{R},$$

is right-continuous, monotone increasing, and differs from f only at the (countably many) points of discontinuity of f. There exists a Borel measure ν on $\mathbb{R}$ such that $\nu((a, b]) = g(b) - g(a)$ for $a < b$ (see Example 5.21). Writing the Lebesgue decomposition $d\nu = u \, d\lambda_1 + \rho$ with $\rho \perp \lambda_1$, we see that

$$g(b) - g(a) = \rho([a, b)) + \int_a^b u \, d\lambda_1, \quad a, b \in \mathbb{R}, a < b.$$

The equality $g' = u$ almost everywhere $[\lambda_1]$ follows as in the proof of part (1), and it is clear that $f' = g'$ at all points of continuity of f where g' exists. The equality

$$f(b) - f(a) = \int_a^b u \, d\lambda_1$$

holds when $\rho((a, b)) = 0$, f is right-continuous at a and left-continuous at b. This implies the final assertion of part (2). □

Example 6.23. Borel measures on $\mathbb{R}$ which assign finite measure to bounded sets can be further decomposed into a discrete and continuous component. A Borel measure μ on $\mathbb{R}$ is said to be continuous if $\mu(\{x\}) = 0$ for every $x \in \mathbb{R}$, and discrete if $\mu(E) = \sum_{x \in E} \mu(\{x\})$ for every $E \in \mathbf{B}_{\mathbb{R}}$. If $\mu((a, b]) = f(b) - f(a)$ for $a, b \in \mathbb{R}$, $a \leq b$, for some right-continuous, monotone increasing function $f : \mathbb{R} \to \mathbb{R}$, it is clear that μ is continuous if and only if f is a continuous function. Indeed,

$$\mu(\{x\}) = \lim_{y \uparrow x}(f(x) - f(y)), \quad x \in \mathbb{R},$$

is precisely the size of the jump discontinuity of f at x.

Given an arbitrary Borel measure μ on $\mathbb{R}$, one defines

$$\mu_{\mathrm{d}}(E) = \sum_{x \in E} \mu(\{x\}), \quad E \in \mathbf{B}_{\mathbb{R}},$$

and $\mu_{\mathrm{c}} = \mu - \mu_{\mathrm{d}}$, which represents μ as the sum of a discrete measure μ_{d} and a continuous one μ_{c}. It is an interesting fact, however, that there exist measures which are continuous but singular $[\lambda_1]$. To construct such a measure it suffices to exhibit a continuous, monotone increasing function $f : \mathbb{R} \to \mathbb{R}$

which is not constant, but $f' = 0$ almost everywhere $[\lambda_1]$. Such a function can be constructed using the Cantor ternary set $C \subset [0, 1]$ which satisfies $\lambda_1(C) = 0$, as follows We recall that $[0, 1] \setminus C$ is the union of open intervals $\{I_{i,j} : i, j \in \mathbb{N}, j \leq 2^{i-1}\}$ such that $\lambda_1(I_{i,j}) = 3^{-i}$, and $I_{i,j}$ is on the left of $I_{i,j+1}$ for $j < 2^{i-1}$. Define a function f on $\mathbb{R} \setminus C$ by setting

$$f(t) = \begin{cases} 0 & \text{for } t < 0, \\ 1 & \text{for } t > 1, \\ \frac{2j-1}{2^i} & \text{for } t \in I_{ij}, i \in \mathbb{N}, j = 1, 2, \dots, 2^{i-1}. \end{cases}$$

The function thus defined is monotone increasing on $\mathbb{R} \setminus C$, and its range is dense in $[0, 1]$. It follows that its right and left limits at each point $t \in C$ must be equal, and therefore f extends to a continuous, monotone increasing function on $\mathbb{R}$, sometimes called the *Cantor function*. Clearly $f'(t) = 0$ on $\mathbb{R} \setminus C$, and thus f determines a continuous singular Borel measure on $\mathbb{R}$. Note also that this function satisfies

$$1 = f(1) - f(0) > \int_0^1 f'(t) \, dt = 0$$

so by Theorem 6.22, f is not absolutely continuous on $[0, 1]$.

Example 6.24. Let $J \subset \mathbb{R}$ be an arbitrary interval, and let $f : J \to \mathbb{R}$ be a function. We say that f is a *convex* function on J if the inequality

$$f(tx + (1 - t)y) \leq tf(x) + (1 - t)f(y) \tag{6.6}$$

holds for all $x, y \in J$ and $t \in (0, 1)$. The function is *strictly convex* if the inequality is strict whenever $x \neq y$. If f is convex on J and $x < y$, the preceding inequality can be rewritten as

$$\frac{f(tx + (1 - t)y) - f(x)}{[tx + (1 - t)y] - x} \leq \frac{f(y) - f(x)}{y - x} \leq \frac{f(y) - f(tx + (1 - t)y)}{y - [tx + (1 - t)y]}.$$

Since $tx + (1 - t)y$ is an arbitrary point in (x, y), we conclude that

$$\frac{f(y) - f(x)}{y - x} \leq \frac{f(y_1) - f(x_1)}{y_1 - x_1} \tag{6.7}$$

whenever $x < y \leq x_1 < y_1$. We deduce that the difference quotients $(f(y) - f(x))/(y-x)$ are bounded above and below when x and y are allowed to vary in a compact interval $I \subset J$. In particular,

$$|f(y) - f(x)| \leq c_I |y - x|, \quad x, y \in I,$$

for some constant $c_I \geq 0$. Obviously, this implies that f is absolutely continuous on the interval I.

Corollary 6.25. *Let $J \subset \mathbb{R}$ be an open (possibly unbounded) interval, and let $f : J \to \mathbb{R}$ be a convex function. Then:*

(1) *There exists a monotone increasing function $u : J \to \mathbb{R}$ such that*

$$f(b) - f(a) = \int_a^b u(t)\,dt, \quad a, b \in J, a < b.$$

(2) *The derivative $f'(t)$ exists and equals $u(t)$ for all but countably many points $t \in J$.*

(3) *The second derivative $f''(t)$ exists and satisfies $f''(t) \geq 0$ almost everywhere $[\lambda_1]$ on J.*

(4) *The function f is strictly convex on J if and only if the function u is strictly increasing on J.*

Conversely, if $f : J \to \mathbb{R}$ is twice differentiable and satisfies $f'' \geq 0$ on J, then f is convex.

Proof. The existence of u and v and the equality $f' = u$ almost everywhere $[\lambda_1]$, follow from the absolute continuity of f on all compact intervals $I \subset J$. Relation (6.7) implies that $f'(a) \leq f'(b)$ if $a < b$ and these two derivatives exist. It follows that the function u can be assumed, after alteration on a $[\lambda_1]$-null set, to be monotone increasing. With this assumption, we see that $f'(a)$ exists for all points of continuity for u, and u only has countably many points of discontinuity. Thus (1) and (2) are proved, and (3) follows from Theorem 6.22. To verify (4), consider $x, y \in J$ and $t \in (0, 1)$ such that $x < y$, and set $z = tx + (1 - t)y$. We have

$$f(z) - tf(x) - (1 - t)f(y) = t[f(z) - f(x)] + (1 - t)[f(z) - f(y)]$$
$$= t \int_x^z u\,d\lambda_1 - (1 - t) \int_z^y u\,d\lambda_1,$$

and

$$t \int_x^z u\,d\lambda_1 \leq tu(z)\lambda_1([x, z]) = t(1 - t)(y - x)u(z).$$

An analogous calculation shows that

$$(1 - t) \int_z^y u\,d\lambda_1 \geq t(1 - t)(y - x)u(z).$$

Thus (6.6) fails to be strict precisely when u is constant on the interval (x, y). This also proves the last statement since, if f'' exists and is nonnegative, it follows that f' is monotone increasing, and the calculation above (with $u = f'$) proves the convexity of f. $\square$

The following extension of (6.6) is due to Jensen.

Theorem 6.26. *Consider an arbitrary (possibly unbounded) interval $J \subset \mathbb{R}$ and a continuous convex function $f : J \to \mathbb{R}$. Let $(X, \mathbf{S}, \mu)$ be a probability space and let $h : X \to J$ be an integrable function. Then $\int_X h \, d\mu \in J$, $\int_X f(h(x)) \, d\mu(x)$ exists, and*

$$f\left(\int_X h \, d\mu \right) \leq \int_X f(h(x)) \, d\mu(x).$$

Proof. If J has a finite left endpoint α, then $h(x) \geq \alpha$ for all $x \in X$, and therefore $\int_X h \, d\mu \geq \int_X \alpha \, d\mu = \alpha$. An analogous argument for the right endpoint shows that $t_0 = \int_X h \, d\mu$ does indeed belong to J. With the notation of Corollary 6.25, we have

$$f(t) - f(t_0) = \int_{t_0}^t u(t) \, dt \geq u(t_0)(t - t_0), \quad t \in J,$$

and therefore

$$f(u(x)) - f(t_0) \geq u(t_0)(u(x) - t_0), \quad x \in X.$$

The conclusion of the theorem follows by integrating this inequality. $\qquad \square$

Example 6.27. Fix $p \in [1, +\infty)$. The function $f(t) = t^p$ satisfies $f''(t) = p(p-1)t^{p-2} > 0$ for $t > 0$ and is therefore convex. Applying (6.6) with $t = 1/2$ yields the inequality

$$(x + y)^p \leq 2^{p-1}(x^p + y^p), \quad x, y \geq 0, p \geq 1. \tag{6.8}$$

On the other hand, if $p \in (0, 1)$, the derivative $f'(t) = pt^{p-1}$ is decreasing on $(0, +\infty)$, and therefore $-f$ is convex. We deduce that

$$(x + y)^p \geq 2^{p-1}(x^p + y^p), \quad x, y \geq 0, p \in (0, 1).$$

For $p \in (0, 1)$ we also have

$$(x + y)^p - x^p = \int_0^y p(x + t)^{p-1} \, dt \leq \int_0^y pt^{p-1} \, dt = y^p,$$

so

$$(x + y)^p \leq x^p + y^p, \quad x, y \geq 0, p \in (0, 1). \tag{6.9}$$

Example 6.28. The function $f(t) = e^t$ is convex on $\mathbb{R}$ since $f'' = f > 0$. Thus

$$e^{tx + (1-t)y} \leq te^x + (1 - t)e^y, \quad x, y \in \mathbb{R}, t \in (0, 1).$$

Setting $a = e^{tx}, b = e^{(1-t)y}$, this can be written as

$$ab \leq ta^{1/t} + (1 - t)b^{1/(1-t)}, \quad a, b \geq 0, t \in (0, 1). \tag{6.10}$$

There is a higher dimensional analog of the Lebesgue differentiation theorem 6.22(1). For the proof we require a different covering lemma, originally due to Vitali. A collection $\mathcal{C}$ of open balls in $\mathbb{R}^d$ is a *Vitali covering* of a set $K \subset \mathbb{R}^d$ if every element of K is contained in balls in $\mathcal{C}$ of arbitrarily small radius.

Lemma 6.29. *Consider a Vitali covering $\mathcal{C}$ of a bounded set $K \subset \mathbb{R}^d$. There exists a sequence of pairwise disjoint balls $\{B_n\} \subset \mathcal{C}$ such that*

$$\lambda_d \left(K \setminus \bigcup_n B_n \right) = 0.$$

Proof. We may assume that all the balls in $\mathcal{C}$ intersect K. Denote by α_1 the least upper bound of the radii of the balls in $\mathcal{C}$. If $\alpha_1 = +\infty$ then a single ball in $\mathcal{C}$ covers K. Otherwise, we construct the sequence B_n inductively. First choose a ball $B_1 = B(x_1, r_1) \in \mathcal{C}$ such that $r_1 > \alpha_1/2$. Assume that balls $B_j = B(x_j, r_j)$ have been constructed for $j = 1, 2, \ldots, n$. If there are no balls in $\mathcal{C}$ which are disjoint from $\bigcup_{j=1}^n B_j$, then $K \subset \bigcup_{j=1}^n \overline{B_j}$ and

$$\lambda_d \left(K \setminus \bigcup_{j=1}^n B_j \right) = 0,$$

so the lemma is proved. Otherwise, denote by α_{n+1} the supremum of the radii of those balls in $\mathcal{C}$ which are disjoint from $\bigcup_{j=1}^n B_j$, and choose $B_{n+1} = B(x_{n+1}, r_{n+1}) \in \mathcal{C}$ such that $B_{n+1} \cap \bigcup_{j=1}^n B_j = \varnothing$ and $r_{n+1} > \alpha_{n+1}/2$. Since the disjoint balls $\{B_n\}_{n=1}^\infty$ are contained in a set of finite measure, we must have

$$\lim_{n \to \infty} \alpha_n = 0.$$

We now show that

$$K \setminus \left(\bigcup_{j=1}^n \overline{B_j} \right) \subset \bigcup_{j=n+1}^\infty B(x_j, 5r_j).$$

Given $x \in K \setminus \left(\bigcup_{j=1}^n \overline{B_j} \right)$, we can choose $B = B(y, r) \in \mathcal{C}$ such that $x \in B$ and

$$B \cap \left(\bigcup_{j=1}^n \overline{B_j} \right) = \varnothing.$$

Observe now that $B \cap B_j \neq \varnothing$ for some $j \geq n+1$. Indeed the hypothesis that $B \cap B_j = \varnothing$ for all j implies $r < \alpha_n$ for all n, and thus $r = 0$. Let N be the first integer such that $B \cap B_N \neq \varnothing$. Then $r \leq \alpha_N < 2r_N$, and this implies that $x \in B \subset B(x_N, 5r_N)$. We conclude that

$$\lambda_d\left(K \setminus \bigcup_j B_j\right) \leq \lambda_d\left(K \setminus \bigcup_{j=1}^n \overline{B_j}\right) \leq 5^d \sum_{j=n+1}^\infty \lambda_d(B(x_j, r_j)),$$

and the desired conclusion is obtained as $n \to +\infty$. $\square$

We record for further use a version of Vitali's lemma for ordinary covers by open balls.

Lemma 6.30. *Let $K \subset \mathbb{R}^d$ be a compact set, and let $\mathcal{C}$ a collection of open balls whose union contains K. There exists a finite collection $\{B(x_j, r_j)\}_{j=1}^N$ of pairwise disjoint balls in $\mathcal{C}$ such that*

$$K \subset \bigcup_{j=1}^N B(x_j, 3r_j).$$

In particular, $\lambda_d(K) \leq 3^d \sum_{j=1}^N \lambda_j(B(x_j, r_j))$.

Proof. Since K is compact, we may assume that $\mathcal{C}$ is finite. Choose the balls $B(x_j, r_j)$ inductively so $B(x_1, r_1)$ is the largest ball in $\mathcal{C}$, and $B(x_{n+1}, r_{n+1})$ is the largest ball in $\mathcal{C}$ which is disjoint from $\bigcup_{j=1}^n B(x_j, r_j)$. The inductive process stops at some integer N such that there are no balls in $\mathcal{C}$ which are disjoint from $\bigcup_{j=1}^N B(x_j, r_j)$. It suffices to show now that any ball $B(x, r) \in \mathcal{C}$ is contained in $\bigcup_{j=1}^N B(x_j, r_j)$. Indeed, let n be the smallest integer such that $B(x, r) \cap B(x_n, r_n) \neq \varnothing$, and note that $r \leq r_n$, and hence $B(x, r) \subset B(x_n, 3r_n)$. $\square$

Given $x \in \mathbb{R}^d$, denote by $\mathcal{B}_x$ the collection of open balls in $\mathbb{R}^d$ that contain x. Given a function $\varphi : \mathcal{B}_x \to \mathbb{R}$ and $\alpha \in \mathbb{R}$, we write

$$\lim_{B \to x} \varphi(B) = \alpha$$

if for every $\varepsilon > 0$ there exists $\delta > 0$ such that $|\varphi(B) - \alpha| < \varepsilon$ for every ball $B \in \mathcal{B}_x$ with radius less than δ. Infinite limits are defined analogously. Setting

$$\limsup_{B \to x} \varphi(B) = \inf_{\delta > 0} \sup\{\varphi(B) : B \in \mathcal{B}_x \text{ with radius } < \delta\}$$

and

$$\liminf_{B \to x} \varphi(B) = \sup_{\delta > 0} \inf\{\varphi(B) : B \in \mathcal{B}_x \text{ with radius } < \delta\},$$

it is clear that the limit $\lim_{B \to x} \varphi(B)$ exists precisely when these two (extended real) numbers coincide. As in the case of functions defined on intervals, $\lim_{B \to x} \varphi(B)$ can be viewed as $\lim_{\mathcal{B}_x} \varphi(B)$, where the limit is taken along the net $\mathcal{B}_x$.

Theorem 6.31. *Let $f : \mathbb{R}^d \to \mathbb{C}$ be integrable $[\lambda_d]$. Then the limit*

$$\lim_{B \to x} \frac{1}{\lambda_d(B)} \int_B f \, d\lambda_d$$

exists and equals $f(x)$ for $[\lambda_d]$-almost every $x \in \mathbb{R}^d$.

Proof. As usual, it suffices to consider the case of positive functions f. As in the proof of Theorem 6.22, we define for each $\alpha > 0$ we define Borel sets

$$E_\alpha = \left\{ x \in \mathbb{R}^d : \liminf_{B \to x} \frac{1}{\lambda_d(B)} \int_B f \, d\lambda_d < \alpha \right\},$$

$$F_\alpha = \left\{ x \in \mathbb{R}^d : \limsup_{B \to x} \frac{1}{\lambda_d(B)} \int_B f \, d\lambda_d > \alpha \right\}.$$

We show that $f(x) \le \alpha$ for $[\lambda_d]$-almost every $x \in E_\alpha$. Assume to the contrary, via regularity, that there exists a compact set $K \subset E_\alpha$ such that $f(x) > \alpha$ on K and $\lambda_d(K) > 0$. For any open set $G \supset K$, the collection $\mathcal{C}$ consisting of those balls in $\mathbb{R}^d$ contained in G and satisfying the inequality

$$\frac{1}{\lambda_d(B)} \int_B f \, d\lambda_d < \alpha$$

is a Vitali covering of K, and therefore a pairwise disjoint subcollection $\{B_n\}$ of $\mathcal{C}$ covers K except for a null set $[\lambda_d]$. We have then

$$\lambda_d(K) \le \sum_{n=1}^{\infty} \lambda_d(B_n) \le \frac{1}{\alpha} \sum_{n=1}^{\infty} \int_{B_n} f \, d\lambda_d$$

$$\le \frac{1}{\alpha} \int_K f \, d\lambda_d + \frac{1}{\alpha} \int_{G \setminus K} f \, d\lambda_d.$$

Since G is an arbitrary open set containing K, the last integral above can be made arbitrarily small, and thus

$$\lambda_d(K) \le \frac{1}{\alpha} \int_K f \, d\lambda_d < \lambda_d(K),$$

which is not possible. An analogous argument shows that $f(x) \ge \alpha$ for $[\lambda_d]$-almost every $x \in F_\alpha$. Thus the intersection $E_\alpha \cap F_\beta$ is λ_d-null when $\alpha < \beta$, and this establishes the existence of the limit in the statement for $[\lambda_d]$-almost every x. The fact that the limit is $f(x)$ follows by noting that, in the contrary case, there would exist a compact set K of positive measure for which one of the following is true for some rational $\alpha > 0$: (a) $K \subset E_\alpha$ and $f(x) \ge \alpha$ for $x \in K$; (b) $K \subset F_\alpha$ and $f(x) \le \alpha$ for $x \in K$. But we have seen that such sets do not exist. $\square$

The conclusion of Theorem 6.31 is still valid if f is only assumed to be integrable on every bounded subset of $\mathbb{R}^d$. This is seen by applying the theorem to the functions $\chi_{B(0,N)} f$ with $N \in \mathbb{N}$.

Example 6.32. If $E \subset \mathbb{R}^d$ is a measurable set, we have

$$\lim_{B \to x} \frac{\lambda_d(B \cap E)}{\lambda_d(B)} = 1 \qquad (6.11)$$

for almost every $x \in E$ $[\lambda_d]$, as can be seen by applying Theorem 6.31 to $f = \chi_E$. Points x for which (6.11) holds are called *density points* of E.

Definition 6.33. Let $f : \mathbb{R}^d \to \mathbb{C}$ be a function which is integrable $[\lambda_d]$. A point $x \in \mathbb{R}^d$ is called a *Lebesgue point* for f if

$$\lim_{B \to x} \frac{1}{\lambda_d(B)} \int_B |f - f(x)| \, d\lambda_d = 0.$$

Corollary 6.34. *Let $f : \mathbb{R}^d \to \mathbb{C}$ be a function which is integrable $[\lambda_d]$. Then $[\lambda_d]$-almost every point in $\mathbb{R}^d$ is a Lebesgue point for f.*

Proof. For each complex number z with rational real and imaginary parts there exists a λ_d-null set E_z such that

$$\lim_{B \to x} \frac{1}{\lambda_d(B)} \int_B |f - z| \, d\lambda_d = |f(x) - z|, \quad x \in \mathbb{R}^d \setminus E_z.$$

Denote by E the union of these sets E_z, so E is also λ_d-null. Every point $x \notin E$ is a Lebesgue point for f. Indeed,   □

Proof.

$$\limsup_{B \to x} \frac{1}{\lambda_d(B)} \int_B |f - f(x)| \, d\lambda_d \leq |f(x) - z| + \lim_{B \to x} \frac{1}{\lambda_d(B)} \int_B |f - z| \, d\lambda_d$$
$$= 2|f(x) - z|$$

for every $z \in \mathbb{C}$ with rational real and imaginary parts, and the right-hand side can be made arbitrarily small.

Problems

6A. Let $f : \mathbb{R} \to \mathbb{R}$ be a function of bounded variation.

(i) Show that f has left and right limits at every point in $\mathbb{R}$. Moreover, if

$$J(x) = \lim_{h \downarrow 0} |f(x + h) - f(x - h)|$$

denotes the jump of J at x, then $\sum_{x \in \mathbb{R}} J(x) < +\infty$. In particular, f has at most
countably many points of discontinuity.

(ii) Assume that f is right-continuous and define $g : \mathbb{R} \to [0, +\infty)$ by the requirement
that $g(t)$ is the variation of f on the interval $(-\infty, t]$ for $t \in \mathbb{R}$, that is,

$$g(t) = \sup \left\{ \sum_{j=1}^{n} |f(b_j) - f(a_j)| \right\},$$

where the supremum is taken over all finite collections of pairwise disjoint intervals $(a_j, b_j] \subset (\infty, t]$. Show that g is monotone increasing, right continuous,
$\lim_{t \to -\infty} g(t) = 0$, and $\lim_{t \to \infty} g(t)$ equals the total variation of f. In addition,
show that $f - g$ is monotone increasing as well. Derive a new proof of Theorem 6.3
for bounded Borel signed measures on $\mathbb{R}$.

(iii) Show that the derivative $f'(x)$ exists almost everywhere $[\lambda_1]$. Moreover, the function
f' is integrable $[\lambda_1]$, and $\int_{\mathbb{R}} |f'| \, d\lambda_1$ is at most the total variation of f.

a. (iv) Are there analogues of the preceding results for complex valued functions with
bounded variation?

6B. (F. Riesz) Consider a continuous function $h : [0, 1] \to \mathbb{R}$, and define an open subset
$G \subset (0, 1)$ to consist of those points $x \in (0, 1)$ for which there exists $y \in (0, x)$ such
that $h(y) < h(x)$. Show that each connected component (a, b) of the set G satisfies
$h(a) \geq (b)$. (In fact $h(a) = h(b)$ unless $a = 0$ or $b = 1$.)

6C. (F. Riesz) Consider a continuous, monotone increasing function $f : [0, 1] \to \mathbb{R}$.

(i) Apply Problem 6B to the function $h(x) = f(x) - \beta x$ for a fixed $\beta > 0$ to show that
the set

$$B = \left\{ x \in (0, 1) : \limsup_{y \uparrow x} \frac{f(x) - f(y)}{x - y} > \beta \right\}$$

satisfies $\lambda_1(B) \leq (f(1) - f(0))/\beta$. Conclude that the set consisting of those points
$x \in (0, 1)$ at which f has an infinite left derivative is null $[\lambda_1]$.

(ii) Fix positive numbers $\alpha < \beta$, define a Borel set by

$$C = \left\{ x \in (0, 1) : \limsup_{y \uparrow x} \frac{f(x) - f(y)}{x - y} > \beta, \liminf_{y \uparrow x} \frac{f(x) - f(y)}{x - y} < \alpha \right\},$$

and let $G \supset C$ be an arbitrary open set. Use Problem 6B find open sets $G_1, G_2 \subset$
$(0, 1)$ such that

(a) $G \supset G_1 \supset G_2 \supset C$,
(b) each connected component (a, b) of G_1 satisfies $b - a \geq (f(b) - f(a))/\alpha$, and
(c) each connected component (c, d) of G_2 satisfies $d - c \leq (f(d) - f(c))/\beta$.
Deduce that

$$\lambda_1(f(G_1)) \geq \alpha \lambda_1(G_1), \quad \lambda_1(f(G_2)) \leq \beta \lambda_1(G_2),$$

and use this fact to show that $\lambda_1(C) = 0$.

(iii) Use parts (i) and (ii), and their analogs for right derivatives, to show that f has a
finite derivative almost everywhere $[\lambda_1]$.

6D. Let $\mathcal{M}$ denote the collection of finite (positive) measures on a measurable space
$(X, \mathbf{S})$. The set $\mathcal{M}$ is a partially ordered set if we define $\mu \leq \nu$ to mean $\mu(E) \leq \nu(E)$
for all $E \in \mathbf{S}$. Show that $\mathcal{M}$ is a lattice, that is, every two elements $\mu, \nu \in \mathcal{M}$ have a
least upper bound $\mu \vee \nu$ and a greatest lower bound $\mu \wedge \nu$. Moreover, every nonempty
subset of $\mathcal{M}$ has a greatest lower bound, and every nonempty subset which has

an upper bound has a least upper bound. (Hint: Fix an upper bound, for instance $\rho = \mu + \nu$, write $d\mu = f \, d\rho$, $d\nu = g \, d\rho$, and consider the measures with densities $\max\{f, g\}$ and $\min\{f, g\}$. Alternatively, define $(\mu \vee \nu)(E)$ and $(\mu \wedge \nu)(E)$ to be, respectively, the supremum and infimum of the set

$$\{\mu(E \cap F) + \nu(E \setminus F) : F \in \mathbf{S}\}.$$

To prove the final assertion, use Theorem 9.15 or extend the above explicit construction to an arbitrary family.)

6E. Consider an interval $I \subset \mathbb{R}$, and a function $f : I \to \mathbb{C}$ which satisfies the Lipschitz condition with constant $c \geq 0$:

$$|f(x) - f(y)| \leq c|x - y|, \quad x, y \in I.$$

Show that there exists a Borel function $h : I \to \mathbb{C}$ such that $|h(x)| \leq c$ for $[\lambda_1]$-almost every $x \in I$, and

$$f(x) - f(y) = \int_x^y h(t) \, d\lambda_1(t), \quad x, y \in I, x < y.$$

Moreover, we have $f' = h$ almost everywhere $[\lambda_1]$. (The converse of this exercise is easily verified.)

6F. Let ξ and ζ be complex measures on measurable spaces $(X, \mathbf{S})$ and $(Y, \mathbf{T})$, respectively.

(i) Show that if G is a set belonging to $\mathbf{S} \times \mathbf{T}$, then the function $\zeta(G_x)$ is measurable $[\mathbf{S}]$ on X, and likewise that the function $\xi(G^y)$ is measurable $[\mathbf{T}]$ on Y. Show also that if f is a measurable complex-valued function on $(X \times Y, \mathbf{S} \times \mathbf{T})$, then the set of all points $x \in X$ such that $\int_Y f(x, y) \, d\xi(x)$ exists is a measurable subset of X and

$$g(x) = \int_Y f(x, y) \, d\zeta(y)$$

is a measurable function on that set. Similarly,

$$h(y) = \int_X f(x, y) \, d\xi(x)$$

defines a measurable function on a measurable subset of Y. (Hint: The collection of all sets $G \in \mathbf{S} \times \mathbf{T}$ with the property that $\zeta(G_x)$ is measurable function of x is a σ-quasiring containing all measurable rectangles. Since $f(x, y)$ is integrable $[\zeta]$ with respect to y if and only if $|f(x, y)|$ is integrable $[|\zeta|]$ with respect to y, the set on which this holds coincides with

$$A = \left\{ x \in X : \int_Y |f(x, y)| \, d|\zeta|(y) < +\infty \right\}.$$

If $\{s_n\}$ is a sequence of measurable simple functions on $X \times Y$ such that $s_n \to f$ and $|s_n| \leq |f|$ pointwise on X, then

$$\lim_n \int_Y s_n(x, y) \, d\zeta(y) = \int_Y f(x, y) \, d\zeta(y), \quad x \in A,$$

by the dominated convergence theorem for complex measures.)

(ii) Verify that there exists a unique complex measure $\xi \times \zeta$ on $(X \times Y, \mathbf{S} \times \mathbf{T})$ satisfying the equation

$$(\xi \times \zeta)(E \times F) = \xi(E)\zeta(F)$$

on measurable rectangles (the measure $\xi \times \zeta$ is called the product of ξ and η), and show that the total variation $|\xi \times \zeta|$ coincides with $|\xi| \times |\zeta|$. (Hint: If $G \in \mathbf{S} \times \mathbf{T}$ then $\zeta(G_x)$ is a measurable function of x, bounded by $|\zeta|(Y)$. Hence one may define $(\xi \times \zeta)(G) = \int_X \zeta(G_x)\, d\xi(x)$. To show that $\xi \times \zeta$ is a complex measure, use the dominated convergence theorem. To obtain the identity $|\xi \times \zeta| = |\xi| \times |\zeta|$, verify it first for measurable rectangles.)

(iii) If f is a complex-valued function on $X \times Y$ that is integrable $[\xi \times \zeta]$, then the function $g(x) = \int_Y f(x, y)\, d\zeta(y)$ is a measurable and integrable function $[|\xi|]$ and

$$\int_X g\, d\xi = \int_X \left[\int_Y\, d\zeta(y) \right] d\xi(x) = \int_{X \times Y} f\, d(\xi \times \zeta).$$

Similarly, the iterated integral

$$\int_Y \left[\int_Y f(x, y)\, d\xi(x) \right] d\zeta(y)$$

exists and is equal to the *product* (or *double*) integral of f. (Hint: Use approximation by simple functions and the dominated convergence theorem.)

6G. Denote by μ counting measure defined on the Borel subsets of $\mathbb{R}$. Verify that $\lambda_1 | \mathbf{B}_{\mathbb{R}} \ll \mu$, but the conclusion of Theorem 6.10 does not hold for $\nu = \lambda_1 | \mathbf{B}_{\mathbb{R}}$.

6H. Let $\mathcal{C}$ denote the collection of all intervals $(\alpha, \beta) \subset (0, 1)$ with irrational endpoints. Thus $\mathcal{C}$ is a Vitali covering of $(0, 1)$. Show that $(0, 1)$ cannot be covered by a collection of pairwise disjoint intervals from $\mathcal{C}$.

6I. (de Guzmán) Under the hypotheses of Proposition 6.21, show that there exists a countable collection $\mathcal{F}$ of pairwise disjoint intervals from $\mathcal{C}$ such that $\mu \left(A \setminus \bigcup_{I \in \mathcal{F}} I \right) = 0$.

6J. Let K be a bounded subset of $\mathbb{R}^d$. Assume that for each $x \in K$ we are given an open cube Q_x centered at x. Show that there exists a subset $A \subset K$ such that $K \subset \bigcup_{x \in A} Q_x$ and each point of K is contained in at most 5^d cubes from the family $\{Q_x\}_{x \in A}$. Show that an analogous result is true if Q_x is replaced by a ball B_x centered at x. (Hint: Choose x_n inductively so $x_n \notin Q_{x_j}$ for $j < n$ and Q_{x_n} is of nearly maximum size.)

6K. Let $\{f_n\}_{n=1}^{\infty}$ be a sequence of nonnegative, monotone increasing functions on $\mathbb{R}$ such that $f(x) = \sum_{n=1}^{\infty} f_n(x) < +\infty$ for every $x \in \mathbb{R}$. Show that $f'(x) = \sum_{n=1}^{\infty} f_n'(x)$ for $[\lambda_1]$-almost every $x \in \mathbb{R}$.

6L. Let ν be a finite Borel measure on $\mathbb{R}^d$ such that $\nu \perp \lambda_d$. Show that

$$\lim_{B \to x} \frac{\nu(B)}{\lambda_d(B)}$$

exists and equals $+\infty$ for $[\nu]$-almost every $x \in \mathbb{R}^d$, where $B \to x$ has the same meaning as in the text.

6M. Assume that a Lebesgue measurable set $E \subset \mathbb{R}^d$ satisfies the condition $\lambda_d(E \cap B) \geq \lambda_d(B)/2$ for every b all B. Show that $\lambda_d(\mathbb{R}^d \setminus E) = 0$.

6N. Given Lebesgue measurable sets $E, F \subset \mathbb{R}^d$, and density points x_0 and y_0 for E and F, respectively, show that $x_0 + y_0$ is an interior point in the sum $E + F = \{x + y : x \in E, y \in F\}$.

6O. Prove that $C + C = [0, 2]$, where C is the ternary Cantor set in $[0, 1]$. Thus $E + F$ can have interior points even when E and F are null sets.

6P. Consider an open set $U \subset \mathbb{R}^d$ and a differentiable function $f : U \to \mathbb{R}^d$ with the property that $\det f'(x) \neq 0$ for every $x \in U$. Denote by $V = f(U)$ the image of f, which is also an open set, and denote by μ the restriction of λ_d to U.

(i) Show that $\mu \circ f^{-1} \ll \lambda_d$.
(ii) Suppose in addition that $f^{-1}(y)$ is a finite set for every $y \in V$. Then the Radon-Nikodým density at $y \in V$ of $\mu \circ f^{-1}$ is given by

$$\sum_{\{x : f(x) = y\}} \frac{1}{|\det f'(x)|}.$$

6Q. Given a continuous function $f : \mathbb{R} \to \mathbb{R}$, define $g : \mathbb{R} \to \mathbb{N}_0 \cup \{+\infty\}$ by setting $g(t)$ equal to the number of solutions x of the equation $f(x) = t$ for $t \in \mathbb{R}$. Prove that the total variation of f is given by $\int_{\mathbb{R}} g \, d\lambda_1$. In particular, if f has finite variation, then the preimage of $[\lambda_1]$-almost every $t \in \mathbb{R}$ is finite.

6R. The product of two functions of finite variation also has finite variation.

6S. Denote by $\mathbf{T}$ the collection of those Borel subsets $E \subset \mathbb{R}$ with the property that $E + \mathbb{Z} = E$. Show that $\mathbf{T}$ is a σ-algebra and that $\lambda_1 | \mathbf{T}$ is not σ-finite. Find a $[\lambda_1]$-integrable function f for which there exists no conditional expectation relative to $\mathbf{T}$.

6T. Consider σ-finite measure spaces $(X_j, \mathbf{S}_j, \mu_j)$, $j = 1, 2$, and let $f : X_1 \times X_2 \to \mathbb{C}$ be integrable $[\mu_1 \times \mu_2]$. Find the conditional expectation $\mathbb{E}[f | \mathbf{T}]$, where $\mathbf{T} = \{E \times X_2 : E \in \mathbf{S}_1\}$.

6U. Let $(X, \mathbf{S}, \mu)$ be a measure space, and let $\mathbf{T} \subset \mathbf{S}$ be a σ-algebra such that $\mu | \mathbf{T}$ is σ-finite. Verify the following properties of conditional expectations.

(i) If $f : X \to [0, +\infty)$ is integrable $[\mu]$, then $\mathbb{E}[f | \mathbf{T}] \geq 0$ almost everywhere $[\mu]$.
(ii) If $\{f_n : X \to [0, +\infty)\}$ is a sequence of functions that are measurable $[\mathbf{S}]$ such that $0 \leq f_n \leq f_{n+1}$ and $f(x) = \lim_{n \to \infty} f_n(x)$ is integrable $[\mu]$, then $\lim_{n \to \infty} \int_X (\mathbb{E}[f | \mathbf{T}] - \mathbb{E}[f_n | \mathbf{T}]) \, d\mu = 0$.
(iii) If $f : X \to \mathbb{C}$ is bounded and measurable $[\mathbf{S}]$ and $h : X \to \mathbb{C}$ is integrable $[\mu]$ and measurable $[\mathbf{T}]$, then $\mathbb{E}[hf | \mathbf{T}] = h \mathbb{E}[f | \mathbf{T}]$.

6V. With the notation of Theorem 6.26, assume that f is strictly convex. Show that the inequality in the statement is strict unless u is constant almost everywhere $[\mu]$.

6W. Consider a probability space $(X, \mathbf{S}, \mu)$ and a measurable function $f : X \to (0, +\infty)$. Show that

$$\frac{1}{\int_X (1/f) \, d\mu} \leq \exp\left[\int_X \log f \, d\mu \right] \leq \int_X f \, d\mu.$$

(The three quantities in this inequality are sometimes called the harmonic, geometric, and arithmetic means of f, respectively.)

6X. Let $(X, \mathbf{S})$ be a measurable space, and let $\mu : \mathbf{S} \to \mathbb{C}$ be a complex measure with total variation equal to 1. Show that there exists a set $E \in \mathbf{S}$ such that $|\mu(E)| \geq 1/\pi$. (Hint: See Problem 8N.)

6Y. Let $(X, \mathbf{S}, \mu)$ be a probability space, and let $\mathbf{T} \subset \mathbf{S}$ be a σ-algebra. Consider also an interval $J \subset \mathbb{R}$ and a continuous function $f : J \to \mathbb{R}$ which is convex on the interior of J, and let $u : X \to J$ be integrable $[\mu]$. Show that $\mathbb{E}[u | \mathbf{T}]$ takes values in J almost everywhere $[\mu]$. Assuming that $f \circ u$ is also integrable $[\mu]$, show that the generalized Jensen inequality

$$f \circ \mathbb{E}[u | \mathbf{T}] \leq \mathbb{E}[f \circ u | \mathbf{T}]$$

holds almost everywhere $[\mu]$.

6Z. Consider a function $f : \mathbb{R} \to \mathbb{R}$ with the property that

$$\limsup_{h \to 0} \frac{f(x + h) - f(x)}{h} \geq 0$$

for every $x \in \mathbb{R}$. Show that f is monotone increasing. (Hint: Replacing $f(x)$ by $f(x) + \varepsilon x$ for some small $\varepsilon > 0$, one may assume that $\limsup_{h \to 0}(f(x + h) - f(x))/h > 0$ for $x \in \mathbb{R}$.)

6AA. Let $E \subset \mathbb{R}$ be a Borel measurable set with $\lambda_1(E) = 0$, and let ε be a positive number. Show that there exists a monotone increasing, absolutely continuous function $g : \mathbb{R} \to \mathbb{R}$ such that $g'(x) = +\infty$ for every $x \in E$ and the total variation of g is less than ε. (Hint: Construct open sets $G_n \supset G_{n+1} \supset E$ with $\lambda_1(G_n) < 2^{-n}$ for $n \in \mathbb{N}$, and choose g such that $g' = \varepsilon \sum_n \chi_{G_n}$.)

6BB. Consider a function $f : \mathbb{R} \to \mathbb{R}$ such that

$$\limsup_{h \to 0} \frac{f(x + h) - f(x)}{h} > -\infty$$

for every $x \in \mathbb{R}$ and $\limsup_{h \to 0}(f(x + h) - f(x))/h \geq 0$ almost everywhere $[\lambda_1]$. Show that f is monotone increasing. (Hint: replace f by $f + g$, where g is monotone increasing and $g'(x) = +\infty$ whenever $\limsup_{h \to 0}(f(x + h) - f(x))/h < 0$.)

6CC. Consider a function $f : \mathbb{R} \to \mathbb{R}$ which has a finite derivative $f'(x)$ everywhere on $\mathbb{R}$ that is integrable $[\lambda_1]$ on every (finite) interval of $\mathbb{R}$.

(i) Define $u_n(x) = \max\{f'(x), -n\}$ and $g_n(x) = \int_0^x u_n(t)\, dt - f(x)$, $x \in \mathbb{R}$. (Here we use the usual convention that $\int_0^x u(t)\, dt = -\int_x^0 u(t)\, dt$ when $x < 0$.) Show that g_n is monotone increasing.

(ii) Let $n \to \infty$ to deduce that

$$f(b) - f(a) \leq \int_a^b f'(t)\, dt$$

for every $a, b \in \mathbb{R}$, $a < b$.

(iii) Use (i) and (ii) to show that

$$f(b) - f(a) = \int_a^b f'(t)\, dt$$

for every $a, b \in \mathbb{R}$, $a < b$.

6DD. Let μ and ν be σ-finite measures on the same measurable space $(X, \mathbf{S})$. Show that there exists a function $f : X \to [0, 1]$ that is measurable $[\mathbf{S}]$ and such that

$$\int_E (1 - f)\, d\nu = \int_E f\, d\mu, \qquad E \in \mathbf{S}. \tag{6.12}$$

Show also that if f satisfies (6.12), and if $A = \{x \in X : f(x) < 1\}$ and $B = \{x \in X : f(x) = 1\}$, then $\mu(B) = 0$, while if $\mu(E) = 0$ for some measurable set $E \subset A$, then $\nu(E) = 0$. Complete the proof of Proposition 6.18. (Hint: Observe that $\nu \ll \mu + \nu$, and also that if two finite measures ν_2 and ν_2' are both singular relative to μ, then there exists a single measurable set C such that $\mu(C) = \nu_2(X \setminus C) = \nu_2'(X \setminus C) = 0$.)

6EE. (Rademacher) Suppose that $d \in \mathbb{N}$ and the function $f : \mathbb{R}^d \to \mathbb{R}$ satisfies the Lipschitz condition $|f(x) - f(y)| \leq c\|x - y\|_2$, $x, y \in \mathbb{R}^d$ for some $c > 0$. Follow the steps below to show that f is differentiable almost everywhere $[\lambda_d]$.

(i) Show that for every nonzero vector $v \in \mathbb{R}^d$, the *directional derivative*

$$D_v f(x) = \lim_{t \to 0} \frac{f(x + tv) - f(x)}{t}$$

exists for all x in a Borel set E_v such that $\lambda_d(\mathbb{R}^d \backslash E_v) = 0$. Moreover, $|D_v f(x)| \le c\|v\|_2$ for all $x \in E_v$. (Hint: The function $t \mapsto f(x + tv)$, $t \in \mathbb{R}$, is absolutely continuous on every bounded interval. Also use the Fubini Theorem 8.4.)

(ii) Suppose that $d = 1$ and $v = 1$ so $D_v f$ is the usual derivative f'. Let $u \in \mathcal{C}_c(\mathbb{R})$ be a continuously differentiable function. Show that

$$\int_{\mathbb{R}} u(x) f'(x) \, dx = -\int_{\mathbb{R}} u'(x) f(x) \, dx.$$

(iii) Denote by $\{e_1, \ldots, e_d\}$ the standard basis in $\mathbb{R}^d$ and set

$$\nabla f(x) = (D_{e_1} f(x), \ldots, D_{e_d} f(x)) \in \mathbb{R}^d$$

for every $x \in \bigcap_{i=1}^d E_{e_i}$. Fix a nonzero vector $v \in \mathbb{R}^d$. Show that, for almost every $x \in E_v$, $D_v f(x)$ can be expressed as a scalar product

$$D_v f(x) = \langle \nabla f(x), v \rangle.$$

(Hint: It suffices to show that

$$\int_{\mathbb{R}^d} u(x) D_v(x) \, dx = \int_{\mathbb{R}^d} \langle \nabla f(x), v \rangle \, dx$$

for every continuously differentiable $u \in \mathcal{C}_c(\mathbb{R}^d)$. Use Theorem 4.35 to write the first integral as

$$\lim_{t \to 0} \int_{\mathbb{R}^d} u(x) \frac{f(x + tv) - f(x)}{t} \, dx$$

$$= \lim_{t \to 0} \int_{\mathbb{R}^d} \frac{u(x - tv) - u(x)}{t} f(x) \, dx$$

$$= -\int_{\mathbb{R}^d} \langle \nabla u(x), v \rangle f(x) \, dx$$

Apply part (b) to each partial derivative $D_{e_i} u$ and Theorem 8.4.)

(iv) Fix a dense sequence $\{v_n\}_{n \in \mathbb{N}} \subset \mathbb{R}^d \backslash \{0\}$ including $e_1, \ldots, e_d$, and let $x \in \bigcap_{n \in \mathbb{N}} E_{v_n}$. Show that the derivative $f'(x)$ exists and $f'(x)y = \langle \nabla f(x), y \rangle$, $y \in \mathbb{R}^d$, that is,

$$\lim_{y \to 0} \frac{\|f(x + y) - f(x) - \langle \nabla f(x), y \rangle\|_2}{\|y\|_2} = 0.$$

(Hint: Given $\delta > 0$, choose $N \in \mathbb{N}$ such that every unit vector $v \in \mathbb{R}^d$ is at distance at most δ from some v_i with $i \le N$. Compare the expressions

$$\frac{f(x + tw) - f(x)}{t} - \langle \nabla f(x), w \rangle$$

for $w = v$ and $w = v_i$.)

6FF. Let $I = [a, b]$ be a compact interval in $\mathbb{R}$ and let $f, g : I \to \mathbb{C}$ be two continuous functions of finite variation on I.

(i) Show that there exists a complex Borel measure μ_f on I such that $\mu_f((s, t]) = f(t) - f(s)$ when $a \le s < t \le b$. Integration of a function h relative to this measure is usually written $\int_I h(t)\, df(t)$.

(ii) (Integration by parts) Prove that

$$\int_I f(t)\, dg(t) = [f(b)g(b) - f(a)g(a)] - \int_I g(t)\, df(t).$$

(Hint: Approximate the integrals by Riemann-like sums and apply summation by parts.)

Chapter 7
Measure and topology

Given a topological space X, there is a natural σ-algebra of subsets of X, namely the σ-algebra $\mathbf{B}_X$ of Borel sets. When X is locally compact (that is, every point has a relatively compact neighborhood) another useful σ-algebra is the σ-algebra generated by the compact G_δ subsets of X. This collection is called the σ-algebra of *Baire sets* and is denoted $\mathbf{Ba}_X$. The σ-algebra $\mathbf{Ba}_X$ is defined the same way for arbitrary topological spaces, but it is not so useful when X is not locally compact. We write $\mathcal{C}_\mathrm{b}(X)$ for the algebra of bounded, continuous complex-valued functions defined on X, and $\mathcal{C}_\mathrm{c}(X)$ for the algebra of complex-valued functions on X which vanish outside some compact subset of X. In other words, a complex-valued function f belongs to $\mathcal{C}_\mathrm{c}(X)$ when $\overline{N_f}$ is compact, where $N_f = E(f \neq 0)$. The set $\overline{N_f}$ is called the *closed support* of f and is denoted $\mathrm{supp}(f)$. The space $\mathcal{C}_\mathrm{c}(X)$ is trivial if X contains no relatively compact open sets. An intermediate space between $\mathcal{C}_\mathrm{c}(X)$ and $\mathcal{C}_\mathrm{b}(X)$ is the algebra $\mathcal{C}_0(X)$ consisting of those continuous functions $f : X \to \mathbb{C}$ that *vanish at infinity*, that is, such that the set $E(|f| \geq \varepsilon)$ is compact for every $\varepsilon > 0$. The space $\mathcal{C}_0(X)$ is easily seen to be the collection of limits of uniformly convergent sequences of functions in $\mathcal{C}_\mathrm{c}(X)$. The subalgebra of real-valued functions in $\mathcal{C}_\mathrm{c}(X)$ is denoted $\mathcal{C}_\mathrm{c}(X, \mathbb{R})$ and analogous notations are used for the other two spaces of continuous functions defined above.

Lemma 7.1. *Assume that X is a locally compact Hausdorff space. Then $\mathbf{Ba}_X$ is the smallest σ-algebra relative to which all functions in $\mathcal{C}_\mathrm{c}(X)$ are measurable.*

Proof. We verify first that each $f \in \mathcal{C}_\mathrm{c}(X)$ is measurable $[\mathbf{Ba}_X]$, and we can restrict ourselves to the collection of nonnegative functions since it generates $\mathcal{C}_\mathrm{c}(X)$ as a vector space. Assume then that $f \in \mathcal{C}_\mathrm{c}(X)$ is nonnegative and observe that

$$E(f \geq \alpha) = \bigcap_{n=1}^{\infty} E\left(f > \alpha - \frac{1}{n}\right)$$

© Springer International Publishing Switzerland 2016
H. Bercovici et al., *Measure and Integration*,
DOI 10.1007/978-3-319-29046-1_7

is a compact G_δ set for $\alpha > 0$, while $E(f \geq \alpha) = X$ for $\alpha < 0$. Thus f is measurable $[\mathbf{Ba}_X]$. In the opposite direction, assume that all the functions in $\mathcal{C}_c(X)$ are measurable $[\mathbf{S}]$ for some σ-algebra $\mathbf{S}$. We need to show that every compact G_δ set belongs to $\mathbf{S}$. Indeed, assume that $K = \bigcap_{n=1}^\infty G_n$ is compact, where G_n is open for every $n \in \mathbb{N}$, and use Urysohn's lemma $[\mathbf{I}, \text{Theorem } 9.32]$ to produce functions $f_n \in \mathcal{C}_c(X)$ such that $\chi_K \leq f_n \leq \chi_{G_n}$. The result follows from the equality $K = \bigcap_n E(f_n > 0)$. $\square$

Example 7.2. Let I be an uncountable set, denote by X_i the discrete topological space $\{0,1\}$ for each $i \in I$, and consider the space $X = \prod_i X_i$ endowed with the product topology. We recall that a base of open sets for this topology consists of products $\prod_i V_i$, where each $V_i \subset X_i$ is open and the set $\{i : V_i \neq X_i\}$ is finite. The space X is Hausdorff and compact; in particular each singleton $\{x\} \subset X$ is closed. The function $\chi_{\{x\}}$ however is not a Baire function, and thus $\{x\} \notin \mathbf{Ba}_X$. Indeed, it can be shown (see Problem 7A) that any function which is measurable $[\mathbf{Ba}_X]$ is countably determined, that is, depends only on countably many variables.

Definition 7.3. Let X be a topological space. A measure on the measurable space $(X, \mathbf{B}_X)$ $[(X, \mathbf{Ba}_X)]$ is called a *Borel* measure $[Baire$ measure$]$ on X.

We have seen in Chapter 3 (Theorems 3.29 and 3.37) that defining a measure on a measurable space $(X, \mathbf{S})$ is equivalent to defining a Lebesgue integral on that space. Under reasonable circumstances, Baire measures are entirely determined by the integrals of compactly supported functions.

Proposition 7.4. *Consider a locally compact Hausdorff space X, and two Baire measures μ, ν on X which assign finite values to all compact sets $K \in \mathbf{Ba}_X$. Then every function in $\mathcal{C}_c(X)$ is integrable $[\mu]$ and $[\nu]$, and we have $\mu = \nu$ if and only if $\int_X f\, d\mu = \int_X f\, d\nu$ for every $f \in \mathcal{C}_c(X)$.*

Proof. Let $f \in \mathcal{C}_c(X)$ be a nonnegative function. There exists a compact set K such that $f(x) = 0$ for $x \notin K$. Local compactness allows us to find an open set U with compact closure such that $K \subset U$, and Urysohn's lemma provides a nonnegative function $g \in \mathcal{C}_c(X)$ such that $\chi_K \leq g \leq \chi_U$. The set $V = E(g > 0)$ is in $\mathbf{Ba}_X$ and has compact closure, so $\mu(V) < +\infty$. Since $0 \leq f \leq (\sup f)\chi_V$, we conclude that f is integrable $[\mu]$. The nonnegative functions span $\mathcal{C}_c(X)$ and therefore every function in this space is integrable $[\mu]$.

Assume now that $\int_X f\, d\mu = \int_X f\, d\nu$ for every $f \in \mathcal{C}_c(X)$. It suffices by Theorem 5.38 to show that $\mu(K) = \nu(K)$ for every compact G_δ set K. Assume that $K = \bigcap_{n=1}^\infty G_n$ is compact, where G_n is open for every $n \in \mathbb{N}$. Since X is locally compact, we can assume that each G_n has compact closure, and we can also assume that $G_{n+1} \subset G_n$. Use Urysohn's lemma to produce functions $f_n \in \mathcal{C}_c(X)$ such that $\chi_K \leq f_n \leq \chi_{G_n}$. The functions f_n converge pointwise to χ_K and $0 \leq f_n \leq f_1$. The dominated convergence theorem (Theorem 4.35) yields

$$\mu(K) = \lim_{n\to\infty} \int_X f_n \, d\mu = \lim_{n\to\infty} \int_X f_n \, d\nu = \nu(K),$$

thus completing the proof, since the opposite implication is trivial. □

Borel measures on a locally compact space are also determined by the integrals of continuous functions, provided that a regularity condition is satisfied.

Definition 7.5. Let X be a topological space, and let $\mu : \mathbf{B}_X \to [0, +\infty]$ be a finitely additive set function. Then μ is said to be *outer regular* provided

$$\mu(A) = \inf\{\mu(G) : G \supset A, G \text{ open}\}$$

for every $A \in \mathbf{B}_X$, and μ is said to be *inner regular* if

$$\mu(A) = \sup\{\mu(K) : K \subset A, K \text{ compact}\}$$

for every $A \in \mathbf{B}_X$.

Theorem 7.6. (Lusin) *Consider a topological space X, an inner regular Borel measure μ on X, a Borel measurable function $f : X \to \mathbb{R}$, and a subset $E \in \mathbf{B}_X$ such that $\mu(E) < +\infty$. For every $\varepsilon > 0$ there exists a compact set $K \subset E$ such that $\mu(E \setminus K) < \varepsilon$ and the restriction of f to K is a continuous function on K.*

Proof. Let $\{I_n\}_{n\in\mathbb{N}}$ be an enumeration of all the open intervals with rational endpoints in $\mathbb{R}$. For each $n \in \mathbb{N}$, the inner regularity of μ implies the existence of compact subsets C_n and D_n of E such that $C_n \subset f^{-1}(I_n)$, $D_n \subset f^{-1}(\mathbb{R} \setminus I_n)$, and

$$\mu(E \setminus (C_n \cup D_n)) < \frac{\varepsilon}{2^n}.$$

We claim that the set

$$K = \bigcap_{n=1}^{\infty} (C_n \cup D_n)$$

satisfies the requirements of the theorem. Indeed, K is compact and

$$\mu(E \setminus K) \le \sum_{n=1}^{\infty} \mu(E \setminus (C_n \cup D_n)) < \varepsilon.$$

If $x \in K$ and V is a neighborhood of $f(x)$, there exists $n \in \mathbb{N}$ such that $x \in I_n \subset V$. The set $W = C_n \cap K = (X \setminus D_n) \cap K$ is a neighborhood of x in K and $f(W) \subset I_n \subset V$. Thus the restriction of f to K is continuous at every point $x \in K$.

Measures with regularity properties on a locally compact space X can be constructed using linear functionals defined on spaces of continuous functions. The following result is known as the Riesz representation theorem.

Theorem 7.7. *Consider a locally compact Hausdorff space X and a positive linear functional $\varphi : \mathcal{C}_c(X, \mathbb{R}) \to \mathbb{R}$. There exists a unique outer regular Borel measure μ on X such that $\mu(C) < +\infty$ for every compact set $C \subset X$,*

$$\varphi(f) = \int_X f \, d\mu, \quad f \in \mathcal{C}_c(X, \mathbb{R}), \tag{7.1}$$

and

$$\mu(A) = \sup\{\mu(K) : K \subset A, K \text{ compact}\} \tag{7.2}$$

whenever A is open. The equality (7.2) is also valid when $\mu(A) < +\infty$. (The measure μ is said to represent *the functional φ.)*

Proof. We prove uniqueness first. Assume therefore that μ and ν are two outer regular measures representing φ. We show first that $\mu(K) = \nu(K)$ for every compact set $K \subset X$. Fix K, $\varepsilon > 0$, choose an open set $G \supset K$ such that $\mu(G) < \mu(K) + \varepsilon$ and $\nu(G) < \nu(K) + \varepsilon$. Consider $f \in \mathcal{C}_c(X)$ such that $\chi_K \leq f \leq \chi_U$. We have then

$$\mu(K) \leq \varphi(f) \leq \mu(K) + \varepsilon, \quad \nu(K) \leq \varphi(f) \leq \nu(K) + \varepsilon,$$

and therefore $|\mu(K) - \nu(K)| \leq \varepsilon$. The equality $\mu(K) = \nu(K)$ follows since ε is arbitrary.

In order to prove existence, we recall from Example 5.37 that the formula

$$\gamma_\varphi(U) = \sup\{\varphi(f) : 0 \leq f \leq \chi_U, f \in \mathcal{C}_c(X, \mathbb{R})\}$$

defines an extendible gauge on the collection of open sets in X, that is, there exists a Borel measure μ on X such that $\mu(U) = \gamma_\varphi(U)$ for each open set $U \subset X$. The definition of μ via an outer measure ensures its outer regularity. If K is compact, $h \in \mathcal{C}_c(X, \mathbb{R})$ satisfies $0 \leq h \leq 1$, and $h(x) = 1$ for x in an open set $U \supset K$, then $\mu(K) \leq \varphi(h)$. Indeed, if $f \in \mathcal{C}_c(X, \mathbb{R})$ is such that $0 \leq f \leq \chi_U$, we have $f \leq h$ and therefore $\mu(K) \leq \mu(U) = \gamma_\varphi(U) \leq \varphi(h)$.

To show that μ represents φ it suffices to prove (7.1) when $0 \leq f \leq 1$. Fix a compact set K_1 containing $E(f \neq 0)$ in its interior, and let $n \in \mathbb{N}$. Define compact sets $K_2 \supset \cdots \supset K_n$ by setting $K_j = E(f \geq (j-1)/n)$ for $j = 2, \ldots, n$. Then K_j is contained in the interior of K_{j-1} for $j = 2, \ldots, n$, and Urysohn's Lemma provides functions $h_j \in \mathcal{C}_c(X)$, $j = 1, \ldots, n$, satisfying $0 \leq h_j \leq \chi_{K_j}$ such that $h_j(x) = 1$ for x in some neighborhood of K_{j+1} for $j < n$. We have then

$$\frac{1}{n}\sum_{j=2}^{n} h_j \leq f \leq \frac{1}{n}\sum_{j=1}^{n} h_j.$$

As noted above, $\varphi(h_j) \geq \mu(K_{j+1})$ for $j < n$, and clearly $\varphi(h_j) \leq \mu(K_j)$. The positivity of φ implies

$$\frac{1}{n}\sum_{j=2}^{n} \mu(K_j) \leq \varphi(f) \leq \frac{1}{n}\sum_{j=1}^{n} \mu(K_j),$$

while

$$\frac{1}{n} \sum_{j=2}^{n} \mu(K_j) \le \int_X \frac{1}{n} \sum_{j=2}^{n} h_j \, d\mu \le \int_X f \, d\mu \le \int_X \frac{1}{n} \sum_{j=1}^{n} h_j \, d\mu \le \frac{1}{n} \sum_{j=1}^{n} \mu(K_j).$$

We conclude that $|\varphi(f) - \int_X f \, d\mu| \le \mu(K_1)/n$ and the equality (7.1) is obtained by letting $n \to +\infty$.

Next we prove (7.2) when A is open, in which case $\mu(A) = \gamma_\varphi(A)$. Fix a number $\alpha \in [0, \mu(A))$ and select $f \in \mathcal{C}_c(X, \mathbb{R})$ such that $0 \le f \le \chi_A$ and $\varphi(f) > \alpha$. There exists $\varepsilon > 0$ such that $\varphi((f - \varepsilon)_+) > \alpha$ as well. Indeed, $\varphi(f) = \lim_{\varepsilon \downarrow 0} \varphi((f - \varepsilon)_+)$ as can be seen from the dominated convergence theorem (Theorem 4.35). The set $K = E(f \ge \varepsilon)$ is a compact subset of A and $\chi_K \ge (f - \varepsilon)_+$. Thus

$$\mu(K) = \int_X \chi_K \, d\mu \ge \int_X (f - \varepsilon)_+ \, d\mu = \varphi((f - \varepsilon)_+) > \alpha,$$

and (7.2) follows by letting $\alpha \to \mu(A)$.

To verify the final assertion, assume that $A \in \mathbf{B}_X$ satisfies $\mu(A) < +\infty$. Fix $\varepsilon > 0$ and choose an open set $U \supset A$ such that $\mu(U) < \mu(A) + \varepsilon$. Now $\mu(U \setminus A) < \varepsilon$ so that $\mu(W) < \varepsilon$ for some open set $W \supset U \setminus A$. Choose a compact set $K \subset U$ such that $\mu(K) > \mu(U) - \varepsilon \ge \mu(A) - \varepsilon$. Then $K \setminus W$ is a compact subset of A and $A \subset (K \setminus W) \cup W \cup (U \setminus K)$, wherefore $\mu(K \setminus W) > \mu(A) - 2\varepsilon$. The equality (7.2) follows in this case because ε is arbitrary. $\quad\square$

Example 7.8. Let Y be an uncountable discrete topological space and consider the locally compact space $X = Y \times [0, 1]$ endowed with the product topology. Every function $f \in \mathcal{C}_c(X, \mathbb{R})$ is supported by a set of the form $Y_0 \times [0, 1]$ with Y_0 a finite set. Define $\varphi : \mathcal{C}_c(X, \mathbb{R}) \to \mathbb{R}$ by

$$\varphi(f) = \sum_{y \in Y} \int_{[0,1]} f(y, t) \, d\lambda_1(t), \quad f \in \mathcal{C}_c(X, \mathbb{R}),$$

and let μ be the outer regular measure representing φ. Then the closed set $A = Y \times \{0\}$ has $\mu(A) = +\infty$, but $\mu(K) = 0$ for every compact set $K \subset A$. Thus μ is not inner regular.

Example 7.9. Denote by X the space $W(\Omega)$ of all countable ordinal numbers endowed with the order topology, that is, the topology generated by open intervals of the form $\{\beta \in W(\Omega) : \beta < \alpha\}$ and $\{\beta \in W(\Omega) : \beta > \alpha\}$ with $\alpha \in W(\Omega)$. Denote by $\mathbf{J}$ the collection consisting of those sets $E \in \mathbf{B}_X$ with the property that the set $X \setminus E$ contains a closed unbounded set. Then $\mathbf{J}$ is a σ-ideal in $\mathbf{B}_X$, and the formula

$$\mu(E) = \begin{cases} 0 & \text{if } E \in \mathbf{J}, \\ 1 & \text{if } E \in \mathbf{B}_X \setminus \mathbf{J}, \end{cases}$$

defines a Borel measure μ on X such that $\mu(K) = 0$ for every compact set $K \subset X$. In particular, μ is not inner regular. The measure μ is outer regular, and since $\int_X f \, d\mu = 0$ for every $f \in \mathcal{C}_c(X)$, this shows that outer regularity alone does not suffice to ensure uniqueness in Theorem 7.7.

One can prove some regularity results for arbitrary measures.

Proposition 7.10. *Let X be a topological space, and let μ be a Borel measure on X. Then*

(1) *If X is a locally compact Hausdorff space and $\mu(C) < +\infty$ for all compact sets $C \subset X$, we have*

$$\mu(A) = \sup\{\mu(C) : C \subset A, C \text{ compact}\} \tag{7.3}$$
$$= \inf\{\mu(G) : G \subset A, G \text{ open}\}$$

for every set A in the σ-algebra $\mathbf{Ba}_X$ generated by the compact G_δ sets. In particular, if X is metrizable and satisfies the second axiom of countability, then μ is regular.

(2) *If X is a complete separable metric space and $\mu(X) < +\infty$, then μ is regular.*

Proof. Denote by $\mathbf{R}$ the collection consisting of those subsets $A \in \mathbf{B}_X$ with the following property: for every $\varepsilon > 0$ there exist a compact set $C \subset A$ and an open set $G \supset A$ such that $\mu(G \setminus C) < \varepsilon$. Every set $A \in \mathbf{R}$ satisfies condition (7.3). Every compact G_δ set that is contained in some open set with finite measure belongs to $\mathbf{R}$. Indeed, assume that $C = \bigcap_{n=1}^{\infty} G_n$ is compact, where $G_1 \supset G_2 \supset \cdots$ are open sets with $\mu(G_1) < +\infty$, and observe that $\lim_{n \to \infty} \mu(G_n \setminus C) = 0$. We verify now that $\mathbf{R}$ is a ring of sets. Indeed, consider $A_1, A_2 \in \mathbf{R}$, fix $\varepsilon > 0$, and choose compact sets C_1, C_2 and open sets G_1, G_2 such that $C_j \subset A_j \subset G_j$ and $\mu(G_j \setminus C_j) < \varepsilon$ for $j = 1, 2$. Then $C_1 \cap C_2 \subset A_1 \cap A_2 \subset G_1 \cap G_2$, $C_1 \cup C_2 \subset A_1 \cup A_2 \subset G_1 \cup G_2$, and $C_1 \setminus G_2 \subset A_1 \setminus A_2 \subset G_1 \setminus C_2$. Moreover, $C_1 \cap C_2$, $C_1 \cup C_2$, and $C_1 \setminus G_2$ are compact, $G_1 \cap G_2$, $G_1 \cup G_2$, and $G_1 \setminus C_2$ are open, and

$$(G_1 \cap G_2) \setminus (C_1 \cap C_2) \subset (G_1 \setminus C_1) \cup (G_2 \setminus C_2),$$
$$(G_1 \cup G_2) \setminus (C_1 \cup C_2) \subset (G_1 \setminus C_1) \cup (G_2 \setminus C_2),$$
$$(G_1 \setminus C_2) \setminus (C_1 \setminus G_2) \subset (G_1 \setminus C_1) \cup (G_2 \setminus C_2).$$

We conclude that each of these sets has measure less than 2ε so that $A_1 \cap A_2$, $A_1 \cup A_2$ and $A_1 \setminus A_2$ belong to $\mathbf{R}$.

Assume next that $(A_n)_{n=1}^{\infty} \subset \mathbf{R}$ is a sequence of pairwise disjoint sets and $A = \bigcup_{n=1}^{\infty} A_n$. We now show that A satisfies (7.3). Given $\varepsilon > 0$, we can find compact sets C_n and open sets G_n such that $C_n \subset A_n \subset G_n$ and $\mu(G_n \setminus C_n) < \varepsilon/2^n$ for $n \in \mathbb{N}$. The set $G = \bigcup_{n=1}^{\infty} G_n$ is open and $\mu(G) \leq \mu(A) + \varepsilon$, while

$$\sup\{\mu(C) : C \subset A, C \text{ compact}\} \geq \mu\left(\bigcup_{n=1}^{N} C_n\right) \geq \sum_{n=1}^{N} \mu(A_n) - \varepsilon.$$

The first equality in (7.3) is obtained by letting first $N \to \infty$ and then $\varepsilon \to 0$.

Assume now that X is locally compact and μ is a Borel measure on X which assigns finite measure to each compact set. The preceding arguments show that $\mathbf{R}_C = \{A \cap C : A \in \mathbf{R}\}$ is a σ-algebra in C whenever C is a compact G_δ set, and therefore $\mathbf{R}$ contains all the Baire sets contained in C. Every element in the σ-algebra generated by the compact G_δ sets is contained in a countable union of compact G_δ sets, and the preceding arguments show that such sets satisfy (7.3). The last assertion in (1) follows from the fact that in a metrizable space every closed set is a G_δ. Moreover, local compactness and the second axiom of countability imply that the complemented σ-algebra generated by the compact sets is precisely $\mathbf{B}_X$.

Finally, assume that X is a separable metric space and μ is a finite measure, in which case $\mathbf{R}$ is a σ-ring. To prove (2), it suffices to show that each open set belongs to $\mathbf{R}$. Assume then that G is an open set and $\varepsilon > 0$. For each $n \in \mathbb{N}$ there exists a sequence $(B_k)_{k=1}^{\infty}$ of closed balls of diameter $< 1/n$ such that $G = \bigcup_k B_k$. We have $\mu(G) = \lim_{N \to \infty} \mu\left(\bigcup_{k=1}^{N} B_k\right)$, so the closed set $C_n = \bigcup_{k=1}^{N_n} B_k$ satisfies $\mu(C_n) > \mu(G) - \varepsilon/2^n$ if N_n is sufficiently large. The set $C = \bigcap_{n=1}^{\infty} C_n$ is closed and totally bounded, and thus C is compact because X was assumed complete. Moreover $\mu(C) > \mu(G) - \varepsilon$, showing that $G \in \mathbf{R}$, as desired. $\qquad \square$

Corollary 7.11. *Consider a complete separable metric space X and a finite Borel measure μ on X. Then there exists a sequence $(C_n)_{n=1}^{\infty}$ of pairwise disjoint compact subsets of X such that $\mu\left(X \setminus \bigcup_{n=1}^{\infty} C_n\right) = 0$.*

Proof. Use the preceding regularity result to choose inductively the sets C_n with the property that $C_n \subset X \setminus \bigcup_{k<n} C_k$ and $\mu\left(X \setminus \bigcup_{k=1}^{n} C_k\right) < 1/n$. $\qquad \square$

A similar uniqueness result is true for more general spaces when the measures are finite. We recall [**I**, Chapter 9] that a topological space X is normal precisely when, given arbitrary sets $C \subset G \subset X$ such that C is closed and G is open, there exists a function $f \in \mathcal{C}_b(X)$ such that $\chi_C \leq f \leq \chi_G$. The notion of integral in the case of finitely additive set function is the one discussed in Problems 3Z and 3AA. We recall that a Hausdorff topological space X is said to be normal if, given disjoint closed sets $F_1, F_2 \subset X$, there exist disjoint open sets $G_1, G_2 \subset X$ such that $F_j \subset G_j$ for $j = 1, 2$.

Proposition 7.12. *Consider a normal topological space X and two finite, outer regular, finitely additive set functions $\mu, \nu : \mathbf{B}_X \to [0, +\infty)$. Then we have $\mu = \nu$ if and only if $\int_X f \, d\mu = \int_X f \, d\nu$ for every $f \in \mathcal{C}_b(X)$.*

Proof. Assume that $\int_X f \, d\mu = \int_X f \, d\nu$ for every $f \in \mathcal{C}_b(X)$, and let $K \subset X$ be a closed set. The argument in the proof of Proposition 7.4 can be repeated, replacing $\mathcal{C}_c$ by $\mathcal{C}_b$, to show that $\mu(K) = \nu(K)$. $\qquad \square$

A version of the Riesz representation theorem holds for normal spaces provided that we do not insist on countable additivity.

Theorem 7.13. *Consider a normal topological space X and a positive linear functional $\varphi : \mathcal{C}_\mathrm{b}(X, \mathbb{R}) \to \mathbb{R}$. There exists a unique finitely additive set function μ defined on the algebra generated by the open sets in X such that*

$$\varphi(f) = \int_X f \, d\mu, \quad f \in \mathcal{C}_\mathrm{b}(X).$$

Proof. The uniqueness assertion is proved like Proposition 7.12. To prove existence we define first

$$\mu_0(F) = \inf\{\varphi(f) : f \in \mathcal{C}_\mathrm{b}(X), f \geq \chi_F\} \tag{7.4}$$

for every closed set $F \subset X$. Then

$$\mu_1(G) = \sup\{\mu_0(F) : F \subset G, F \text{ closed}\} \tag{7.5}$$

for every open set $G \subset X$, and finally

$$\mu^*(A) = \inf\{\mu_1(G) : G \supset A, G \text{ open}\} \tag{7.6}$$

for every set $A \subset X$. We observe immediately that the functions μ_0 and μ_1 are increasing, and thus $\mu^*(G) = \mu_1(G)$ if G is open. Therefore (7.6) proves the outer regularity of μ^*. It is also immediate (since $\chi_\varnothing = 0$) that $\mu_0(\varnothing) = \mu_1(\varnothing) = \mu(\varnothing) = 0$. We verify next that all of these set functions are subadditive. Indeed, let $F_1, F_2 \subset X$ be closed, and let $f_1, f_2 \in \mathcal{C}_\mathrm{b}(X)$ satisfy $f_1 \geq \chi_{F_1}$ and $f_2 \geq \chi_{F_2}$. Then $f_1 + f_2 \geq \chi_{F_1} + \chi_{F_2} \geq \chi_{F_1 \cup F_2}$ so that

$$\mu_0(F_1 \cup F_2) \leq \varphi(f_1 + f_2) = \varphi(f_1) + \varphi(f_2).$$

Taking the infimum over all functions f_1 and f_2 yields

$$\mu_0(F_1 \cup F_2) \leq \mu_0(F_1) + \mu_0(F_2).$$

Suppose that $G_1, G_2 \subset X$ are open and $F \subset G_1 \cup G_2$ is a closed set. Use the fact that X is normal to find an open set $O \supset F \setminus G_2$ such that $\overline{O} \subset G_1$, and set $F_1 = F \cap \overline{O}$ and $F_2 = F \setminus O$. The sets F_1 and F_2 are closed, $F_1 \cup F_2 = F$, and $F_j \subset G_j$ for $j = 1, 2$. Thus

$$\mu_0(F) \leq \mu_0(F_1) + \mu_0(F_2) \leq \mu_1(G_1) + \mu_1(G_2),$$

and we obtain $\mu_1(G_1 \cup G_2) \leq \mu_1(G_1) + \mu_1(G_2)$ by taking the supremum over closed $F \subset G_1 \cup G_2$. The subadditivity of μ^* is obtained in a similar manner. The set function μ_0 is, in fact, additive; that is,

$$\mu_0(F_1 \cup F_2) = \mu_0(F_1) + \mu_0(F_2) \text{ if } F_1 \cap F_2 = \varnothing.$$

To see this, note that in this case there exists a function $h \in \mathcal{C}_b(X, \mathbb{R})$ such that $\chi_{F_1} \leq h \leq \chi_{X \backslash F_2}$, and it follows that every function satisfying $f \geq \chi_{F_1 \cup F_2}$ can be written as $f = f_1 + f_2$, where $f_1 = hf \geq \chi_{F_1}$ and $f_2 = (1 - h)f \geq \chi_{F_2}$. The desired equality follows easily from this observation.

The next observation is that μ extends μ_0, that is, $\mu^*(F) = \mu_0(F)$ if F is closed. Indeed, assume that F is closed. The definition of μ_1 implies that $\mu_0(F) \leq \mu_1(G) = \mu^*(G)$ for any open set $G \supset F$, and therefore $\mu_0(F) \leq \mu(F)$. On the other hand, assume that $f \in \mathcal{C}_c(X, \mathbb{R})$ satisfies $f \geq \chi_F$, and fix $\varepsilon > 0$. The set $G = E(f > 1/(1 + \varepsilon))$ is open, $F \subset G$, and $(1 + \varepsilon)f \geq \chi_{\overline{G}}$. Any closed set $F_1 \subset G$ satisfies then $\mu_0(F_1) \leq (1 + \varepsilon)\varphi(f)$, and therefore $\mu_1(G) \leq (1 + \varepsilon)\varphi(f)$. We have then $\mu(F) \leq (1 + \varepsilon)\varphi(f)$, and the inequality $\mu^*(F) \leq \mu_0(F)$ is obtained by letting ε tend to zero and taking the infimum over all functions $f \geq \chi_F$.

We prove now a measurability property of open sets, namely the identity

$$\mu^*(A) = \mu^*(A \cap G) + \mu^*(A \backslash G)$$

holds for every set $A \subset X$ when G is open. Only one inequality needs to be proved because μ is subadditive, and this amounts to showing that

$$\mu_1(O) \geq \mu^*(A \cap G) + \mu^*(A \backslash G)$$

for every open set $O \supset A$. It will then suffice to prove the stronger inequality

$$\mu_1(O) \geq \mu_1(O \cap G) + \mu^*(O \backslash G).$$

This in turn will follow if we can prove that

$$\mu_1(O) \geq \mu_0(F_1) + \mu^*(O \backslash G)$$

for every closed $F_1 \subset O \cap G$. Fix then such a set F_1 and observe that $O \backslash F_1$ is an open set containing $O \backslash G$; thus $\mu_1(O \backslash F_1) \geq \mu^*(O \backslash G)$. We show that in fact

$$\mu_1(O) \geq \mu_0(F_1) + \mu_1(O \backslash F_1).$$

Indeed, if $F_2 \subset O \backslash F_1$ is closed we have $F_1 \cup F_2 \subset O$ so

$$\mu_1(O) \geq \mu_0(F_1 \cup F_2) = \mu(F_1) + \mu(F_2).$$

Take the supremum over F_2 to obtain the desired inequality.

The first part of the proof of Proposition 5.8 shows that the restriction μ of μ^* to the algebra generated by the open sets in X is a finitely additive. The proof of the fact that μ represents φ is similar to the corresponding proof in Theorem 7.7 and is not repeated here. $\qquad \square$

We conclude this chapter with an application of the Riesz representation theorem.

Definition 7.14. Let μ be a Borel measure on $\mathbb{R}$ with the property that $\int_{\mathbb{R}} |t|^k \, d\mu(t) < +\infty$ for every $k \in \mathbb{N}_0$. The number

$$c_k(\mu) = \int_{\mathbb{R}} t^k \, d\mu(t), \quad k \in \mathbb{N}_0,$$

is called the *moment* of order k of μ.

The *moment problem* on the real line (also known as the Hamburger moment problem) asks under what conditions a given sequence $(c_n)_{n=0}^{\infty}$ of real numbers is the sequence of moments of some Borel measure on $\mathbb{R}$.

Example 7.15. The concentration of λ_1 on the unit interval $[0, 1]$ has moments

$$\int_{[0,1]} t^k \, dt = \frac{1}{k+1}, \quad k \in \mathbb{N}_0.$$

Example 7.16. The Gaussian measure on $\mathbb{R}$ with density $e^{-t^2/2}/\sqrt{2\pi}$ has zero moments of odd order and

$$\frac{1}{\sqrt{2\pi}} \int_{\mathbb{R}} t^{2k} e^{-t^2/2} \, dt = (2k-1)!!,$$

where $(2k-1)!!$ is the product of the first k odd integers or, equivalently, $(2k-1)!! = (2k)!/(2^k k!)$.

Example 7.17. The semicircle measure with density $(1/2\pi)\sqrt{4-t^2}\chi_{[-2,2]}$ has zero moments of odd order and

$$\frac{1}{2\pi} \int_{[-2,2]} t^{2k} \sqrt{4-t^2} \, dt = \frac{1}{k+1}\binom{2k}{k}, \quad k \in \mathbb{N}_0,$$

is the kth Catalan number for $k \geq 0$. The moments in this example and the preceding one can be evaluated using the methods in Examples 8.11 and 8.12.

Necessary conditions for a sequence $(c_k)_{k=0}^{\infty}$ to be a moment sequence can be obtained by integrating positive polynomials.

Example 7.18. Suppose that $c_k = c_k(\mu)$ for $k \in \mathbb{N}_0$, and note that

$$\int_{\mathbb{R}} (t - c_1)^2 \, d\mu(t) \geq 0.$$

Expanding the square we obtain $c_1^2 \leq c_0 c_2$ which is essentially the Cauchy-Schwarz inequality. Equality holds precisely when μ is a point mass. More generally, consider an arbitrary polynomial $p(t) = \sum_{j=0}^{N} \alpha_j t^j$ with complex coefficients. The inequality $\int_{\mathbb{R}} |p(t)|^2 \, d\mu(t) \geq 0$ yields

$$\sum_{i,j=0}^{N} \alpha_i \overline{\alpha_j} c_{i+j} \geq 0.$$

Since the coefficients are arbitrary complex numbers, we conclude that the *Hankel matrix* $[c_{i+j}]_{i,j=0}^{N}$ is nonnegative definite for every $N \in \mathbb{N}_0$. By an elementary result of linear algebra this property of the Hankel matrices is equivalent to

$$\det[c_{i+j}]_{i,j=0}^{N} \geq 0, \quad N \in \mathbb{N}_0. \tag{7.7}$$

The necessary conditions just described do in fact characterize moment sequences.

Theorem 7.19. *Let $(c_n)_{n=0}^{\infty}$ be a sequence of real numbers. There exists a Borel measure μ on $\mathbb{R}$ such that $c_n = c_n(\mu)$ for all n if and only if the inequalities (7.7) are satisfied.*

The proof requires a preliminary result about polynomials.

Lemma 7.20. *Suppose p is a polynomial with real coefficients such that $p(t) \geq 0$ for every $t \in \mathbb{R}$. Then there exist polynomials q, r with real coefficients such that $p(t) = q(t)^2 + r(t)^2$ for every $t \in \mathbb{R}$.*

Proof. The leading coefficient c of p is clearly positive. We claim that the roots of p can be listed as $\alpha_1, \ldots, \alpha_k, \overline{\alpha_1}, \ldots, \overline{\alpha_k}$ for some $\alpha_1, \ldots, \alpha_k \in \mathbb{C}$. Indeed, the complex roots of p come in conjugate pairs because p has real coefficients, while the real roots of p must have even multiplicity because p does not change sign. Write now

$$c^{1/2} \prod_{j=1}^{k} (t - \alpha_j) = q(t) + ir(t),$$

so

$$c^{1/2} \prod_{j=1}^{k} (t - \overline{\alpha_j}) = q(t) - ir(t),$$

and observe that $p(t) = (q(t) + ir(t))(q(t) - ir(t))$. □

We can now proceed with the proof of Theorem 7.19, which can now be viewed as a form of the Riesz representation theorem for positive functionals defined on the space of polynomials (but without a uniqueness assertion).

Proof. The necessity of (7.7) was proved in Example 7.18. Conversely, assume that (7.7) holds, so the Hankel matrices $[c_{i+j}]_{i,j=1}^{N}$ are nonnegative definite for all $N \in \mathbb{N}_0$. Denote by X_0 the (real) vector space of all polynomials with real coefficients, and let X be the (real) vector space generated by X_0 and $C_c(\mathbb{R}, \mathbb{R})$. The elements of X can be (uniquely) written as $u + v$, where u is a polynomial and v is a continuous function with compact support. Define a linear functional $\varphi_0 : X_0 \to \mathbb{R}$ by setting

$$\varphi_0(p) = \sum_{k=0}^{N} c_k a_k \text{ for } p(t) = \sum_{k=0}^{N} a_k t^k \in X_0.$$

We show that φ_0 is a positive functional, that is, $\varphi_0(p) \geq 0$ when p takes only nonnegative values. By Lemma 7.20, it suffices to prove that $\varphi_0(q^2) \geq 0$ when q is a polynomial with real coefficients, and this inequality follows immediately from that fact that the matrices $[c_{i+j}]_{i,j=1}^N$ are nonnegative definite. The remainder of the proof is organized as follows.

(1) We show that there exists a positive functional $\varphi : X \to \mathbb{R}$ such that $\varphi | X_0 = \varphi_0$.
(2) We apply Theorem 7.7 to $\varphi | C_c(\mathbb{R}, \mathbb{R})$ to obtain a measure μ.
(3) We show that $c_k(\mu) = c_k$ for $k \in \mathbb{N}_0$.

To prove the existence of φ in (1) we use Zorn's lemma. Denote by $\mathcal{F}$ the collection of all pairs (Y, ψ), where $Y \subset X$ is a vector space, $X_0 \subset Y$, and $\psi : Y \to \mathbb{R}$ is a positive linear functional, that is, $\psi(y) \geq 0$ when $y \in Y$ and y is a nonnegative function. If $(Y, \psi), (Y', \psi') \in \mathcal{F}$, we write $(Y, \psi) \leq (Y', \psi')$ when $Y \subset Y'$ and $\psi' | Y = \psi$. Then $\mathcal{F}$ becomes an ordered set, and this order is inductive, that is, every totally ordered subset $\mathcal{G} \subset \mathcal{F}$ has an upper bound. Indeed, we can define an upper bound (Y_0, ψ_0) for $\mathcal{G}$ by setting $Y_0 = \bigcup Y$, where the union is extended over all spaces Y such that $(Y, \psi) \in \mathcal{G}$ for some ψ, and defining $\psi_0(y) = \psi(y)$ for $y \in Y$ and $(Y, \psi) \in \mathcal{G}$. The fact that $\mathcal{G}$ is totally ordered implies that ψ_0 is well defined, and clearly $(Y_0, \psi_0) \geq (Y, \psi)$ for $(Y, \psi) \in \mathcal{G}$. Zorn's lemma implies that $\mathcal{F}$ contains a maximal element (Y_m, ψ_m), and (1) follows if we show that $Y_m = X$. Assume to the contrary that there exists a function $f_0 \in X \setminus Y_m$ and denote by

$$Y' = \{tf_0 + g : t \in \mathbb{R}, g \in Y_m\}$$

the vector space generated by Y_m and f_0. We show that there exists an element $(Y', \psi') \in \mathcal{F}$ such that $(Y', \psi') > (Y_m, \psi_m)$, thus contradicting the maximality of (Y_m, ψ_m). The linear extensions of ψ to Y' are of the form

$$\psi_\alpha(tf_0 + g) = t\alpha + \psi_m(g), \quad t \in \mathbb{R}, g \in Y_m,$$

for some $\alpha \in \mathbb{R}$. We have $\psi_\alpha(tf_0 + g) = |t|\psi_\alpha((t/|t|)f_0 + (1/|t|)g)$ if $t \neq 0$. Writing $g' = (1/|t|)g$ when $t < 0$ and $g'' = -(1/|t|)g$ when $t > 0$, we see that ψ_α is a positive functional precisely when

$$\sup\{\psi_m(g') : g' \in Y_m, g' \leq f_0\} \leq \alpha \leq \inf\{\psi_m(g'') : g'' \in Y_m, g'' \geq f_0\}. \tag{7.8}$$

Note now that the two sets in this inequality are nonempty and $\psi_m(g') \leq \psi_m(g'')$ whenever $g', g'' \in Y_m$ and $g' \leq f_0 \leq g''$. Hence there exists at least one number α satisfying (7.8), and this establishes the desired contradiction.

Let now $\varphi : X \to \mathbb{R}$ be one of the positive extensions whose existence was just proved, and let μ be a Borel measure on $\mathbb{R}$ such that $\varphi(f) = \int_\mathbb{R} f \, d\mu$ for every $f \in C_c(\mathbb{R}, \mathbb{R})$. Denote by $f_N \in C_c(\mathbb{R}, \mathbb{R})$ a function such that $\chi_{[-N,N]} \leq f_N \leq \chi_{[-N-1,N+1]}$, $N \in \mathbb{N}$. For any $k \in \mathbb{N}_0$ we have $0 \leq f_N(t)t^{2k} \leq t^{2k}, t \in \mathbb{R}$, and the positivity of φ implies

$$\int_{[-N,N]} t^{2k}\, d\mu(t) \le \int_{\mathbb{R}} f_N(t) t^{2k}\, d\mu(t) = \varphi(f_N t^{2k}) \le \varphi(t^{2k}) = c_{2k}.$$

Letting $N \to +\infty$, we conclude from the monotone convergence theorem that $\int_{\mathbb{R}} t^{2k}\, d\mu(t) < +\infty$. In particular, the measure μ has finite moments of all

$$t^k - \varepsilon t^{2k} \le f_N(t) t^k \le t^k + \varepsilon t^{2k}, \quad t \in \mathbb{R},$$

provided that N is sufficiently large. The positivity of φ yields now

$$c_k - \varepsilon c_{2k} = \varphi(t^k - \varepsilon t^{2k}) \le \int_{\mathbb{R}} f_N(t) t^k\, d\mu(t) \le c_k + \varepsilon c_{2k}$$

for N large. The dominated convergence theorem allows us to let $N \to +\infty$ and conclude that $c_k - \varepsilon c_{2k} \le \int_{\mathbb{R}} t^k\, d\mu(t) \le c_k + \varepsilon c_{2k}$. The desired conclusion $c_k = \int_{\mathbb{R}} t^k\, d\mu(t)$ is reached by letting $\varepsilon \to 0$. $\qquad \square$

The preceding theorem is due to Hamburger. The idea of extending a positive linear functional was first applied by F. Riesz in his original proof of the representation theorem. The application of this idea to the proof of Hamburger's theorem is due to M. Riesz.

Problems

7A. Let I be an uncountable set, let K be a compact Hausdorff space, and set $X = K^I$. That is, X consists of families $(x_i)_{i \in I}$ of elements of K, indexed by I. Endow X with the product topology so X is a compact Hausdorff space.

    a. Denote by $\mathcal{A}$ the algebra consisting of all countably determined functions $f : X \to \mathbb{C}$, that is, functions which only depend on countably many coordinates. Show that $\mathcal{A}$ is closed under the formation of pointwise sequential limits.

    b. Denote by $\mathcal{C} \subset \mathcal{A}$ the subalgebra consisting of the continuous functions in $\mathcal{A}$. Show that $\mathcal{A}$ separates the points of X. Conclude from Problem 4BB that $\mathcal{C} = \mathcal{C}(X)$.

    c. Verify the last assertion of Example 7.2.

7B. Let X be a locally compact Hausdorff space. Show that every function in $\mathcal{C}_0(X)$ is a uniform limit of a sequence of functions in $\mathcal{C}_c(X)$.

7C. Consider the space $W(\Omega)$ of Example 7.9.

    (i) Given a subset $A \subset W(\Omega)$ and an element $\alpha \in W(\Omega) \setminus A$, show that α is in the closure of A (in the order topology) if and only if $(\beta, \alpha) \cap A \ne \varnothing$ for every $\beta < \alpha$. In particular, $\overline{A} \setminus A$ consists entirely of limit ordinals.

    (ii) Let $A, B \subset W(\Omega)$ be two unbounded sets, and let $\alpha \in W(\Omega)$. Show that there exist elements $\{\alpha_n\}_{n=1}^{\infty}$ in A and $\{\beta_n\}_{n=1}^{\infty}$ in B such that

$$\alpha < \alpha_1 < \beta_1 < \cdots < \alpha_n < \beta_n < \cdots.$$

    Deduce that $\overline{A} \cap \overline{B}$ is also unbounded.

    (iii) Verify the assertions of Example 7.9.

7D. Consider a locally compact Hausdorff space X and a positive linear functional $\varphi :$ $\mathcal{C}_0(X) \to \mathbb{C}$. Theorem 7.7 yields a Borel measure μ on X such that

$$(\dagger) \qquad \varphi(f) = \int_X f \, d\mu$$

for every $f \in \mathcal{C}_c(X)$.

(i) Show that $\mu(X) < +\infty$. (Hint: If $\mu(X) = +\infty$ choose compact sets K_n with $\mu(K_n) > 2^n$ and functions $u_n \in \mathcal{C}_c(X)$ such that $\chi_{K_n} \le u_n \le 1$. Estimate $\varphi(f)$, where $f = \sum_{n=1}^{\infty} 2^{-n} u_n$.)

(ii) Show that the equality ($\dagger$) holds for all $f \in \mathcal{C}_0(X)$. (Hint: Assuming f is positive, construct functions $u_n \in \mathcal{C}_c(X)$ such that $0 \le u_n \le 1$, $(1 - u_{n+1})u_n = 0$, and $f \le \sum_{n=1}^{\infty} 2^{-n} u_n$. Set $g = \sum_{n=1}^{\infty} n 2^{-n} u_n$ and note that $f - \varepsilon g \le u_n \le f + \varepsilon g$ if $\varepsilon > 0$ and $n = n(\varepsilon)$ is large.)

7E. Consider the discrete space $\mathbb{N}$ and a positive linear functional $\varphi : \mathcal{C}_b(\mathbb{N}) = \ell^\infty \to \mathbb{C}$ such that $\varphi((x_n)_{n=1}^{\infty}) = \lim_{n \to \infty} x_n$ when the limit exists.

(i) Show that there is no measure μ on $\mathbb{N}$ such that $\varphi(f) = \int_\mathbb{N} f \, d\mu$ for every $f \in \mathcal{C}_b(\mathbb{N})$.

(ii) More generally, show that every positive functional $\varphi : \mathcal{C}_b(\mathbb{N}) = \ell^\infty \to \mathbb{C}$ can be written as $\varphi = \varphi_1 + \varphi_2$, where $\varphi_1 | \mathcal{C}_0(\mathbb{N}) = 0$, while $\varphi_2((x_n)_{n=1}^{\infty}) = \sum_{n=1}^{\infty} a_n x_n$ for some summable sequence $a_n \ge 0$.

(iii) Let μ be a finitely additive positive function defined on all the subsets of $\mathbb{N}$. Show that $\mu = \mu_1 + \mu_2$, where μ_2 is a positive measure, and μ_1 is a finitely additive positive set function such that $\mu_1(\{n\}) = 0$ for every $n \in \mathbb{N}$.

7F. The first moment problem to be solved concerns measures on $[0, +\infty)$, and it was solved by Stieltjes.

(i) Let p be a polynomial such that $p(t) \ge 0$ for all $t \in [0, +\infty)$. Show that there exist real polynomials u and v such that $p(t) = u(t)^2 + t v(t)^2$ for all $t \in \mathbb{R}$. (Hint:

$$(a^2 + tb^2)(c^2 + td^2) = (ac + tbd)^2 + t(ad - bc)^2 .)$$

(ii) Consider a sequence $(c_n)_{n=0}^{\infty}$ of real numbers. Show that there exists a positive Borel measure μ on $[0, +\infty)$ such that $c_n = c_n(\mu)$ for all n if and only if the determinants $\det[c_{i+j}]_{i,j=0}^{N}$ and $\det[c_{i+j+1}]_{i,j=0}^{N}$ are nonnegative for every $N \in \mathbb{N}_0$.

7G. Consider a positive Borel measure μ on $\mathbb{R}$ with finite moments $c_n = c_n(\mu)$. Assume that $\det[c_{i+j}]_{i,j=0}^{N} = 0$ for some $N \in \mathbb{N}_0$. Show that μ has finite support, that is, there exists a finite set $F \subset \mathbb{R}$ such that $\mu(\mathbb{R} \setminus F) = 0$. The smallest cardinality of such a set F does not exceed N.

7H. The moment problem for measures on the unit circle $\mathbb{T} = \{\lambda \in \mathbb{C} : |\lambda| = 1\}$ is called the trigonometric moment problem.

(i) Let $p(\lambda) = \sum_{n=-N}^{N} \alpha_n \lambda^n$ be a Laurent polynomial such that $p(\lambda) \ge 0$ for all $\lambda \in \mathbb{T}$. Show that there exists an analytic polynomial $q(\lambda) = \sum_{n=0}^{N} \beta_n \lambda^n$ such that $p(\lambda) = |q(\lambda)|^2$ for all $\lambda \in \mathbb{T}$. (Hint: note that $\overline{p(\lambda)} = p(1/\overline{\lambda})$ and show that the roots of the equation $p(\lambda) = 0, \lambda \in \mathbb{C} \setminus \{0\}$, can be listed as $z_1, \ldots, z_N, 1/\overline{z_1}, \ldots 1/\overline{z_N}$.)

(ii) Consider a sequence $(c_n)_{n=0}^\infty$ of real numbers. Show that there exists a positive Borel measure μ on $\mathbb{T}$ such that $c_n = c_n(\mu)$ for all n if and only if the matrices $[c_{i-j}]_{i,j=0}^N$ are nonnegative definite for every $N \in \mathbb{N}_0$.

7I. The moment problem for measures on a bounded interval was solved by Hausdorff. A complete description of positive polynomials on such an interval is replaced by the Bernstein approximation used in the proof of Theorem 2.41.

(i) Let p be a polynomial of degree $m \geq 2$ with complex coefficients. Show that there exist polynomials $q_1, q_2, \ldots, q_{m-1}$ such that

$$B_n(p, t) - p(t) = \sum_{k=1}^{m-1} \frac{q_k(t)}{n^k}$$

for every $n \in \mathbb{N}$, where $B_n(p, t) = \sum_{i=0}^n \binom{n}{i} p(i/n) t^i (1 - t)^{n-i}$ is the Bernstein polynomial of degree n associated with p. (Hint: It suffices to consider $p(t) = t^m$. The cases $m \leq 2$ are done in Problem 2K. Proceed by induction.)

(ii) Show that a linear functional $\varphi : \mathcal{C}([0, 1]) \to \mathbb{C}$ is positive if and only if $\varphi(p_{m,n}) \geq 0$ for all $m, n \in \mathbb{N}_0$, where $p_{m,n}(t) = t^m (1 - t)^n$.

(iii) Consider a sequence $(c_n)_{n=0}^\infty$ of real numbers. Show that there exists a Borel measure μ on $[0, 1]$ such that $c_n = c_n(\mu)$ for all n if and only if

$$\sum_{k=0}^n (-1)^k \binom{n}{k} c_{m+k} \geq 0$$

for all $m, n \in \mathbb{N}_0$.

7J. Let X be a locally compact Hausdorff space. A linear functional $\varphi : \mathcal{C}_c(X) \to \mathbb{C}$ is said to be *self-conjugate* if $\varphi(\overline{f}) = \overline{\varphi(f)}$ for every $f \in \mathcal{C}_c(X)$, where the bar indicates complex conjugation. Show that an arbitrary linear functional $\varphi : \mathcal{C}_c(X) \to \mathbb{C}$ can be written in a unique way as $\varphi = \varphi_1 + i\varphi_2$, where φ_1 and φ_2 are self-conjugate linear functionals on $\mathcal{C}_c(X)$. Similar statements hold for linear functionals defined on $\mathcal{C}(X)$ and $\mathcal{C}_0(X)$. (Hint: $\varphi_1(f) = (\varphi(f) + \overline{\varphi(\overline{f})})/2$.)

7K. Let X be a locally compact Hausdorff space and let $\varphi : \mathcal{C}_c(X) \to \mathbb{C}$ be a self-conjugate linear functional. Suppose that for every compact set $K \subset X$ there exists a constant c_K such that

$$|\varphi(f)| \leq c_K \sup_{x \in X} |f(x)|$$

for every function $f \in \mathcal{C}_c(X)$ that vanishes on $X \setminus K$.

(i) Let $f \in \mathcal{C}_c(X)$ be a nonnegative function. Show that

$$\sup\{\varphi(g) : g \in \mathcal{C}_c(X), 0 \leq g \leq f\}$$

is finite.

(ii) Show that there exists a positive linear functional $\psi : \mathcal{C}_c(X) \to \mathbb{C}$ such that

$$\psi(f) = \sup\{\varphi(g) : g \in \mathcal{C}_c(X), 0 \leq g \leq f\}$$

for all nonnegative functions $f \in \mathcal{C}_c(X)$.

(iii) Prove that the functional $\psi - \varphi$ is positive.

(iv) Deduce the existence of positive outer regular Borel measures μ and ν on X such that

$$\varphi(f) = \int_X f \, d\mu - \int_X f \, d\nu, \quad f \in \mathcal{C}_c(X).$$

(v) Prove an analogous representation theorem for linear functionals $\varphi : \mathcal{C}_c(X) \to \mathbb{C}$ which are not assumed to be self-conjugate.

7L. Let X be a compact Hausdorff space. A complex Borel measure μ on X is said to be *regular* if, given an arbitrary Borel set $E \subset X$ and $\varepsilon > 0$, there exists an open set G containing A such that $|\mu|(G \setminus A) < \varepsilon$. Let $\varphi : \mathcal{C}(X) \to \mathbb{C}$ be a linear functional. Suppose that there exists a constant $c \geq 0$ such that

$$\varphi(f)| \leq c_K \sup_{x \in X} |f(x)|, \quad f \in \mathcal{C}(X). \tag{7.9}$$

Show that there exists a unique regular complex Borel measure μ on X such that

$$\varphi(f) = \int_X f \, d\mu, \quad f \in : \mathcal{C}(X).$$

Moreover, the smallest constant c satisfying (7.9) equals the total variation $|\mu|(X)$ of μ.

Chapter 8
Product measures

In this chapter we treat an additional important topic in the theory of measure and integration concerning one way that new measures can be constructed from old ones. If $(X, \mathbf{S})$ and $(Y, \mathbf{T})$ are measurable spaces, then a set of the form $E \times F$, where $E \in \mathbf{S}$ and $F \in \mathbf{T}$, is called a *measurable rectangle* in $X \times Y$. The σ-algebra of subsets of $X \times Y$ generated by the collection of all measurable rectangles is denoted by $\mathbf{S} \times \mathbf{T}$, and the space $X \times Y$ equipped with the *product σ-algebra* $\mathbf{S} \times \mathbf{T}$ is a measurable space called the *product of* $(X, \mathbf{S})$ and $(Y, \mathbf{T})$.

Example 8.1. Consider topological spaces X and Y endowed with their Borel σ-algebras $\mathbf{B}_X$ and $\mathbf{B}_Y$. The product σ-algebra $\mathbf{B}_X \times \mathbf{B}_Y$ is always a subalgebra of $\mathbf{B}_{X \times Y}$. If X and Y both satisfy the second axiom of countability, then $\mathbf{B}_X \times \mathbf{B}_Y = \mathbf{B}_{X \times Y}$. (See Problem 8A.)

Assume now that $(X, \mathbf{S}, \mu)$ and $(Y, \mathbf{T}, \nu)$ are measure spaces. A measure $\lambda : \mathbf{S} \times \mathbf{T} \to [0, +\infty]$ is called *a product measure* of μ and ν if the identity

$$\lambda(E \times F) = \mu(E)\nu(F)$$

holds for all measurable rectangles $E \times F \in \mathbf{S} \times \mathbf{T}$.

Example 8.2. Given $p, q \in \mathbb{N}$, we have seen in Example 8.1 that $\mathbf{B}_{\mathbb{R}^{p+q}} = \mathbf{B}_{\mathbb{R}^p} \times \mathbf{B}_{\mathbb{R}^q}$. We claim that the Lebesgue-Borel measure $\lambda_{p+q}|\mathbf{B}_{\mathbb{R}^{p+q}}$ is a product of $\lambda_p|\mathbf{B}_{\mathbb{R}^p}$ and $\lambda_q|\mathbf{B}_{\mathbb{R}^q}$. Indeed, the equality

$$\lambda_{p+q}(E \times F) = \lambda_p(E)\lambda_q(F), \quad E \in \mathbf{B}_{\mathbb{R}^p}, F \in \mathbf{B}_{\mathbb{R}^q}, \tag{8.1}$$

is true by definition if E and F are cells. The general case of Borel sets is obtained by applying the Theorem 1.23 twice. For instance, for fixed $F \in \mathbf{B}_{\mathbb{R}^q}$, the collection of those sets $E \in \mathbf{B}_{\mathbb{R}^p}$ for which the equality (8.1) holds is seen to be a σ-algebra.

© Springer International Publishing Switzerland 2016
H. Bercovici et al., *Measure and Integration*,
DOI 10.1007/978-3-319-29046-1_8

Product measures always exist (see Problem 8C), but they are not generally unique. The most amenable situations arise when there is only one product measure, in which case we refer to it as *the* product measure and denote it $\lambda = \mu \times \nu$. One such situation occurs when the measures μ and ν are σ-finite. To explore this, we need some additional notation.

Given a function f defined on a product $X \times Y$ and a point $x \in X$, we denote by

$$f_x(y) = f(x, y), \quad y \in Y,$$

the *X-section* of f determined by x. If $f = \chi_E$ is the characteristic function of a set $E \subset X \times Y$, the function f_x is the characteristic function of the *X-section* E_x of E determined by x, where

$$E_x = \{y \in Y : (x, y) \in E\}.$$

One defines analogously the *Y*-sections, denoted f^y and E^y for $y \in Y$.

Theorem 8.3. *Let $(X, \mathbf{S}, \mu)$ and $(Y, \mathbf{T}, \nu)$ be two σ-finite measure spaces, and suppose $G \in \mathbf{S} \times \mathbf{T}$. Then:*

(1) *The set G_x is measurable $[\mathbf{T}]$ for every $x \in X$.*
(2) *The map $x \mapsto \nu(G_x)$ is measurable $[\mathbf{S}]$.*
(3) *The set G^y is measurable $[\mathbf{S}]$ for every $y \in Y$.*
(4) *The map $y \mapsto \mu(G^y)$ is measurable $[\mathbf{T}]$.*
(5) *We have*

$$\int_X \nu(G_x) \, d\mu(x) = \int_Y \mu(G^y) \, d\nu(y).$$

(6) *The formula*

$$\lambda(G) = \int_X \nu(G_x) \, d\mu(x), \quad G \in G \in \mathbf{S} \times \mathbf{T}$$

defines a σ-finite measure λ on $(X \times Y, \mathbf{S} \times \mathbf{T})$ which is the unique product measure of μ and ν.

Proof. It suffices to prove the theorem under the additional assumption that μ and ν are finite measures. Indeed, there exist increasing sequences $(X_n)_{n=1}^{\infty} \subset \mathbf{S}$ and $(Y_n)_{n=1}^{\infty} \subset \mathbf{T}$ such that $\mu(X_n) + \nu(Y_n) < +\infty$. The set G is then the increasing union of the sets $G_n = G \cap (X_n \times Y_n)$, and assertions (1–5) follow from the corresponding assertions with G replaced by G_n using the countable additivity of μ, ν and the monotone convergence theorem 4.24. Similarly, $\lambda(G) = \lim_{n \to \infty} \lambda(G_n)$, so λ is uniquely determined by its restriction to measurable sets contained in $X_n \times Y_n$. Assume therefore that $\mu(X) + \nu(Y) < +\infty$, and denote by $\mathbf{C}$ the collection of all sets $G \in \mathbf{S} \times \mathbf{T}$ which satisfy conditions (1–5). We first observe that measurable rectangles belong to $\mathbf{C}$. Indeed, given $E \in \mathbf{S}$ and $F \in \mathbf{T}$, we have

$$G_x = \begin{cases} F, & x \in E, \\ \varnothing, & x \in X \setminus E, \end{cases}$$

so the map $x \mapsto \nu(G_x)$ is the simple measurable function $\nu(F)\chi_E$ and

$$\int_X \nu(G_x)\, d\mu(x) = \mu(E)\nu(F).$$

A similar calculation holds for the Y-sections of G. The collection of measurable rectangles is closed under the formation of finite intersections, and therefore Theorem 1.23 can be applied. In order to show that $\mathbf{C} = \mathbf{S} \times \mathbf{T}$ it suffices to show that $\mathbf{C}$ is a σ-quasiring. Indeed, if $P, Q \in \mathbf{C}$ satisfy $P \supset Q$ and $G = P \setminus Q$, we have

$$G_x = P_x \setminus Q_x, G^y = P^y \setminus Q^y, \nu(G_x) = \nu(P_x) - \nu(Q_x), \mu(G^y) = \mu(P^y) - \mu(Q^y),$$

and properties (1–5) follow from the corresponding properties for P and Q using the additivity of measures and integrals. Similarly, if $(P(n))_{n=1}^{\infty}$ is a sequence of pairwise disjoint sets in $\mathbf{C}$ and $G = \bigcup_{n=1}^{\infty} P(n)$, then

$$G_x = \bigcup_{n=1}^{\infty} P(n)_x, \quad \mu(G_x) = \sum_{n=1}^{\infty} \mu(P(n)_x)$$

by the countable additivity of μ. A similar calculation for Y-sections and an application of the theorem of Beppo-Levi (Corollary 4.25) show that $G \in \mathbf{C}$. Thus we have $\mathbf{C} = \mathbf{S} \times \mathbf{T}$. The usual properties of measures and integrals show then that the function λ defined in (6) is a finite measure, and the calculation done earlier for measurable rectangles shows that λ is a product measure of μ and ν. It remains to show that λ is the unique product measure. Assume indeed that λ' is another product measure of μ and ν. The measure λ' is finite since $\lambda'(X \times Y) = \mu(X)\nu(Y)$. It follows immediately that $\{G \in \mathbf{S} \times \mathbf{T} : \lambda'(G) = \lambda(G)\}$ is a σ-quasiring containing all measurable rectangles. The equality $\lambda' = \lambda$ follows now from Theorem 1.23. $\square$

One important property of product measures is that integrals relative to such measures can be calculated as *iterated* integrals. This is Fubini's theorem, which is part (2) of the following result. The first part, due to Tonelli, applies to positive functions, and it can be viewed as a test of integrability. Tonelli's theorem is a generalization of the theorem of Beppo-Levi which is recovered by taking μ to be the counting measure on $\mathbb{N}$.

Theorem 8.4. *Consider σ-finite measure spaces $(X, \mathbf{S}, \mu)$ and $(Y, \mathbf{T}, \nu)$ and their product $(X \times Y, \mathbf{S} \times \mathbf{T}, \mu \times \nu)$.*

(1) *Let $f : X \times Y \to [0, +\infty]$ be measurable $[\mu \times \nu]$. Then the function f_x is measurable $[\nu]$ for every $x \in X$, the function $x \mapsto \int_Y f_x\, d\nu$ is measurable $[\mu]$, and*

$$\int_{X \times Y} f \, d(\mu \times \nu) = \int_X \left[\int_Y f_x(y) \, d\nu(y) \right] d\mu(x) \qquad (8.2)$$

$$= \int_Y \left[\int_X f^y(x) \, d\mu(x) \right] d\nu(y),$$

meaning that if any of the three above integrals is finite, then all three are, and equality holds.

(2) *Let $f : X \times Y \to \mathbb{C}$ be integrable $[\mu \times \nu]$. Then the function f_x is integrable $[\nu]$ for $[\mu]$-almost every $x \in X$, the function $x \mapsto \int_Y f_x \, d\nu$ is integrable $[\mu]$, and (8.2) holds.*

Proof. Part (2) of the theorem follows easily from (1). Indeed, if f is complex-valued, we can write

$$f = u_+ - u_- + i(v_+ - v_-),$$

with $u_\pm, v_\pm$ the positive and negative parts of $\Re f, \Im f$, respectively. Then the conclusion of (1) applies to these four nonnegative measurable functions and (2) follows by linearity. Assume therefore that $f : X \times Y \to [0, +\infty]$ is measurable. Theorem 8.3 shows that the conclusions of (1) are verified in case f is the characteristic function of a set in $\mathbf{S} \times \mathbf{T}$. By linearity, (1) also follows when f is a simple function. For the general case we choose, according to Proposition 2.36, a sequence of simple functions

$$0 \le s_1 \le s_2 \le \cdots \le f,$$

which converges to f at all points in $X \times Y$. Then the x-sections of s_n also converge pointwise to f_x, thus proving the measurability of these functions. An application of the monotone convergence theorem 4.24 shows that the map $x \mapsto \int_Y f_x \, d\nu$ is measurable, and two more applications of that theorem allow us to deduce (8.2) from the corresponding identity for the functions s_n.   □

Example 8.5. Consider $X = Y = [0, 1]$ and $\mathbf{S} = \mathbf{T} = \mathbf{B}_{[0,1]}$, so $(X \times Y, \mathbf{S} \times \mathbf{T})$ is simply the unit square Q equipped with its σ-algebra of Borel sets $\mathbf{B}_Q$ (see Example 8.1). Let $\mu = \lambda_1 | [0, 1]$ be ordinary Lebesgue-Borel measure on $(X, \mathbf{S})$, and let ν be the counting measure on $(Y, \mathbf{T})$. Fix a Borel set $A \subset Q$. The proof of Theorem 8.3 shows that the function $h(y) = \mu(A^y)$ is Borel measurable on $[0, 1]$. Irrespective of this fact, we can certainly define

$$\lambda(A) = \int_Y h(y) \, d\nu(y) = \sum_{0 \le y \le 1} \mu(A^y),$$

and it follows from the theorem of Beppo-Levi (Corollary 4.25) that the set function λ is a measure on $(Q, \mathbf{B}_Q)$. Indeed, λ satisfies the identity $\lambda(E \times F) = \mu(E)\nu(F)$ for measurable rectangles. Consider the Borel set $\Delta = \{(x, x) : x \in [0, 1]\} \subset Q$. Obviously, $\lambda(\Delta) = 0$, while

$$\int_X \nu(\Delta_x)\,d\mu = 1.$$

This shows that the conclusion of Theorem 8.3 may fail if ν is not σ-finite. There is a more subtle, set-theoretical, failure as well. There exist Borel sets $A \subset Q$ for which the function

$$g(x) = \nu(A_x), \quad x \in [0,1],$$

is not Borel measurable. (This comes from the fact that every analytic subset of $[0,1]$ is the projection onto the first factor of a Borel set in $[0,1] \times [0,1]$, and there are analytic sets which are not Borel. These facts are somewhat outside the scope of this exposition.)

Returning to the example of Euclidean space, we observed that the Borel-Lebesgue measure $\lambda_{m+n}|\mathbf{B}_{\mathbb{R}^{m+n}}$ is a product measure: using the standard identification of $\mathbb{R}^{m+n}$ with $\mathbb{R}^m \times \mathbb{R}^n$,

$$\lambda_{m+n}|\mathbf{B}_{\mathbb{R}^{m+n}} = (\lambda_m|\mathbf{B}_{\mathbb{R}^m}) \times (\lambda_n|\mathbf{B}_{\mathbb{R}^n}).$$

It is not the case however that $\lambda_{m+n} = \lambda_m \times \lambda_n$ (see Problem 8G), and therefore the Fubini-Tonelli theorem as stated above does not apply to all Lebesgue integrable functions. Note however that λ_n is the completion of $\lambda_n|\mathbf{B}_{\mathbb{R}^n}$, and this makes it fairly easy to extend Theorem 8.4 to cover these more general integrable functions. We state only the extension of the Tonelli theorem, using the convention that $\overline{\mu}$ denotes the completion of a measure μ.

Theorem 8.6. *Consider σ-finite measure spaces $(X, \mathbf{S}, \mu)$ and $(Y, \mathbf{T}, \nu)$ and their product $(X \times Y, \mathbf{S} \times \mathbf{T}, \mu \times \nu)$. Let $f : X \times Y \to [0, +\infty]$ be measurable $\overline{[\mu \times \nu]}$. Then the function f_x is measurable $[\overline{\nu}]$ for $[\overline{\mu}]$-almost every $x \in X$ and the function $x \mapsto \int_Y f_x\,d\overline{\nu}$ defined for such x is measurable $[\overline{\mu}]$. Similarly, the function f^y is measurable $[\overline{\mu}]$ for $[\overline{\nu}]$-almost every and the function $x \mapsto \int_X f^y\,d\overline{\mu}$ defined for such y is measurable $[\overline{\nu}]$. Moreover, we have*

$$\int_{X \times Y} f\,d(\overline{\mu \times \nu}) = \int_X \left[\int_Y f_x(y)\,d\overline{\nu}(y) \right] d\overline{\mu}(x) \int_Y \left[\int_X f^y(x)\,d\overline{\mu}(x) \right] d\overline{\nu}(y).$$

Proof. It suffices to prove the theorem in case $f = \chi_A$ for some $\overline{[\mu \times \nu]}$-measurable $A \subset X \times Y$. There exist sets $B, C \in \mathbf{S} \times \mathbf{T}$ such that $A \triangle B \subset C$ and $(\mu \times \nu)(C) = 0$. By Theorem 8.3, the set $D = \{x \in X : \nu(C_x) \neq 0\}$ belongs to $\mathbf{S}$ and $\mu(D) = 0$. Since $A_x \triangle B_x \subset C_x$ for each $x \in X$, we conclude that A_x is $[\overline{\nu}]$-measurable for each $x \in X \setminus D$ and $\overline{\nu}(A_x) = \nu(B_x)$. Thus the map $x \mapsto \overline{\nu}(A_x)$ is measurable $\overline{\mu}$, and we have

$$\overline{\mu \times \nu}(A) = (\mu \times \nu)(B) = \int_X \nu(B_x)\,d\mu(x) = \int_X \overline{\nu}(A_x)\,d\overline{\mu}(x),$$

so the theorem follows.   □

The Fubini and Tonelli theorems extend, naturally, to arbitrary finite products $X_1 \times \cdots \times X_n$ of σ-finite measure spaces (see Problem 8U). For instance, $\lambda_n | \mathbf{B}_{\mathbb{R}^n}$ is easily seen to equal the product of n copies of $\lambda_1 | \mathbf{B}_{\mathbb{R}}$.

Remark 8.7. The assumption that the measure spaces $(X, \mathbf{S}, \mu)$ and $(Y, \mathbf{T}, \nu)$ are both σ-finite is certainly sufficient to assure the validity of the Fubini and Tonelli theorems, but this condition has seemed to many to be unjustifiably restrictive. In this connection it is appropriate to remark that if $(X, \mathbf{S}, \mu)$ and $(Y, \mathbf{T}, \nu)$ are two absolutely arbitrary measure spaces, then there exists a product measure λ of μ and ν. This measure, which is constructed in a completely standard fashion, possesses certain rather appealing properties. For example, the Fubini theorem is valid for the measure λ, again without any restrictions whatever on X and Y (Problem 8C). A serious difficulty is that the Tonelli theorem is *not* valid in general for the measure λ, and since this is ordinarily the only feasible way to determine when the Fubini theorem is applicable, we have relegated this measure to Problems 7C and 7D.

Example 8.8. Fix a real number α, an integer $d \geq 2$, and consider the linear transformation $T_\alpha : \mathbb{R}^d \to \mathbb{R}^d$ defined by

$$T_\alpha(x_1, x_2, \ldots, x_d) = (x_1 + \alpha x_2, x_2, \ldots, x_d).$$

We use Fubini's theorem to show that a set $E \subset \mathbb{R}^d$ is Lebesgue measurable if and only if $T_0(E)$ is Lebesgue measurable, and $\lambda_d(T_0(E)) = \lambda_d(E)$ when E is measurable. Consider first a compact set of the form

$$E = [a_1, b_1] \times [a_2, b_2] \times \cdots \times [a_d, b_d]$$

and note that $T_\alpha(E)$ is also compact, hence measurable. Setting $y = (x_2, \ldots, x_d)$, observe that

$$T_0(E^y) = \begin{cases} [a_1 + \alpha x_2, b_1 + \alpha x_2], & y \in [a_2, b_2] \times \cdots \times [a_d, b_d] \\ \varnothing, & \text{otherwise,} \end{cases}$$

so $\lambda_1(T_0(E^y)) = (b_1 - a_1)\chi_{[a_2, b_2] \times \cdots \times [a_d, b_d]}(y)$. Therefore

$$\lambda_d(T_\alpha(E)) = \int_{[a_2, b_2] \times \cdots \times [a_d, b_d]} \lambda_1(T_\alpha(E^y)) \, d\lambda_{d-1}(y) = \prod_{j=1}^{d} (b_j - a_j) = \lambda_d(E).$$

The monotone class theorem implies then the equality $\lambda_d(T_\alpha(E)) = \lambda_d(E)$ for all Borel sets $E \subset \mathbb{R}^d$. We deduce that, given a measurable set E with $\lambda_d(E) = 0$, the set $T_\alpha(E)$ has Lebesgue outer measure equal to zero. Thus the equality extends to arbitrary Lebesgue measurable sets. The same argument can be applied to the inverse transformation $T_\alpha^{-1} = T_{-\alpha}$ to see that E is measurable if $T_\alpha(E)$ is measurable. This observation extends as follows.

Proposition 8.9. *Fix an integer $d \geq 2$ and an invertible linear transformation $T : \mathbb{R}^d \to \mathbb{R}^d$. Then a set $E \subset \mathbb{R}^d$ is Lebesgue measurable if and only if $T(E)$ is Lebesgue measurable, and $\lambda_d(T(E)) = |\det(T)|\lambda_d(E)$ when E is measurable.*

Proof. It is known that we can write T as a product

$$T = X_1 X_2 \cdots X_N$$

of transformations where each X_j is of one of the following three types:

(1) $X_j = T_\alpha$, where T_α is the transformation of Example 8.8.
(2) $X_j(x_1, x_2, \ldots, x_d) = (x_{\sigma(1)}, x_{\sigma(2)}, \ldots, x_{\sigma(d)})$ for some permutation σ of $\{1, 2, \ldots, d\}$.
(3) $X_j(x_1, x_2, \ldots, x_d) = (\lambda_1 x_1, \lambda_2 x_2, \ldots, \lambda_d x_d)$ for some nonzero real numbers $\lambda_1, \lambda_2, \ldots, \lambda_d$.

(A proof of this result from linear algebra is outlined in Problem 8Z.) Because the determinant is a multiplicative function, it suffices to prove the proposition when T equals one of these factors X_j. The transformation T_0 was already treated above. For the other cases we restrict ourselves, as in Example 8.8, to sets E which are products of intervals. When T is of the form (2) or (3), the set $T(E)$ is also a product of intervals, and the identity $\lambda_d(T(E)) = |\det(T)|\lambda_d(E)$ is trivially verified.

$\square$

The preceding result is the simplest case of the change of variables formula. To state the general result, we begin with some notation. Fix $d \in \mathbb{N}$, two open sets $U, V \subset \mathbb{R}^d$, and a map $f : U \to V$ which is bijective, continuously differentiable, and is such that the inverse $f^{-1} : V \to U$ is continuously differentiable as well. Such a map is called a C^1-*diffeomorphism* of U onto V. We recall that for each $x \in U$, the derivative $f'(x) : \mathbb{R}^d \to \mathbb{R}^d$ is a linear transformation with the property that

$$\lim_{h \to 0} \frac{\|f(x+h) - f(x) - f'(x)h\|_2}{\|h\|_2} = 0,$$

where $\|h\|_2$ denotes the Euclidean length of the vector h in $\mathbb{R}^n$, that is,

$$\|(h_1, h_2, \ldots, h_d)\|_2 = (h_1^2 + h_2^2 + \cdots + h_d^2)^{1/2}.$$

Theorem 8.10. *Fix $d \geq 1$, two open sets $U, V \subset \mathbb{R}^d$, and let $f : U \to V$ be a C^1-diffeomorphism. A function $h : V \to \mathbb{C}$ is Lebesgue integrable if and only if the function $x \mapsto h(f(x))|\det f'(x)|$ is Lebesgue integrable on U, and*

$$\int_V h(y) \, d\lambda_d(y) = \int_U h(f(x))|\det f'(x)| \, d\lambda_d(x)$$

when h is integrable.

Proof. We prove first that $f(E)$ is measurable and that

$$\lambda_d(f(E)) \le \int_E |\det f'(x)| \, d\lambda_d(x)$$

for every Lebesgue measurable set $E \subset U$. As in the proof of Proposition 8.9, we can restrict ourselves to the case in which E is a product of compact intervals. Assume then that $E = [a_1, b_1] \times [a_2, b_2] \times \cdots \times [a_d, b_d] \subset U$. Since E is compact and f is continuous, $f(E)$ is also compact and thus a Borel set in $\mathbb{R}^n$. The fact that $y \mapsto (f^{-1})'(y)$ is continuous, hence bounded, on $f(E)$ implies the existence of $\gamma > 0$ with the following property: for every $\eta \in \mathbb{R}^d$ and every $x \in E$, there exists $\xi \in \mathbb{R}^d$ such that $f'(x)\xi = \eta$ and $\|\xi\| \le \gamma\|\eta\|_2$. Denote by δ the diameter of E, set

$$\varepsilon_E = \sup_{x,y \in E, x \ne y} \frac{\|f(y) - f(x) - f'(x)(y - x)\|_2}{\|y - x\|_2}$$

and fix a point $x \in E$. For every $y \in E$ we have

$$f(y) = f(x) + f'(x)(y - x) + \eta_y,$$

with $\|\eta_y\|_2 \le \varepsilon_E \|y - x\|_2$. Write then $\eta_y = f'(x)\xi_y$ with $\|\xi_y\|_2 \le \gamma\varepsilon_E\|y - x\|_2$, so

$$f(y) - f(x) = f'(x)(y - x + \xi_y).$$

Since the absolute value of each coordinate of ξ_y is at most $\|\xi_y\|_2$, we conclude that $\{f(y) - f(x) : y \in E\}$ is contained in $f'(x)(E_1)$, where E_1 is a product of intervals of length

$$b_j - a_j + \gamma\varepsilon_E\delta \le (1 + k\varepsilon_E)(b_j - a_j)$$

and $k = \max\{\gamma\delta/(b_j - a_j) : j = 1, 2, \ldots, n\}$. Thus

$$\lambda_d(f(E)) \le (1 + k\varepsilon_E)^d |\det f'(x)|\lambda_d(E) \qquad (8.3)$$

by Proposition 8.9. This estimate can be improved as follows. Divide E into N^d parts $E_1, \ldots, E_{N^d}$ of equal volume by dividing each side of E into N intervals of equal length and apply (8.3) to each E_j to obtain

$$\lambda_d(f(E)) \le \sum_{j=1}^{N^d} (1 + k\varepsilon_{E_j})^d |\det f'(x_j)|\lambda_d(E_j),$$

with x_j an arbitrary point in E_j. Note that the constant k need not be changed, but the fact that f' is continuous implies that $\delta_N = \max\{\varepsilon_{E_j} : j = 1, 2, \ldots, N^d\}$ tends to 0 as $N \to +\infty$. We have therefore

$$\lambda_d(f(E)) \le (1 + k\delta_N)^d \int_E u_N \, d\lambda_d(x), \qquad (8.4)$$

where $u_N = \sum_{j=1}^{N^d} |\det f'(x_j)| \chi_{E_J}$. Clearly, $\lim_{N \to +\infty} u_N(x) = |\det f'(x)|$ for $[\lambda_d]$-almost every $x \in E$, and the functions u_N are uniformly bounded. Letting $N \to \infty$, (8.4) implies

$$\lambda_n(f(E)) \le \int_E |\det f'(x)| \, d\lambda_d(x). \tag{8.5}$$

Standard arguments based on Theorem 1.23 imply now that (8.5) holds for all Lebesgue measurable subsets $E \subset U$. Writing $F = f(E)$, this inequality can be rewritten as

$$\int_V \chi_F(y) \, d\lambda_d(y) \le \int_U \chi_F(f(x)) |\det f'(x)| \, d\lambda_d(x),$$

which shows, in particular, that f maps sets of measure zero to sets of measure zero. Using approximations by simple functions and the monotone convergence theorem, we conclude that

$$\int_V h(y) \, d\lambda_d(y) \le \int_U h(f(x)) |\det f'(x)| \, d\lambda_d(x) \tag{8.6}$$

for every Lebesgue measurable function $h : V \to [0, +\infty]$. The same argument applied to f^{-1} implies the inequality

$$\int_U k(x) \, d\lambda_d(x) \le \int_V k(f^{-1}(y)) |\det(f^{-1})'(y)| \, d\lambda_d(y)$$

for all Lebesgue measurable functions $k : U \to [0, +\infty]$. We apply these inequalities to $k = \chi_E$ for some measurable set $E \subset U$ to obtain

$$\begin{aligned}
\lambda_d(E) &= \int_U \chi_E(x) \, d\lambda_d(x) \\
&\le \int_V \chi_E(f^{-1}(y)) |\det(f^{-1})'(y)| \, d\lambda_d(y) \\
&\le \int_U \chi_E(f^{-1}(f(x))) |\det(f^{-1})'(f(x))| |\det f'(x)| \, d\lambda_d(x) = \lambda_d(E).
\end{aligned}$$

Here we used the fact that $\det(f^{-1})'(f(x)) = 1/\det f'(x)$. It follows that

$$\lambda_d(E) = \int_{f(E)} |\det(f^{-1})'(y)| \, d\lambda_d(y).$$

By symmetry we also have equality in (8.5), and therefore in (8.6). The equality in the theorem follows by the linearity of the integral.   □

Example 8.11. A frequently used change of variables occurs when using polar coordinates. Consider the C^1-diffeomorphism $f : (0, +\infty) \times (-\pi, \pi) \to \mathbb{R}^2 \setminus \{(x, 0) : x \leq 0\}$ given by $f(r, t) = (r \cos t, r \sin t)$. The range of f differs from $\mathbb{R}^2$ by a set of measure zero, so Theorem 8.10 implies the validity of the formula

$$\int_{-\infty}^{+\infty} \int_{-\infty}^{+\infty} f(x, y)\, dx\, dy = \int_{-\pi}^{\pi} \int_{0}^{+\infty} f(r \cos t, r \sin t) r\, dr\, dt$$

for any Lebesgue integrable function f on $\mathbb{R}^2$. Here is a simple application:

$$\left[\int_{\mathbb{R}} e^{-x^2}\, dx \right]^2 = \int_{\mathbb{R}} \int_{\mathbb{R}} e^{-x^2} e^{-y^2}\, dx\, dy$$

$$= \int_{-\pi}^{\pi} \int_{0}^{+\infty} e^{-r^2} r\, dr\, dt$$

$$= 2\pi \int_{0}^{+\infty} e^{-r^2} r\, dr = \pi,$$

thus yielding

$$\int_{\mathbb{R}} e^{-x^2}\, dx = \sqrt{\pi}.$$

A more general instance of this calculation relates the Euler gamma and beta functions defined by

$$\Gamma(p) = \int_{0}^{+\infty} t^{p-1} e^{-t}\, dt, \quad B(p, q) = \int_{0}^{1} t^{p-1}(1-t)^{q-1}\, dt,$$

for complex numbers p, q satisfying $\Re p > 0$ and $\Re q > 0$. The following formula

$$B(p, q) = \frac{\Gamma(p)\Gamma(q)}{\Gamma(p+q)}$$

is proved by writing the numerator as a double integral

$$\int_{0}^{+\infty} \int_{0}^{+\infty} t^{p-1} s^{q-1} e^{-t-q}\, dt\, ds,$$

replacing t by x^2, s by y^2, and then using polar coordinates. The preceding calculation corresponds to the values $p = q = 1/2$ because $\Gamma(1/2) = \int_{\mathbb{R}} e^{-x^2}\, dx$.

Example 8.12. Denote by $B_d = \{x \in \mathbb{R}^d : \|x\|_2 \leq 1\}$ the Euclidean unit ball in $\mathbb{R}^d$, and set $\omega_d = \lambda_d(B_d)$. The change of variable $x \mapsto rx$ for any fixed $r > 0$ shows that the volume of a ball of radius r in $\mathbb{R}^d$ equals $\omega_d r^d$. We know that $\omega_1 = 2$ and $\omega_2 = \pi$. We calculate other values of ω_d by applying

Fubini's theorem. The x_1 section of B_d is a ball with radius $(1 - x_1^2)^{1/2}$ in $\mathbb{R}^{d-1}$ provided that $x_1 \in [-1, 1]$. Thus

$$\omega_d = \int_{-1}^{1} \omega_{d-1}(1 - x^2)^{(d-1)/2}\, dx$$

$$= 2\omega_{d-1} \int_0^1 (1 - x^2)^{(d-1)/2}\, dx$$

$$= \omega_{d-1} \int_0^1 (1 - t)^{(d-1)/2} t^{-1/2} dt = \omega_{d-1} B\left(\frac{1}{2}, \frac{d+1}{2}\right),$$

so

$$\omega_d = \omega_{d-1} \frac{\Gamma\left(\frac{1}{2}\right) \Gamma\left(\frac{d+1}{2}\right)}{\Gamma\left(\frac{d+2}{2}\right)},$$

and repeated application of this formula yields

$$\omega_d = \omega_1 \frac{\Gamma\left(\frac{1}{2}\right)^{d-1} \Gamma\left(\frac{3}{2}\right)}{\Gamma\left(\frac{d}{2}+1\right)} = \frac{\Gamma\left(\frac{1}{2}\right)^d}{\Gamma\left(\frac{d}{2}+1\right)} = \frac{\pi^{d/2}}{\Gamma\left(\frac{d}{2}+1\right)}.$$

Further manipulations of Γ (see Problem 8F) show that

$$\Gamma\left(\frac{d}{2}+1\right) = \begin{cases} k!, & d = 2k, k \in \mathbb{N}, \\ \frac{(2k+1)!!}{2^{k+1}} \sqrt{\pi}, & d = 2k+1, k \in \mathbb{N}, \end{cases}$$

where $(2k+1)!! = \prod_{\ell=1}^{k}(2\ell + 1)$.

For the next example, we return to general measure spaces.

Example 8.13. Let $(Y, \mathbf{T}, \nu)$ be a σ-finite measure space, let $\mu = \lambda_1 | \mathbf{B}_\mathbb{R}$ denote Lebesgue-Borel measure on $(\mathbb{R}, \mathbf{B}_\mathbb{R})$, and consider a function $f : Y \to [0, +\infty]$. Then f is measurable $[\mathbf{T}]$ precisely when the set

$$R_f = \{(t, y) : 0 \le t < f(y)\}$$

is measurable $[\mathbf{B}_\mathbb{R} \times \mathbf{T}]$. Likewise, f is integrable $[\nu]$ precisely when $(\mu \times \nu)$ $(R_f) < +\infty$, and we have $(\mu \times \nu)(R_f) = \int_Y f\, d\nu$. Changing the order of integration we obtain

$$\int_Y f\, d\nu = (\mu \times \nu)(R_f) = \int_0^{+\infty} \nu(\{y : f(y) > t\})\, dt,$$

a formula which expresses the integral of f in terms of its distribution function $t \mapsto \nu(\{y : f(y) > t\})$. More generally, assume that $f : Y \to \mathbb{C}$ is a measurable function, $p \in (0, +\infty)$ is such that $|f|^p$ is integrable, and set $g(t) = \nu(\{y : |f(y)| > t\})$ for $t > 0$. Then a change of variables shows that

$$\int_Y |f|^p \, d\nu = \int_0^{+\infty} \nu(\{y : |f(y)|^p > t\}) \, dt$$

$$= \int_0^{+\infty} g(t^{1/p}) \, dt = p \int_0^{+\infty} g(t) t^{p-1} \, dt.$$

We conclude this chapter with a discussion of products of infinitely many measure spaces. It is convenient to restrict ourselves to probability spaces. Assume then that we are given a family $\{(X_i, \mathbf{S}_i, \mu_i) : i \in I\}$ of measure spaces with the property that $\mu_i(X_i) = 1$ for every $i \in I$, and denote by $X = \prod_{i \in I} X_i$ the product of the sets X_i. Thus X consists of all functions $x : I \to \bigcup_{i \in I} X_i$ with the property that $x(i) \in X_i$ for all $i \in I$. A *measurable cylinder* in X is a set of the form

$$E = \prod_{i \in I} E_i,$$

where $E_i \in \mathbf{S}_i$ for every $i \in I$ and the set $\{i \in I : E_i \neq X_i\}$ is finite. Denote by $\mathbf{T}_0$ the ring generated by all measurable cylinders, and by $\mathbf{T}$ the σ-ring generated by $\mathbf{T}_0$. It is clear that both $\mathbf{T}_0$ and $\mathbf{T}$ are complemented. Note that the infinite numerical product in the following statement only contains a finite number of factors which are different from 1, and therefore its definition does not pose any difficulties.

Theorem 8.14. *There exists a unique measure ν on $(X, \mathbf{T})$ with the property that*

$$\nu(E) = \prod_{i \in I} \mu_i(E_i)$$

for every measurable cylinder $E = \prod_{i \in I} E_i$.

Proof. It is clear from our discussion of finite products of measures that there exists a finitely additive set function $\nu_0 : \mathbf{T}_0 \to [0, 1]$ such that

$$\nu_0(E) = \prod_{i \in I} \mu_i(E_i)$$

for every measurable cylinder $E = \prod_{i \in I} E_i$. To prove the existence of ν, it suffices to show that ν_0 is countably additive (Proposition 5.19). This follows from the following statement: Given a sequence $F_1 \supset F_2 \supset \cdots$ of sets in $\mathbf{T}_0$ with the property that $\bigcap_{n=1}^{\infty} F_n = \varnothing$, it follows that $\lim_{n \to \infty} \nu_0(F_n) = 0$. Indeed, if $A_1, A_2, \ldots$ are pairwise disjoint subsets in $\mathbf{T}_0$, we have

$$\nu_0 \left(\bigcup_{k=1}^{\infty} A_k \right) = \sum_{k=1}^{n} \nu_0(A_k) + \nu_0(F_n),$$

where $F_n = \bigcup_{k > n} A_k$. Assume to the contrary that the sets $F_n \in \mathbf{T}_0$ form a decreasing sequence, $\bigcap_{n=1}^{\infty} F_n = \varnothing$, but

$$\alpha = \lim_{n \to \infty} \nu_0(F_n) > 0. \qquad (8.7)$$

Observe first that for each n there exists a finite set $I_n \subset I$ such that, upon identifying X with

$$\left[\prod_{i \in I_n} X_i \right] \times \left[\prod_{i \notin I_n} X_i \right],$$

we have

$$F_n = G_n \times \left[\prod_{i \notin I_n} X_i \right]$$

for some set $G_n \subset \prod_{i \in I_n} X_i$ which is measurable in this finite product. Since $\bigcup_n I_n$ is at most countable, there is no loss of generality in assuming for this part of the argument that $I = \mathbb{N}$ and that $I_n = \{1, 2, \ldots, m_n\}$ for some integers $m_1 < m_2 < \cdots$. Given an integer $N \in \mathbb{N}$, denote by $\mathbf{T}_N$ the ring generated by the measurable cylinders E in $\prod_{i > N} X_i$, and denote by $\nu_N : \mathbf{T}_N \to [0, 1]$ the additive set function satisfying

$$\nu_N(E) = \prod_{i > N} \mu_i(E_i)$$

for all measurable cylinders $E = \prod_{i > N} E_i \in \mathbf{T}_N$. Using the notation

$$F_{x_1, x_2, \ldots, x_N} = \{x \in X : x(i) = x_i \text{ for } i \leq N\},$$

where $x_i \in X_i$ for $i = 1, 2, \ldots, N$, we observe that $(F_n)_{x_1, x_2, \ldots, x_N} \in \mathbf{T}_N$ for all $n, N \in \mathbb{N}$, and therefore we can define functions $h_{n,N} : \prod_{i=1}^N X_i \to [0, 1]$ by setting

$$h_{n,N}(x_1, x_2, \ldots, x_N) = \nu_N((F_n)_{x_1, x_2, \ldots, x_N}).$$

The extension of Fubini's theorem to finitely many factors implies that $h_{n,N}$ is measurable $[\mu_1 \times \mu_2 \times \cdots \times \mu_N]$, and

$$\nu_0(F_n) = \int_{\prod_{i \leq N} X_i} h_{n,N} \, d(\mu_1 \times \mu_2 \times \cdots \times \mu_N), \quad n, N \in \mathbb{N}. \qquad (8.8)$$

Note now that we have $h_{n,N} \geq h_{n+1,N}$ for all $n, N \in \mathbb{N}$, and therefore the pointwise limit

$$h_N(x_1, x_2, \ldots, x_N) = \lim_{n \to \infty} h_{n,N}(x_1, x_2, \ldots, x_N)$$

exists for all $N \in \mathbb{N}$ and is measurable $[\mu_1 \times \mu_2 \times \cdots \times \mu_N]$. Apply now the dominated convergence theorem to (8.8), and use (8.7) to deduce that

$$\int_{\prod_{i \leq N} X_i} h_N \, d(\mu_1 \times \mu_2 \times \cdots \times \mu_N) = \alpha.$$

Another application of Fubini's theorem shows that

$$h_{n,N}(x_1, x_2, \ldots, x_N) = \int_{X_{N+1}} h_{n,N+1}(x_1, x_2, \ldots, x_N, x_{N+1}) \, d\mu_{N+1}(x_{N+1}),$$

and the dominated convergence theorem implies that

$$h_N(x_1, x_2, \ldots, x_N) = \int_{X_{N+1}} h_{N+1}(x_1, x_2, \ldots, x_N, x_{N+1}) \, d\mu_{N+1}(x_{N+1}).$$

Since $\alpha > 0$, there exists $x_1 \in X_1$ such that $h_1(x_1) > 0$. Use then the preceding equation to find inductively points $x_N \in X_N$ such that

$$h_N(x_1, x_2, \ldots, x_N) > 0, \quad N \in \mathbb{N}.$$

We derive a contradiction by observing that the element $x \in X$ defined by $x(i) = x_i$ belongs to F_n for every $n \in \mathbb{N}$. Indeed, fix $n \in \mathbb{N}$ and choose $N = k_n$ to deduce that $(x_1, x_2, \ldots, x_{k_n}) \in G_n$.

The contradiction just derived shows that ν_0 is indeed countably additive, and the existence of ν is established. Uniqueness follows in the usual way from Theorem 1.23. $\qquad\qquad\qquad\qquad\qquad\qquad\qquad\qquad\qquad\qquad\qquad\qquad\qquad\square$

The measure space $(X, \mathbf{T}, \lambda)$ is called the product of the probability spaces $\{(X_i, \mathbf{S}_i, \mu_i) : i \in I\}$, and we write $\nu = \prod_{i \in I} \mu_i$.

Example 8.15. Denote $X_n = \{0, 1\}$ for $n \in \mathbb{N}$, and define probability measures $\mu_n = \mu$ by setting $\mu(\{0\}) = \mu(\{1\}) = 1/2$. The measure $\nu = \prod_{n \in I} \mu_n$ on $X = \prod_{n=1}^{\infty} X_n$ can be understood in terms of Lebesgue measure on $[0, 1]$. Indeed, the map $f : X \to [0, 1]$ defined by

$$f(x) = \sum_{n=1}^{\infty} \frac{x(n)}{2^n}, \quad x \in X,$$

is a bijection when restricted to the set

$$\{x \in X : x(n) = 1 \text{ for infinitely many values of } n\}$$

whose complement in X is countable. It is easy to see that f maps a cylinder of the form $E = \{x \in X : x(i) = x_i, i = 1, 2, \ldots n\}$ onto an interval of length $2^{-n} = \nu(E)$. It follows easily that a set $E \subset X$ belongs to the product σ-algebra $\mathbf{T}$ if and only if $f(E)$ is a Borel set, and $\lambda_1(f(E)) = \nu(E)$.

Problems

8A. Let W denote the topological space $W(\Omega)$ of countable ordinal numbers in the order topology (Example 7.9 and Problem 7C), and let $\mathbf{B}$ denote the σ-algebra of Borel subsets of W. Show that the diagonal $\Delta = \{(\alpha, \alpha) : \alpha \in W\}$ is a Borel set in the topological space $W \times W$, but that Δ does not belong to the σ-algebra $\mathbf{B} \times \mathbf{B}$. (Hint: Every set A in the σ-algebra $\mathbf{B} \times \mathbf{B}$ has the property that either A or its complement $(W \times W) \setminus A$ contains a set of the form $F \times F$, where F is a closed unbounded subset of W.)

8B. Denote by μ and ν counting measure on the measurable spaces $(X, \mathbf{S})$ and $(Y, \mathbf{T})$ respectively. Show that there is a unique product of the measures μ and ν, namely, counting measure on $(X \times Y, \mathbf{S} \times \mathbf{T})$.

8C. Let $(X, \mathbf{S}, \mu)$ and $(Y, \mathbf{T}, \nu)$ be arbitrary measure spaces, and let $\mathbf{R}$ denote the collection of all measurable rectangles $E \times F \in \mathbf{S} \times \mathbf{T}$, so the σ-algebra $\mathbf{S} \times \mathbf{T}$ coincides with $\mathbf{S}(\mathbf{R})$.

   (i) Show that the gauge $\gamma : \mathbf{R} \to [0, +\infty]$ defined by $\gamma(E \times F) = \mu(E)\nu(F)$ for $E \times F \in \mathbf{R}$ is countably additive. (Hint: Given a partition $E \times F = \bigcup_{n \in \mathbb{N}} E_n \times F_n$, show that

$$\nu(E)\chi_F = \sum_n \nu(F_n)\chi_{E_n}$$

and then apply the theorem of Beppo-Levi.)
   (ii) Conclude that the outer measure λ^* generated by the gauge γ extends γ, and therefore $\lambda = \lambda^* | \mathbf{S} \times \mathbf{T}$ is a product measure of μ and ν.
   (iii) Show that a set $A \in \mathbf{S} \times \mathbf{T}$ is σ-finite with respect to λ if and only if there exist sets E and F, σ-finite with respect to μ and ν, respectively, such that $A \subset E \times F$. Use this fact to show that any product measure of μ and ν on $(X \times Y, \mathbf{S} \times \mathbf{T})$ must agree with λ on every set that is σ-finite with respect to λ. Conclude from this that any product measure of μ and ν is necessarily dominated (setwise) by λ and agrees with λ on any set in $\mathbf{S} \times \mathbf{T}$ that is not an infinite atom with respect to λ.
   (iv) Assume that $X = Y = [0, 1]$, $\mu = \lambda_1 | \mathbf{B}_{[0,1]}$, and ν is the counting measure on $\mathbf{B}_{[0,1]}$. Show that the diagonal $\Delta = \{(x, x) : x \in [0, 1]\}$ is an infinite atom for $\lambda = \mu \times \nu$. (Note that the measure λ in this exercise is different from the one used in Example 8.5.)

8D. Consider measure spaces $(X, \mathbf{S}, \mu)$ and $(Y, \mathbf{T}, \nu)$, and denote by λ the product measure constructed in the preceding problem.

   (i) Verify that the Fubini theorem is valid for every function on $X \times Y$ that is measurable $[\mathbf{S} \times \mathbf{T}]$ and integrable $[\lambda]$. (Hint: An integrable function always vanishes outside some σ-finite set.)
   (ii) (Mukherjea) Assume that λ is semi-finite (Definition 3.22). Show that Tonelli's theorem 8.4 is also valid for an arbitrary function $f : X \times Y \to [0, +\infty]$ that is measurable $[\mathbf{S} \times \mathbf{T}]$, in the sense that if either of the iterated integrals

$$\int_X \left[\int_Y f(x, y) \, d\nu(y) \right] d\mu(x) \text{ or } \int_Y \left[\int_X f(x, y) \, d\mu(x) \right] d\nu(y)$$

exists and is finite, then f is integrable $[\lambda]$ on $X \times Y$. (Hint: It suffices to prove that f vanishes outside some set that is σ-finite with respect to λ. Assume to the contrary that $\{(x, y) : f(x, y) > \varepsilon\}$ has infinite measure for some $\varepsilon > 0$, and use part (i) to obtain a contradiction.)

Remark 8.16. The last two problems show that, provided λ is semifinite, both the Fubini and Tonelli theorems hold for measurable functions on $(X \times Y, \mathbf{S} \times \mathbf{T}, \lambda)$ without any further restriction on X or Y. It should be noted, however, that in this more general version of the Tonelli theorem, it is a part of the hypothesis of the theorem that the function

$g(x) = \int_Y f(x, y)\, d\mu(y)$ be measurable $[\mu]$ (or that the function $h(y) = \int_X f(x, y)\, d\mu(x)$ be measurable $[\nu]$).

8E. With the notation of the preceding two problems, define $\lambda' : \mathbf{S} \times \mathbf{T} \to [0, +\infty]$ by setting

$$\lambda'(G) = \sup\{\lambda(G \cap (E \times F)) : E \in \mathbf{S}, F \in \mathbf{T}, \mu(E) < +\infty, \nu(F) < +\infty\}$$

for $G \in \mathbf{S} \times \mathbf{T}$.

   (i) Show that λ' is a measure such that $\lambda'(E \times F) = \mu(E)\nu(F)$ for measurable rectangles satisfying $\mu(E)\nu(F) < +\infty$, but possibly not for all rectangles.
   (ii) The following assertions are equivalent:
      (a) $\lambda' = \lambda$.
      (b) λ' is semi-finite.
      (c) $\lambda'(E) = 0$ if and only $\lambda(E) = 0$ for every $E \in \mathbf{S} \times \mathbf{T}$.
   (iii) If ρ is an arbitrary product measure of μ and ν on $(X \times Y, \mathbf{S} \times \mathbf{T})$, then $\lambda'(E) \le \rho(E) \le \lambda(E)$ for every $E \in \mathbf{S} \times \mathbf{T}$.

8F. (i) Verify that the integrals defining $B(p, q)$ and $\Gamma(p)$ exist when $\Re p > 0$ and $\Re q > 0$.
   (ii) Show that $\Gamma(1/2) = \int_{\mathbb{R}} e^{-x^2}\, dx$. More generally, verify the moment formulas from Examples 7.16 and 7.17.
   (iii) Verify the identity $\Gamma(p+1) = p\Gamma(p)$ for $\Re p > 0$. Deduce the formula for $\Gamma(1+d/2)$ given in Example 8.12.
   (iv) Derive the formula for ω_d by calculating $\int_{\mathbb{R}^d} e^{-\|x\|_2^2}\, d\lambda_d(x) = \pi^{d/2}$ via distribution functions, as in Example 8.13.

8G. Consider a subset $E \subset \mathbb{R}$ which is not measurable $[\lambda_1]$. Show that the set $E \times \{0\}$ is not measurable $[\lambda_1 \times \lambda_1]$ but it is measurable $[\lambda_2]$. Thus, λ_2 is a proper extension of $\lambda_1 \times \lambda_1$.

8H. Consider the measurable space $(X, \mathbf{S})$ where $X = W(\Omega)$ and $\mathbf{S} = 2^X$ is the σ-algebra consisting of all subsets of X.

   (i) Let $f : X \to \mathbb{R}$ be a measurable function. Show that the graph $G_f = \{(x, f(x)) : x \in X\}$ of f is measurable $[\mathbf{S} \times \mathbf{B}_{\mathbb{R}}]$. (This part of the problem actually applies to an arbitrary measurable space $(X, \mathbf{S})$.)
   (ii) For an arbitrary function $f : X \to X$, show that G_f is measurable $[\mathbf{S} \times \mathbf{S}]$. (Hint: X has cardinality $\aleph_1 \le 2^{\aleph_0}$, and hence it is in one-to-one correspondence with some subset of $\mathbb{R}$.)
   (iii) Let $E \subset X \times X$ be such that E_x is at most countable for each $x \in X$. Show that E is measurable $[\mathbf{S} \times \mathbf{S}]$. In particular, every subset of $\{(x, y) : y \le x\}$ is measurable $[\mathbf{S} \times \mathbf{S}]$.
   (iv) Show that every subset $E \subset X \times X$ is measurable $[\mathbf{S} \times \mathbf{S}]$.
   (v) Let $\mu : \mathbf{S} \to [0, 1]$ be a measure such that $\mu(\{x\}) = 0$ for all $x \in X$. Show that $\mu = 0$. (Hint: Use Fubini's theorem to show that $(\mu \times \mu)(\{(x, y) : y \le x\}) = 0$. Deduce that $\mu(\mathbf{S})^2 = (\mu \times \mu)(\mathbf{S} \times \mathbf{S}) = 0$.)

Remark 8.17. The argument suggested in (i–iv) is due to B. V. Rao. The *measure problem* asks whether there exists an infinite set X and a probability measure μ on 2^X such that $\mu(\{x\}) = 0$ for all $x \in X$. Part (v) of the preceding problem shows that such a measure does not exist if the cardinality of X is $\aleph_1$. This fact was known to Banach and Kuratowski. The existence of such probability measures is independent of the standard axioms of set theory.

8I. Consider a measurable space $(X, \mathbf{S})$.

(i) If $G \subset X \times X$ is measurable $[\mathbf{S} \times \mathbf{S}]$, then there exist sequences of sets $\{E_n\}_{n \in \mathbb{N}}$ and $\{F_n\}_{n \in \mathbb{N}}$ in $\mathbf{S}$ such that G belongs to the σ-algebra generated by $\{E_n \times F_n : n \in \mathbb{N}\}$. In particular, the collection of sections $\{G_x : x \in X\}$ is contained in the σ-algebra generated by $\{F_n : n \in \mathbb{N}\}$. Conclude that there are at most $2^{\aleph_0}$ distinct sets of the form G_x with $x \in X$.

(ii) Show that the diagonal $\Delta = \{(x, x) : x \in X\}$ does not belong to $\mathbf{S} \times \mathbf{S}$ if the cardinality of X is strictly greater than $2^{\aleph_0}$. In particular, the σ-algebra $2^X \times 2^X$ does not equal $2^{X \times X}$ for such sets X.

8J. Let $(X, \mathbf{S}, \mu)$ be a σ-finite, atom free, measure space, and let $f : X \to \mathbb{R}^d$ be a $[\mathbf{S}, \mathbf{B}_{\mathbb{R}^d}]$ measurable function. Show that the graph G_f is null $[\mu \times \lambda_d]$. (The set G_f is measurable $[\mathbf{S} \times \mathbf{B}_{\mathbb{R}^d}]$ according to Problem 8H(1).)

8K. Let $(X, \mathbf{S}, \mu)$ and $(Y, \mathbf{T}, \nu)$ be σ-finite measure spaces, and let $f : X \times Y \to \mathbb{R}$ be a measurable function.

(i) Assume that the function f_x is $[\nu]$-almost everywhere constant for $[\mu]$-almost every $x \in X$, and the function f^y is $[\mu]$-almost everywhere constant for $[\nu]$-almost every $y \in Y$. Show that f is $[\mu \times \nu]$-almost everywhere constant.

(ii) Assume that f is integrable $[\mu \times \nu]$ and $\int_{E \times F} f \, d(\mu \times \nu) = 0$ whenever $E \in \mathbf{S}$, $F \in \mathbf{T}$, and $\mu(E \times F) < +\infty$. Show that $f = 0$ $[\mu \times \nu]$-almost everywhere.

8L. Define $f : \mathbb{R}^2 \to \mathbb{R}$ by setting

$$f(x, y) = \begin{cases} \sin(x - y), & 0 \le y \le x \le y + 2\pi, \\ 0, & \text{otherwise.} \end{cases}$$

Then the integrals $\int_{\mathbb{R}} \left[\int_{\mathbb{R}} f(x, y) \, dx \right] dy$ and $\int_{\mathbb{R}} \left[\int_{\mathbb{R}} f(x, y) \, dy \right] dx$ exist and they are different.

8M. Consider a nonconstant polynomial function $p : \mathbb{R}^d \to \mathbb{R}$. Show that $\lambda_d(\{x \in \mathbb{R}^d : p(x) = 0\}) = 0$. (Hint: Proceed by induction on n.)

8N. Let f be an integrable function on the σ-finite measure space $(X, \mathbf{S}, \mu)$.

(i) Show that the set $F = \{(t, x) \in (-\pi, \pi) \times X : \Re(e^{it} f(x)) > 0\}$ is measurable $[\lambda_1 \times \mu]$.

(ii) Show that $\int_F \Re(e^{it} f(x)) \, d(\lambda_1 \times \mu)(t, x) = 2 \int_X |f| \, d\mu$. Deduce that there exists a set $E \in \mathbf{S}$ such that $\left| \int_E f \, d\mu \right| \ge (1/\pi) \int_X |f| \, d\mu$.

(iii) Use the function $f(t) = e^{it}, t \in (-\pi, \pi)$, to show that the constant $1/\pi$ in (ii) cannot be improved.

8O. (Sard) Consider an open set $U \subset \mathbb{R}^d$ and a continuously differentiable function $f : U \to \mathbb{R}^d$. Show that

$$\lambda_d(f(E)) \le \int_E |\det f'(x)| \, d\lambda_d(x)$$

for every Lebesgue measurable set $E \subset U$. Deduce that $f(E_0)$ is $[\lambda_d]$-null where the set E_0 consists of those points $x \in U$ for which $f'(x)$ is not invertible.

8P. Let $(X_n, \mathbf{S}_n, \mu_n)$ be a probability space for each $n \in \mathbb{N}$, and let $(X, \mathbf{S}, \mu)$ be the product of these probability spaces. Show that $\mu(E) = \prod_{n=1}^{\infty} \mu_n(E_n)$ if $E = \prod_{n=1}^{\infty} E_n$ and $E_n \in \mathbf{S}_n$ for $n \in \mathbb{N}$. In particular, $\mu(E) \ne 0$ precisely when $\sum_{n=1}^{\infty} (1 - \mu_n(E_n)) < +\infty$.

8Q. Assume that $(X_n, \mathbf{S}_n, \mu_n) = ([0, 1], \mathbf{B}_{[0,1]}, \lambda_1 | \mathbf{B}_{[0,1]})$, $n \in \mathbb{N}$, and let $(X, \mathbf{S}, \mu)$ be the product of these probability spaces. Show that the set $\{x \in X : x(i) = x(j) \text{ for some } i \ne j\}$ is $[\mu]$-null.

8R. Let I be an uncountable set. Endow the set $X = \{0, 1\}^I$ with the σ-algebra $\mathbf{S}$ generated by cylinders. Show that singletons are not measurable $[\mathbf{S}]$.

8S. Consider a probability space $(X, \mathbf{S}, \mu)$, and let $Y = (X^{\mathbb{N}}, \mathbf{T}, \nu)$ be the product of countably many copies of $(X, \mathbf{S}, \mu)$. Given a bijection $\sigma : \mathbb{N} \to \mathbb{N}$, we define a transformation $f_\sigma : Y \to Y$ by setting $f_\sigma(y) = y \circ \sigma^{-1}$ for $y \in Y$.

(i) Show that f_σ is measurable, and $\nu(f_\sigma(E)) = \nu(E)$ for every $E \in \mathbf{T}$.

(ii) (Savage) Let $E \in \mathbf{S}$ satisfy $\nu(E \triangle f_\sigma(E)) = 0$ for every permutation σ of $\mathbb{N}$ for which the set $\{n \in \mathbb{N} : \sigma(n) \neq n\}$ is finite. Show that $\nu(E)$ equals either 0 or 1. (Hint: If F is a cylinder, $\nu(F \triangle f_\sigma(F)) = 2\nu(F)(1 - \nu(F))$ for some permutations σ.)

8T. (Kolmogorov) Let $(X, \mathbf{S}, \mu) = \prod_{n \in \mathbb{N}}(X_n, \mathbf{S}_n, \mu_n)$ be an infinite product of probability spaces. Denote by $\mathbf{T}_n$ the σ-algebra generated by cylinders $\prod_{j \in \mathbb{N}} E_j$ with the property that $E_j = X_j$ for $j \leq n$, and denote by $\mathbf{T} = \bigcap_{n \in \mathbb{N}} \mathbf{T}_n$ the *tail σ-algebra*. Show that $\mu(E) \in \{0, 1\}$ for every $E \in \mathbf{T}$. (Hint: We have $\mu(E \cap F) = \mu(E)\mu(F)$ if $E \in \mathbf{T}$ and F is a cylinder.)

8U. Consider three σ-finite measure spaces $(X_j, \mathbf{S}_j, \mu_j)$. Show that, after making the natural identification of the sets $X_1 \times (X_2 \times X_3)$ and $(X_1 \times X_2) \times X_3$, we have $\mathbf{S}_1 \times (\mathbf{S}_2 \times \mathbf{S}_3) = (\mathbf{S}_1 \times \mathbf{S}_2) \times \mathbf{S}_3$ and $\mu_1 \times (\mu_2 \times \mu_3) = (\mu_1 \times \mu_2) \times \mu_3$. Use this fact to define the product of finitely many measure spaces.

8V. Define an appropriate notion of a *direct sum* of two measure spaces and prove (in the σ-finite case) that the product of measures is distributive relative to direct sums. (Distributivity always occurs at the level of sets, that is, $(X_1 \cup X_2) \times X_3 = (X_1 \times X_3) \cup (X_2 \times X_3)$.)

8W. Endow the space $\mathbb{C}^{\mathbb{R}}$ of all complex-valued functions on $\mathbb{R}$ with the product σ-algebra, where each factor $\mathbb{C}$ is given its usual Borel σ-algebra $\mathbf{B}_{\mathbb{C}}$. Show that $\varnothing$ is the only measurable set contained in the space $\mathcal{C}(\mathbb{R}) \subset \mathbb{C}^{\mathbb{R}}$ of continuous functions.

8X. Let $(X, \mathbf{S}, \mu)$ and $(Y, \mathbf{T}, \nu)$ be probability spaces, and let the measurable set $E \subset X \times Y$ satisfy $(\mu \times \nu)(E) > 0$. Show that there exists a subset $A \in \mathbf{S}$ such that $\mu(A) > 0$ and for each $n \in \mathbb{N}$ we have $\inf\{\nu(E_{x_1} \cap E_{x_2} \cap \cdots \cap E_{x_n}) : x_1, x_2, \ldots, x_n \in A\} > 0$.

8Y. (Steiner) Let $K \subset \mathbb{R}^d$ be a compact convex set for some $d \geq 1$, and denote by

$$\delta = \delta(K) = \sup\{\|x - y\|_2 : x, y \in K\}$$

the diameter of K.

(i) Denote by K' the set consisting of those points $x' = (x_2, x_3, \ldots, x_d) \in \mathbb{R}^{d-1}$ for which there exists $x_1 \in \mathbb{R}$ such that $(x_1, x') \in K$, and define

$$\ell(x') = \sup\{|x_1 - x_2| : x_1, x_2 \in \mathbb{R}, (x_1, x'), (y_1, y') \in K\}.$$

Show that the set

$$K_1 = \{(x_1, x') : x' \in K', |x_1| \leq \ell(x')/2\}$$

is also a compact convex set, $\lambda_d(K_1) = \lambda_d(K)$, and $\delta(K_1) \leq \delta$.

(ii) Show that there exists a compact convex set K_d satisfying $\lambda_d(K_d) = \lambda_d(K)$, $\delta(K_d) \leq \delta$, and such that for every $(x_1, x_2, \ldots, x_d) \in K_d$ and for every choice of signs $\varepsilon_j \in \{-1, +1\}, j = 1, 2, \ldots, d$, we have $(\varepsilon_1 x_1, \varepsilon_2 x_2, \ldots, \varepsilon_d x_d) \in K_d$.

(iii) Show that the set K_d is contained in the ball centered at zero with radius $\delta/2$, and deduce that $\lambda_d(K) \leq \omega_d \delta^d / 2^d$.

8Z. Let T be an invertible linear transformation on $\mathbb{R}^d$ for some $d \geq 1$. Denote by $e_j, j = 1, 2, \ldots, d$ the standard basis vectors in $\mathbb{R}^d$, that is, the only nonzero component of e_j is the jth component which is equal to 1.

(i) Show that there exists a permutation matrix X such that the first component of XTe_1 is different from zero.

(ii) Assume that the first component of Te_1 is different from zero. Show that there exist scalars $\alpha_j, \beta_j \in \mathbb{R}$, $j = 1, 2, \ldots, d-1$ and $\gamma \in \mathbb{R} \setminus \{0\}$ such that

$$T_{\alpha_1} T_{\alpha_2} \cdots T_{\alpha_{d-1}} T T_{\beta_1} T_{\beta_2} \cdots T_{\beta_{d-1}} e_1 = \gamma e_1.$$

The transformations T_α were defined in Example 8.8.

(iii) Use induction to prove that T can be written as a product of the type described in the proof of Proposition 8.9.

Chapter 9
The L^p spaces

So far in this book we have considered only pointwise, or almost everywhere, convergence of sequences (and nets) of measurable functions. In many problems where Lebesgue integration occurs, several different kinds of convergence appear naturally. In most cases, different kinds of convergence are associated with different vector spaces of (equivalence classes of) measurable functions, and it is the purpose of this chapter to describe some of these spaces and discuss the corresponding types of convergence. One of the most basic requirements for a mode of convergence is that the usual operations of addition and multiplication by scalars preserve convergence. Because of this, we will be dealing with vector spaces endowed with a topology that is compatible with the linear structure. We start therefore with a brief discussion of the concept of a topological vector space. The discussion is carried out for complex vector spaces, but the analogous concepts make sense for real vector spaces as well.

Definition 9.1. Suppose given a complex vector space $\mathcal{V}$ and a topology τ on $\mathcal{V}$. We say that $(\mathcal{V}, \tau)$ (or simply $\mathcal{V}$ if no confusion will result) is a *topological vector space* if the vector operations

$$(x, y) \mapsto x + y, \quad x, y \in \mathcal{V},$$
$$(\lambda, x) \mapsto \lambda x, \quad \lambda \in \mathbb{C}, x \in \mathcal{V},$$

are continuous when $\mathcal{V} \times \mathcal{V}$ and $\mathbb{C} \times \mathcal{V}$ are endowed with their product topologies.

The topology of a topological vector space is often generated by a family of quasinorms.

Definition 9.2. Consider a complex vector space $\mathcal{V}$ and a function $q : \mathcal{V} \to [0, +\infty)$. We say that q is a *quasi-seminorm* if it satisfies the following three conditions:

© Springer International Publishing Switzerland 2016
H. Bercovici et al., *Measure and Integration*,
DOI 10.1007/978-3-319-29046-1_9

(1) $q(x + y) \leq q(x) + q(y)$, $x, y \in \mathcal{V}$,
(2) $q(\lambda x) \leq q(x)$, $|\lambda| \leq 1, x \in \mathcal{V}$, and
(3) $\lim_{\lambda \to 0} q(\lambda x) = 0$, $x \in \mathcal{V}$.

We say that a quasi-seminorm q is a *seminorm* if it satisfies

(4) $q(\lambda x) = |\lambda| q(x)$, $(\lambda, x) \in \mathbb{C} \times \mathcal{V}$.

A (quasi-)seminorm q is called a *(quasi)norm* if $q(x) \neq 0$ whenever $x \neq 0$.

Condition (4) in the preceding definition is certainly stronger than (2) and (3). Given a collection $\mathcal{Q}$ of quasi-seminorms on a vector space $\mathcal{V}$, a topological vector space $(\mathcal{V}, \tau_\mathcal{Q})$ is defined by specifying that the sets

$$\{y \in \mathcal{V} : q(x - y) < \varepsilon, q \in \mathcal{F}\}$$

form a neighborhood base at x as ε ranges over $(0, +\infty)$ and $\mathcal{F}$ ranges over the finite subsets of $\mathcal{Q}$. (This topology may also be defined by saying that a net $\{x_\lambda\}$ in $\mathcal{V}$ converges to an element $x_0 \in \mathcal{V}$ if and only if the net $\{q(x_\lambda - x_0)\}$ converges to 0 for every $q \in \mathcal{F}$.) The topological vector space $(\mathcal{V}, \tau_\mathcal{Q})$ is Hausdorff when for every $x \in \mathcal{V} \setminus \{0\}$ we have $q(x) \neq 0$ for some $q \in \mathcal{Q}$. A topological vector space defined by a single (quasi)norm q is called a *(quasi)normed space*. Clearly, a (quasi)normed space is also a metric space relative to the metric $d(x, y) = q(x - y)$. A normed space which is complete relative to this metric is called a *Banach space*. A quasinormed space which is complete is called an *F-space*.

The norm of an element x of a generic Banach space $\mathcal{V}$ is denoted by $\|x\|$ or $\|x\|_\mathcal{V}$.

Example 9.3. The space $\mathbb{C}^d$ endowed with the usual Euclidean norm $\|\cdot\|_2$ is a (complex) Banach space. The space $\mathbb{C}^\mathbb{N}$ of all sequences of complex numbers, endowed with the sequence of seminorms

$$p_n((\lambda_i)_{i \in \mathbb{N}}) = |\lambda_n|, \quad n \in \mathbb{N},$$

is a topological vector space. Convergence in this space is equivalent to componentwise convergence. This topology on $\mathbb{C}^\mathbb{N}$ can also be defined by a single quasinorm

$$q((\lambda_i)_{i \in \mathbb{N}}) = \sum_{n=1}^{\infty} 2^{-n} \min\{1, |\lambda_n|\},$$

and $\mathbb{C}^\mathbb{N}$ is an F-space. If I is an uncountable set, the topological vector space $\mathbb{C}^I$, endowed with the topology of componentwise convergence, is not metrizable.

We recall that a map $T : \mathcal{V} \to \mathcal{W}$ between two complex vector spaces is said to be *linear* if

$$T(\lambda x + y) = \lambda T(x) + T(y), \quad \lambda \in \mathbb{C}, x, y \in \mathcal{V}.$$

We use the term *linear functional* for a linear map $T : \mathcal{V} \to \mathbb{C}$.

Definition 9.4. Let $\mathcal{V}$ and $\mathcal{W}$ be topological vector spaces. We denote by $\mathcal{B}(\mathcal{V}, \mathcal{W})$ the vector space consisting of all *continuous* linear maps $T : \mathcal{V} \to \mathcal{W}$. The *topological dual* of a topological vector space $\mathcal{V}$ is the vector space $\mathcal{V}^* = \mathcal{B}(\mathcal{V}, \mathbb{C})$.

We remark here that a topological vector space $\mathcal{V}$ may have a trivial dual space, that is $\mathcal{V}^* = \{0\}$; see Problem 9E.

Note that $\mathcal{B}(\mathcal{V}, \mathcal{W})$ is simply a vector space, although it can be endowed with various natural topologies. For example, when $\mathcal{V}$ and $\mathcal{W}$ are normed spaces, a linear map $T : \mathcal{V} \to \mathcal{W}$ is continuous if and only if *the norm of T*, defined by

$$\|T\| = \sup\{\|Tx\|_{\mathcal{W}} : x \in \mathcal{V}, \|x\| \leq 1\} \tag{9.1}$$

is finite. The necessity of (9.1) is seen because the set

$$\{x \in \mathcal{V} : \|Tx\|_{\mathcal{W}} < 1\}$$

is a neighborhood of 0 if T is continuous, say

$$\{x \in \mathcal{V} : \|Tx\|_{\mathcal{W}} < 1\} \supset \{x \in \mathcal{V} : \|x\|_{\mathcal{V}} < \delta\}$$

for some $\delta > 0$. It follows that $\|T\| \leq 1/\delta$. It is immediate that $\|T\| < +\infty$ implies the continuity of T. The space $\mathcal{B}(\mathcal{V}, \mathcal{W})$ becomes a normed space with the norm defined by (9.1). A sequence of linear maps in $\mathcal{B}(\mathcal{V}, \mathcal{W})$ converges in this norm if and only if it converges uniformly on the closed unit ball $\{x \in \mathcal{V} : \|x\|_{\mathcal{V}} \leq 1\}$ of $\mathcal{V}$. It is thus easily seen that $\mathcal{B}(\mathcal{V}, \mathcal{W})$ is a Banach space if $\mathcal{W}$ is complete. In particular, the norm turns the dual $\mathcal{V}^*$ of a normed space into a Banach space.

The following result, the *uniform boundedness theorem*, is due to Banach and Steinhaus.

Theorem 9.5. *Let $\mathcal{V}$ be a Banach space, let $\mathcal{W}$ be a normed space, and let $\mathcal{F} \subset \mathcal{B}(\mathcal{V}, \mathcal{W})$ satisfy*

$$\sup_{T \in \mathcal{F}} \|T(x)\|_{\mathcal{W}} < +\infty, \quad x \in \mathcal{V}.$$

Then $\sup\{\|T\| : T \in \mathcal{F}\} < +\infty$.

Proof. Assume, to get a contradiction, that $\sup\{\|T\| : T \in \mathcal{F}\} = +\infty$. We construct inductively vectors $x_n \in \mathcal{V}$ and linear maps $T_n \in \mathcal{F}$ such that

$$\|T_n(x_n)\|_{\mathcal{W}} > n + \sum_{j<n} \sup_{\varphi \in \mathcal{F}} \|T(x_j)\|_{\mathcal{W}}$$

and

$$\|x_n\|_{\mathcal{V}} < \frac{2^{-n}}{1 + \sum_{j<n} \|T_j\|}$$

for all $n \in \mathbb{N}$. The vector $x = \sum_{n=1}^{\infty} x_n$ exists because $\mathcal{V}$ is complete, and

$$\|T_n(x)\|_{\mathcal{W}} \geq \|T_n(x_n)\|_{\mathcal{W}} - \sum_{j<n} \|T_n(x_j)\|_{\mathcal{W}} - \sum_{j>n} \|T_n(x_j)\|_{\mathcal{W}}$$

$$> n - \sum_{j>n} \|T_n\| \|x_j\|_{\mathcal{V}}$$

$$> n - \sum_{j>n} 2^{-j} > n - 1$$

is an unbounded sequence, contradicting the hypothesis of the theorem. $\quad\square$

Uniform boundedness is relevant in the study of pointwise convergence for sequences of linear functionals. The reader will have already encountered the following general principle, for instance in Problem 4EE. We record it here for later use.

Proposition 9.6. *Let $\mathcal{V}$ be a normed space, let $\mathcal{W}$ be a Banach space, and let $\{T_n\}_{n=1}^{\infty} \subset \mathcal{B}(\mathcal{V}, \mathcal{W})$ be a bounded sequence, that is, $\sup_{n \in \mathbb{N}} \|T_n\| < +\infty$. Then the set*

$$S = \{x \in \mathcal{V} : \lim_{n \to \infty} T_n(x) \ exists\}$$

is a closed linear manifold in $\mathcal{V}$. The limit $T(x) = \lim_{n \to \infty} T_n(x)$ defines a continuous linear map T on S.

Proof. Set $M = \sup_{n \in \mathbb{N}} \|T_n\|$. Given $x \in S$ and $y \in \mathcal{V}$, we have

$$\|T_n(y) - T_m(y)\|_{\mathcal{W}} \leq \|T_n(x) - T_m(x)\|_{\mathcal{W}} + 2M\|x - y\|_{\mathcal{V}}, \quad m, n \in \mathbb{N},$$

and therefore

$$\limsup_{n,m \to +\infty} \|T_n(y) - T_m(y)\|_{\mathcal{W}} \leq 2M\|x - y\|.$$

Thus the sequence $\{T_n(y)\}_{n=1}^{\infty}$ is convergent if y belongs to the closure of S. The fact that S is a vector space follows from the linearity of limits and the linearity of each T_n. Finally, we have $\|T\| \leq \sup_{n \in \mathbb{N}} \|T_n\|$, so T is continuous. $\quad\square$

Definition 9.7. Given a topological vector space $\mathcal{V}$, consider the seminorms $q_{\varphi} : \mathcal{V} \to [0, +\infty)$ given by

$$q_{\varphi}(x) = |\varphi(x)|, \quad x \in \mathcal{V}, \varphi \in \mathcal{V}^*.$$

The topology defined on $\mathcal{V}$ by these seminorms is called the *weak topology* and, as noted above, may be defined by specifying that a net $\{x_{\lambda}\}$ converges to x_0 if and only if the net $\{\varphi(x_{\lambda})\}$ converges to $\varphi(x_0)$ for every $\varphi \in \mathcal{V}^*$. Analogously, the seminorms $q_x : \mathcal{V}^* \to [0, +\infty)$ given by

$$q_x(\varphi) = |\varphi(x)|, \quad \varphi \in \mathcal{V}^*, x \in \mathcal{V},$$

define the *weak* topology* on $\mathcal{V}^*$. We observe that the weak* topology on $\mathcal{V}^*$ is always Hausdorff, but the weak topology on $\mathcal{V}$ is Hausdorff only when

$$\{x \in \mathcal{V} : \varphi(x) = 0 \text{ for every } \varphi \in \mathcal{V}^*\} = \{0\}.$$

One important feature of weak* topologies is that they provide useful examples of compactness in an infinite dimensional setting. The following result is due to Banach and Alaoglu.

Theorem 9.8. *For every normed space* $\mathcal{V}$*, the closed unit ball*

$$\mathcal{V}_1^* = \{\varphi \in \mathcal{V}^* : \|\varphi\| \leq 1\}$$

is compact in the weak* topology. If $\mathcal{V}$ *is separable then* $\mathcal{V}_1^*$ *is, in addition, metrizable.*

Proof. The weak* topology on $\mathcal{V}^*$ is simply the restriction of the product topology on $\mathbb{C}^{\mathcal{V}}$ to the subset of continuous linear maps. The condition $\|\varphi\| \leq 1$ is equivalent to saying $|\varphi(x)| \leq \|x\|$, $x \in \mathcal{V}$, and therefore $\mathcal{V}_1^*$ is a subset of the compact product space $K = \prod_{x \in \mathcal{V}} K_x$, where $K_x = \{\lambda \in \mathbb{C} : |\lambda| \leq \|x\|\}$ (see [**I**, Corollary 9.57]). To show that $\mathcal{V}_1^*$ is compact, it therefore suffices to verify that it is a closed set. This is obvious because an element $(\alpha_x)_{x \in X} \in K$ belongs to $\mathcal{V}_1^*$ if and only if it satisfies the equations

$$\alpha_{x+y} = \alpha_x + \alpha_y, \alpha_{\lambda x} = \lambda \alpha_x, \quad x, y \in \mathcal{V}, \lambda \in \mathbb{C},$$

and each of these equations defines a closed subset of K. Assume now that X is separable, and fix a dense sequence $\{x_n\}_{n=1}^{\infty}$ in the unit ball of $\mathcal{V}$. The formula

$$d(\varphi, \psi) = \sum_{n=1}^{\infty} 2^{-n} |\varphi(x_n) - \psi(x_n)|, \quad \varphi, \psi \in \mathcal{V}_1^*,$$

defines a metric on $\mathcal{V}_1^*$ which is continuous in the weak* topology. Since X_1^* is compact, it follows that d is defines the weak* topology on this set. $\square$

Remark 9.9. The metric d defined in the preceding proof defines a topology on the entire dual $\mathcal{V}^*$. This topology is generally weaker than the weak* topology. In fact, d is compatible with the weak* topology precisely when $\mathcal{V}$ has a countable Hamel basis as a vector space.

Remark 9.10. The reader who wishes to inquire more deeply into the topic of topological vector spaces might consult [**I**, Chapters 11–18].

We next explore some concrete and very important topological vector spaces which occur naturally in the study of measure and integration. Historically, these were among the first infinite dimensional topological vector spaces to be studied.

Definition 9.11. Let $(X, \mathbf{S}, \mu)$ be a measure space. We denote by $\mathcal{L}^0(X, \mathbf{S}, \mu)$ or, when no confusion can result, $\mathcal{L}^0(\mu)$, the collection of functions $f : Y \to \mathbb{C}$, where $Y \subset X$ is such that there exists a set $E \in \mathbf{S}$ with $\mu(E) = 0$ that satisfies $X \setminus E \subset Y$ and $f|(X \setminus E)$ is measurable $[\mu]$.

The functions in $\mathcal{L}^0(\mu)$ which are defined on the entire set X form a complex vector space. Generally, however, the set $\mathcal{L}^0(\mu)$ is not a vector space, though the sum $f + g$ of two elements (defined on the common domain of f and g) coincides almost everywhere $[\mu]$ with a function in $\mathcal{L}^0(\mu)$. We can construct a vector space from such functions as follows.

Definition 9.12. Two functions $f, g \in \mathcal{L}^0(\mu)$ are said to be *equivalent* if the set $\{x : f(x) \neq g(x)\}$ is contained in a set $E \in \mathbf{S}$ with $\mu(E) = 0$. Denote by $[f]$ the equivalence class of a function f, and write

$$L^0(\mu) = \{[f] : f \in \mathcal{L}^0(\mu)\}.$$

Every equivalence class in $L^0(\mu)$ contains a function which is measurable and defined everywhere on X. This is seen as follows. If f, Y, and E are as in Definition 9.11, the function

$$g(x) = \begin{cases} f(x), & x \in X \setminus E, \\ 0, & x \in E, \end{cases}$$

is equivalent to f. One reason for the definition of the set $\mathcal{L}^0(\mu)$ is that in many cases a function may be defined by a limit process such that the limit exists almost everywhere $[\mu]$. In such cases it may be tedious to verify that the set on which convergence occurs is measurable $[\mu]$ and to redefine the function on a null set so as to be measurable on X.

It is immediate that the operations

$$[f] + [g] = [f + g], \lambda[f] = [\lambda f], \quad f, g \in \mathcal{L}^0(\mu), \lambda \in \mathbb{C},$$

are well defined, and therefore that $L^0(\mu)$ is a vector space. Multiplication is defined by

$$[f][g] = [fg], \quad f, g \in \mathcal{L}^0(\mu),$$

and this turns $L^0(\mu)$ into an associative algebra with unit $[1]$ determined by the constant function 1. Complex conjugation can be defined by

$$\overline{[f]} = [\bar{f}], \quad f \in \mathcal{L}^0(\mu).$$

An element $u \in L^0(\mu)$ is said to be *real* [*positive*] if it can be written as $u = [f]$ for some $f \in \mathcal{L}^0(\mu)$ which is real-valued [nonnegative]. For real elements $u, v \in L^0(\mu)$ one defines $\max\{u, v\}$ and $\min\{u, v\}$ in the obvious way.

A natural topology on the space $L^0(\mu)$ is that of *convergence in measure* which we now define. For every subset $E \in \mathbf{S}$ with finite measure, define a quasi-seminorm on $L^0(\mu)$ by

$$q_E([f]) = q_E(f) = \int_E \min\{1, |f|\}\, d\mu, \quad f \in \mathcal{L}^0(\mu). \tag{9.2}$$

The topology of *convergence in measure* is the topology determined by this family of quasi-seminorms on $L^0(\mu)$. Thus a net $\{f_\lambda\} \subset \mathcal{L}^0(\mu)$ converges to $f \in \mathcal{L}^0(\mu)$ if and only if $\lim_\lambda q_E(f_\lambda - f) = 0$ for every $E \in \mathbf{S}$ with finite measure.

Let us note at this point that replacing the measure μ by its completion, that is, declaring all subsets of null sets to be measurable, does not change the space $L^0(\mu)$ or the quasi-seminorms q_E.

Lemma 9.13. *Let $(X, \mathbf{S}, \mu)$ be a complete measure space. Then a function $f \in \mathcal{L}^0(\mu)$ satisfies $q_E([f]) = 0$ for every set $E \in \mathbf{S}$ with finite measure precisely when $N_f = \{x \in X : f(x) \neq 0\}$ is either $[\mu]$-null or is an infinite atom of μ. In particular, the topology of convergence in measure is Hausdorff if μ is σ-finite or when μ is the counting measure on X. A net $\{[f_\lambda]\} \subset L^0(\mu)$ converges to zero in measure if and only if*

$$\lim_\lambda \mu(\{x \in E : |f_\lambda(x)| > \varepsilon\}) = 0$$

for every set E with finite measure and every $\varepsilon > 0$.

Proof. Assume that $q_E([f]) = 0$ for every E with $\mu(E) < +\infty$. If $E \subset N_f$ is any fixed measurable set with $\mu(E) < +\infty$, we deduce from (9.2) that $\mu(E) = 0$. Thus, if $\mu(N_f) > 0$, it follows that N_f is an infinite atom.

If the measure μ is semifinite (see Proposition 3.24), it has no infinite atoms. Hence any function $f \in \mathcal{L}^0(\mu)$ such that $q_E(f) = 0$ whenever $E \in \mathbf{S}$ has finite measure satisfies $\mu(N_f) = 0$, so $[f] = 0$. It follows that $L^0(\mu)$ is Hausdorff if μ is semifinite. Clearly, σ-finite measures and the counting measure on X are semifinite.

To prove the last assertion we observe that for every $\varepsilon \in (0, 1)$ and every $E \in \mathbf{S}$ with finite measure we have

$$\varepsilon \chi_{\{x \in E : |f_\lambda(x)| > \varepsilon\}} \leq \min\{1, |f_\lambda|\} \chi_E \leq \varepsilon \chi_{\{x \in E : |f_\lambda(x)| \leq \varepsilon\}} + \chi_{\{x \in E : |f_\lambda(x)| > \varepsilon\}},$$

and therefore

$$\varepsilon \mu(\{x \in E : |f_\lambda(x)| > \varepsilon\}) \leq q_E([f_\lambda]) \leq \varepsilon \mu(E) + \mu(\{x \in E : |f_\lambda(x)| > \varepsilon\}).$$

The desired conclusion is obtained by using the fact that ε is arbitrary. $\square$

Convergence in measure is closely related to almost everywhere convergence.

Proposition 9.14. *Let $(X, \mathbf{S}, \mu)$ be a measure space, let $\{f_n\}_{n=1}^{\infty} \subset \mathcal{L}^0(\mu)$ be a sequence, and let $f \in \mathcal{L}^0(\mu)$.*

(1) *If $\lim_{n\to\infty} f_n = f$ almost everywhere $[\mu]$, then $\lim_{n\to\infty} q_E([f - f_n]) = 0$ for every $E \in \mathbf{S}$ with $\mu(E) < +\infty$, that is, $\{f_n\}$ converges to f in measure.*
(2) *Assume, conversely, that $\{f_n\}$ converges to f in measure, and μ is σ-finite. Then $\{f_n\}_{n=1}^{\infty}$ has a subsequence which converges to f almost everywhere $[\mu]$.*
(3) *Assume that μ is σ-finite and the sequence $\{f_n\}_{n=1}^{\infty}$ is Cauchy in measure, that is, $\lim_{m,n\to\infty} q_E([f_n - f_m]) = 0$ for every $E \in \mathbf{S}$ with $\mu(E) < +\infty$. Then $\{f_n\}_{n=1}^{\infty}$ has a subsequence which converges to a function $f \in \mathcal{L}^0(\mu)$ almost everywhere $[\mu]$.*

Proof. The first statement follows immediately from the bounded convergence theorem 4.36. Clearly, (2) follows from (3). To prove (3), choose pairwise disjoint sets $\{E_n\}_{n\in\mathbb{N}} \subset \mathbf{S}$ with finite measure such that $X = \bigcup_{n\in\mathbb{N}} E_n$. It suffices to show that there exist sequences $\{f_{m,n}\}_{n\in\mathbb{N}_0}$, $m \in \mathbb{N}$, with the following properties:

(1) $f_{0,n} = f_n$, $n \in \mathbb{N}$,
(2) $\{f_{m+1,n}\}_n$ is a subsequence of $\{f_{m,n}\}_n$, $m \in \mathbb{N}$,
(3) $\{f_{m,n}|E_m\}_n$ converges almost everywhere $[\mu]$ to a measurable function.

Then the diagonal sequence $\{f_{n,n}\}_{n\in\mathbb{N}}$ satisfies the required properties. The construction of these sequences is inductive. Assume that the sequence $\{f_{m,n}\}_{n\in\mathbb{N}}$ has been constructed and use the fact that

$$\lim_{n_1,n_2\to\infty} q_{E_m}(f_{n_1} - f_{n_2}) = 0, \quad m \in \mathbb{N},$$

to choose natural numbers $n_1 < n_2 < \cdots$ such that $\mu(F_k) < 2^{-k}$, where

$$F_k = \{x \in E_{m+1} : |f_{m,n_k}(x) - f_{m,n_{k+1}}(x)| > 2^{-k}\}, \quad k \in \mathbb{N}.$$

Clearly the sequence $\{f_{m,n_k}(x)\}_{k=1}^{\infty}$ is Cauchy if x belongs to only finitely many of the sets F_k. According to Example 4.28, this situation occurs for $[\mu]$-almost every $x \in E$. We can thus define $f_{m+1,k} = f_{m,n_k}$, $k \in \mathbb{N}$, and thereby conclude the proof. $\qquad\square$

Suppose that $(X, \mathbf{S}, \mu)$ is a σ-finite measure space and $\{E_n\}_{n\in\mathbb{N}} \subset \mathbf{S}$ is a sequence satisfying $X = \bigcup_{n\in\mathbb{N}} E_n$ and $0 < \mu(E_n) < +\infty$, $n \in \mathbb{N}$. Then

$$q(u) = \sum_{n=1}^{\infty} \frac{q_{E_n}(u)}{2^n \mu(E_n)}, \quad u \in L^0(\mu),$$

is a single quasinorm on $L^0(\mu)$ which defines the topology of convergence in measure. Part (3) of the preceding proposition shows in this case that the space $L^0(\mu)$ is an F-space when μ is σ-finite, that is, it is complete

relative to this quasinorm. It is interesting to notice that this space also
has a completeness property in the sense of its order structure. A collection
$\mathcal{F} \subset L^0(\mu)$ consisting of real elements is said to have an upper bound u if u
is real, and $u - v$ is positive (or $v \leq u$) for every $v \in \mathcal{F}$. An upper bound u
is the least upper bound of $\mathcal{F}$ if every other upper bound u_1 for $\mathcal{F}$ satisfies
$u \leq u_1$.

Theorem 9.15. *Assume that μ is a σ-finite measure. Then every nonempty
set of real elements $\mathcal{F} \subset L^0(\mu)$ which has an upper bound also has a least
upper bound.*

Proof. We may and do restrict ourselves to finite measures μ. The σ-finite
case follows easily. Assume that $\mathcal{F}$ is a collection of real elements of $L^0(\mu)$
with an upper bound u. Note that $\max\{v_1, v_2\} \leq u$ for every $v_1, v_2 \in \mathcal{F}$.
Thus there is no loss of generality in assuming that $\max\{v_1, v_2\} \in \mathcal{F}$ for all
$v_1, v_2 \in \mathcal{F}$. Replacing $\mathcal{F}$ and u by $\{v - v_0 : v \in \mathcal{F}\}$ and $u - v_0$, respectively, for
some fixed $v_0 \in \mathcal{F}$ we may also assume that $0 \in \mathcal{F}$. Under this assumption,
w is an upper bound for $\mathcal{F}$ if and only if $w \geq v$ for every positive element
$v \in \mathcal{F}$. Therefore we also assume that every $v \in \mathcal{F}$ is positive and that $u - v_0$
is an upper bound of $\mathcal{F}$. Define now a number $\alpha \geq 0$ by setting

$$\alpha = \sup_{[f] \in \mathcal{F}} \int_X \frac{f}{1+f} \, d\mu,$$

and choose a sequence $\{f_n\}_{n=1}^\infty \in \mathcal{L}^0(\mu)$ such that each $[f_n]$ belongs to $\mathcal{F}$ and

$$\alpha = \lim_{n \to \infty} \int_X \frac{f_n}{1+f_n} \, d\mu.$$

Replacing f_n by $\max\{f_1, f_2, \ldots, f_n\}$, we may assume that $f_n \leq f_{n+1}$, and we
can define now

$$g(x) = \lim_{n \to \infty} f_n(x) (\leq u(x) - v_0(x) \text{ almost everywhere } [\mu]).$$

Clearly $g \in \mathcal{L}^0(\mu)$, and we claim that $[g]$ is the least upper bound of $\mathcal{F}$. Indeed,
assume that $[h]$ is an upper bound of $\mathcal{F}$, so $f_n \leq h$ almost everywhere $[\mu]$
for every $n \in \mathbb{N}$. Thus $g \leq h$ almost everywhere $[\mu]$. Once we show that $[g]$
is an upper bound of $\mathcal{F}$, the preceding argument shows that it is the least
upper bound. We conclude by showing that $[g]$ is indeed an upper bound
of $\mathcal{F}$. Indeed, consider an arbitrary element $[f] \in \mathcal{F}$, and set

$$g_1(x) = \lim_{n \to \infty} \max\{f_n(x), f(x)\} (\leq u(x) - v_0(x) \text{ almost everywhere } [\mu]).$$

We see as before that $g_1 \in \mathcal{L}^0(\mu)$, and clearly $g \leq g_1$ almost everywhere $[\mu]$.
Next we observe that

$$\frac{f_n(x)}{1+f_n(x)} \leq \frac{\max\{f_n(x), f(x)\}}{1+\max\{f_n(x), f(x)\}},$$

and the bounded convergence theorem implies that

$$\alpha = \int_X \frac{g(x)}{1+g(x)}\, d\mu(x) \leq \int \frac{g_1(x)}{1+g_1(x)}\, d\mu(x) \leq \alpha,$$

where the last inequality follows because $\max\{[f_n],[f]\} \in \mathcal{F}$. Since $g \leq g_1$ almost everywhere $[\mu]$, we also have $g/(1+g) \leq g_1/(1+g_1)$ almost everywhere $[\mu]$, and the equality

$$\int_X \frac{g(x)}{1+g(x)}\, d\mu(x) = \int \frac{g_1(x)}{1+g_1(x)}\, d\mu(x) = \alpha$$

implies that $[g] = [g_1]$. In particular, $[f] \leq [g]$, and the proposition follows.
□

We turn now to an important family of subspaces of $L^0(\mu)$ for an arbitrary measure space $(X, \mathbf{S}, \mu)$, first introduced by F. Riesz. Fix a number $p \in (0, +\infty)$, and denote by $\mathcal{L}^p(\mu)$ the collection of those functions $f \in \mathcal{L}^0(\mu)$ for which $\int_X |f|^p\, d\mu < +\infty$. Also define

$$L^p(\mu) = \{[f] \in L^0(\mu) : f \in \mathcal{L}^p(\mu)\}.$$

We set

$$\|[f]\|_p = \|f\|_p = \left(\int_X |f|^p\, d\mu \right)^{1/p} < +\infty, \quad f \in \mathcal{L}^p(\mu),$$

and it is convenient to set

$$\|[f]\|_p = \|f\|_p = +\infty, \quad f \in \mathcal{L}^0(\mu) \setminus \mathcal{L}^p(\mu).$$

We observe immediately that $\|f\|_p = 0$ if and only if f vanishes almost everywhere $[\mu]$, and $\|\lambda f\|_p = |\lambda| \|f\|_p$, $\lambda \in \mathbb{C}$, $[f] \in L^p(\mu)$. The number $\|f\|_p$ is called the *p-norm* of f if $p \geq 1$; we show later that this appellation is appropriate. When $p \in (0, 1)$, the map $u \mapsto \|u\|_p$ is not a norm, but $u \mapsto \|u\|_p^p$ is a quasinorm on $L^p(\mu)$. The inequalities (6.8) and (6.9) imply that $L^p(\mu)$ is a vector space for every $p \in (0, +\infty)$. We also define the space $L^\infty(\mu)$ to consist of all elements in $L^0(\mu)$ which can be represented as $[f]$ for some bounded function f. A function $f \in \mathcal{L}^0(\mu)$ is said to be *essentially bounded* if the class $[f]$ belongs to $L^\infty(\mu)$. We denote by $\mathcal{L}^\infty(\mu)$ the collection of all essentially bounded functions $f \in \mathcal{L}^0(\mu)$. Note that f is essentially bounded if and only if

$$\mu(\{x : |f(x)| > M\}) = 0$$

for some $M \in [0, +\infty)$. The set $S = \{M \geq 0 : \mu(\{x : |f(x)| > M\}) = 0\}$ has a minimum element. Indeed, let $M_n \geq 0$ be a decreasing sequence in S converging to $M = \inf S$, and observe that

$$\{x : |f(x)| > M\} = \bigcup_{n=1}^{\infty} \{x : |f(x)| > M_n\}.$$

The minimum element of S is denoted by $\|f\|_\infty = \|[f]\|_\infty$, and it is called the *essential supremum* of $|f|$. If f is not essentially bounded, we set $\|f\|_\infty = +\infty$.

The study of the spaces $L^p(\mu)$ for $p \geq 1$ requires an inequality due to Hölder. For every $p \in (1, +\infty)$ we define the *conjugate exponent* p' by

$$p' = \frac{p}{p-1}.$$

Note that

$$\frac{1}{p} + \frac{1}{p'} = 1,$$

so the conjugate exponent of p' is p. It is also convenient to agree that 1 and $+\infty$ are conjugate exponents.

Lemma 9.16. *Let $(X, \mathbf{S}, \mu)$ be a measure space. For every $f, g \in \mathcal{L}^0(\mu)$ and for every $p \in [1, +\infty]$ we have*

$$\|fg\|_1 \leq \|f\|_p \|g\|_{p'}. \tag{9.3}$$

If $f \in \mathcal{L}^p(\mu)$ and $p \in [1, +\infty)$, then

$$\|f\|_p = \max \left\{ \left| \int_X fg \, d\mu \right| : g \in \mathcal{L}^{p'}(\mu), \|g\|_{p'} \leq 1 \right\}. \tag{9.4}$$

If μ is semifinite and $f \in \mathcal{L}^\infty(\mu)$, we also have

$$\|f\|_\infty = \sup \left\{ \left| \int_X fg \, d\mu \right| : g \in \mathcal{L}^1(\mu), \|g\|_1 \leq 1 \right\}.$$

Proof. We deal first with the inequality (9.3) in the simple case $p = 1$. It is clear that

$$|f(x)g(x)| \leq |f(x)| \|g\|_\infty$$

almost everywhere $[\mu]$, and integration of this inequality yields

$$\|fg\|_1 \leq \|f\|_1 \|g\|_\infty.$$

Assume now that $p \in (1, +\infty)$. The inequality (9.3) is trivial if $\|f\|_p = 0$ or $\|g\|_{p'} = 0$. In the case in which both of these numbers are different from zero, the inequality (9.3) is also trivial if $\|f\|_p$ or $\|g\|_{p'}$ is infinite. Hence we suppose that $\|f\|_p$ and $\|g\|_{p'}$ are finite and different from zero. Replacing f and g by $f/\|f\|_p$ and $g/\|g\|_{p'}$, respectively, we see that it suffices to consider the case in which $\|f\|_p = \|g\|_{p'} = 1$. Here we use the inequality

$$ab \leq \frac{a^p}{p} + \frac{b^{p'}}{p'}, \quad a, b \geq 0,$$

which follows from (6.10) with $t = 1/p$, to see that in this case

$$\|fg\|_1 = \int_X |f(x)||g(x)| \, d\mu(x)$$

$$\leq \frac{1}{p} \int_X |f(x)|^p \, d\mu + \frac{1}{p'} \int_X |g(x)|^{p'} \, d\mu = \frac{1}{p} + \frac{1}{p'} = 1,$$

and this is precisely the stated inequality.

Turning to the proof of (9.4) we observe that (9.3) implies that

$$\|f\|_p \geq \sup \left\{ \left| \int_X fg \, d\mu \right| : g \in \mathcal{L}^{p'}(\mu), \|g\|_{p'} \leq 1 \right\}.$$

So it suffices to show that, given $f \in \mathcal{L}^p(\mu)$, there exists $g \in \mathcal{L}^{p'}(\mu)$ such that $\|g\|_{p'} \leq 1$ and $\int_X fg \, d\mu = \|f\|_p$. When $[f] = 0$ we can take $g = 0$. If $[f] \neq 0$, the function

$$g(x) = \begin{cases} \dfrac{|f(x)|^{p-2}\overline{f(x)}}{\|f\|_p^{p-1}}, & f(x) \neq 0, \\ 0, & \text{otherwise}, \end{cases}$$

is easily seen to do the job.

Finally, assume that μ is semifinite (that is, μ has no infinite atoms) and $\|f\|_\infty > \alpha$. Then there is a set $E \in \mathbf{S}$ such that $0 < \mu(E) < +\infty$ and $|f(x)| > \alpha$ for every $x \in E$. The function $g = \mu(E)^{-1}\chi_E \overline{f}/|f|$ satisfies $\|h\|_1 = 1$ and $\int_X fg \, d\mu > \alpha$. This proves that $\|f\|_\infty$ is at most equal to the supremum in the statement. The opposite inequality follows from (9.3). $\square$

Henceforth we use $u, v, \ldots$ in place of $[f], [g], \ldots$ to denote elements of $L^0(\mu)$ when it is convenient.

Theorem 9.17. *Let $(X, \mathbf{S}, \mu)$ be a measure space. For each $p \in [1, +\infty]$, the map $u \mapsto \|u\|_p$ is a norm on $L^p(\mu)$, and $L^p(\mu)$ endowed with this norm is a Banach space.*

Proof. The fact that $\|u\|_p = 0$ implies $u = 0$ for $u \in L^p(\mu)$ is immediate. The inequality

$$\|u + v\|_1 \leq \|u\|_1 + \|v\|_1, \quad u, v \in L^0(\mu),$$

follows from the standard properties of integrals, while

$$\|[f] + [g]\|_\infty \leq \|[f]\|_\infty + \|[g]\|_\infty, \quad [f], [g] \in L^\infty(\mu),$$

follows because the relations $|f(x)| \leq M$ almost everywhere $[\mu]$ and $|g(x)| \leq N$ almost everywhere $[\mu]$ imply that

$$|f(x) + g(x)| \leq |f(x)| + |g(x)| \leq M + N$$

almost everywhere $[\mu]$. For $p \in (1, +\infty)$, the inequality

$$\|[f] + [g]\|_p \leq \|[f]\|_p + \|[g]\|_p, \quad [f], [g] \in L^p(\mu),$$

follows from (9.4) because

$$\left| \int_X (f + g)h \, d\mu \right| \leq \left| \int_X fh \, d\mu \right| + \left| \int_X gh \, d\mu \right|$$

for every $[h] \in L^{p'}(\mu)$. It remains to prove that $L^p(\mu)$, $p \in [1, +\infty]$, is complete, and for this purpose it suffices to show that a sequence $\{[f_n]\}_{n=1}^\infty \subset L^p(\mu)$ satisfying

$$\|f_n - f_{n+1}\|_p < 2^{-n}, \quad n \in \mathbb{N},$$

converges in the p-norm to an element of $L^p(\mu)$. Indeed, any Cauchy sequence in $L^p(\mu)$ has a subsequence with this stronger property, and the limit of this subsequence is also the limit of the original sequence. We consider first the case $p = +\infty$ when the inequality above implies that

$$|f_n(x) - f_{n+1}(x)| < 2^{-n}, \quad n \in \mathbb{N},$$

almost everywhere $[\mu]$. It follows that the series

$$f_1(x) + \sum_{n=1}^\infty (f_{n+1}(x) - f(x)) \tag{9.5}$$

converges almost everywhere $[\mu]$ to a function $f \in \mathcal{L}^0(\mu)$ satisfying $|f(x)| \leq |f_1(x)| + \sum_{n=1}^\infty 2^{-n} \leq \|f_1\|_\infty + 1$ almost everywhere $[\mu]$, and hence $[f] \in L^\infty(\mu)$. The same calculation shows that $\|f - f_n\|_\infty \leq 2^{-n+1} \to 0$ as $n \to \infty$, which shows that $L^\infty(\mu)$ is complete. Assume now that $p \in [1, +\infty)$ and, returning to the sequence $\{[f_n]\}$, we set

$$h_N(x) = \sum_{n=1}^N |f_{n+1}(x) - f_n(x)|.$$

We have

$$\|h_N\|_p \leq \sum_{n=1}^N \|f_{n+1} - f_n\|_p < 1,$$

and the monotone convergence theorem implies that $h(x) = \sup_N h_N(x) = \lim_N h_N(x)$ is finite almost everywhere $[\mu]$ and $\int_X h^p \, d\mu \leq 1$. Since the series (9.5) is absolutely convergent almost everywhere $[\mu]$, its sum is a function $f \in \mathcal{L}^p(\mu)$ with $\|f\|_p \leq \|f_1\|_p + 1$. Finally we note that $|f - f_n|^p \leq h^p$, $n \in \mathbb{N}$, and $\lim_{n\to\infty}(f(x) - f_n(x)) = 0$ almost everywhere $[\mu]$. The dominated convergence theorem now implies that $\lim_{n\to\infty} \|f - f_n\|_p = 0$, so $L^p(\mu)$ is complete. $\qquad \square$

The triangle inequality $\|f+g\|_p \leq \|f\|_p + \|g\|_p$ for $L^p(\mu)$, $p \geq 1$, is usually called the Minkowski inequality. One can replace addition by integration in this inequality; see Problem 9G. The fact that $L^p(\mu)$, $p \in [1,+\infty]$, is complete is due to Riesz and Fisher.

There is no analog of the Hölder inequality for $p \in (0,1)$, and in fact the map $[f] \mapsto \|f\|_p$ is not even a quasinorm for such values of p. Instead, we have the following result.

Proposition 9.18. *Let $(X,\mathbf{S},\mu)$ be a measure space. Then for each $p \in (0,1)$, the map $u \mapsto \|u\|_p^p$ is a quasinorm on $L^p(\mu)$, and $L^p(\mu)$ endowed with this quasinorm is an F-space.*

Proof. Once we have verified the triangle inequality $\|u+v\|_p^p \leq \|u\|_p^p + \|v\|_p^p$, $u,v \in L^p(\mu)$, $p \in (0,1)$, the proof of completeness proceeds as in Theorem 9.17. The triangle inequality follows from (6.9) by integration, and the relation $\lim_{\lambda \to 0} \|\lambda u\| = 0$ is immediate for $u \in L^p(\mu)$. $\qquad\square$

Example 9.19. Consider an arbitrary index set Γ endowed with the counting measure μ on Γ. The space $L^p(\mu)$ is usually denoted by $\ell^p(\Gamma)$ or, more simply, ℓ^p if $\Gamma = \mathbb{N}$. Since there are no nonempty sets of measure zero, $\ell^0(\Gamma)$ consists of all families $\alpha = (\alpha_\gamma)_{\gamma \in \Gamma}$ of complex numbers, while $\ell^p(\Gamma)$, $p \in (0,+\infty)$, consists of those families $\alpha = (\alpha_\gamma)_{\gamma \in \Gamma}$ for which

$$\|\alpha\|_p = \left(\sum_{\gamma \in \Gamma} |\alpha_\gamma|^p \right)^{1/p},$$

is finite. The space $\ell^\infty(\Gamma)$ consists of the bounded families $\alpha = (\alpha_\gamma)_{\gamma \in \Gamma}$, and the ∞-norm is given by

$$\|\alpha\|_\infty = \sup_{\gamma \in \Gamma} |\alpha_\gamma|.$$

Since summable families must be bounded, we have $\ell^p(\Gamma) \subset \ell^\infty(\Gamma)$ for all $p \in (0,+\infty)$. In fact, more is true. If $0 < p < q < +\infty$ and $\alpha \in \ell^p(\Gamma)$, it follows that $\alpha \in \ell^q(\Gamma)$ and

$$\|\alpha\|_q \leq \|\alpha\|_p.$$

Indeed, by homogeneity, it suffices to prove the inequality when $\|\alpha\|_p = 1$. In this case we must have $|\alpha_\gamma| \leq 1$, $\gamma \in \Gamma$, and therefore $|\alpha_\gamma|^q \leq |\alpha_\gamma|^p$, for all $\gamma \in \Gamma$, so $\|\alpha\|_q \leq 1$.

Example 9.20. Consider now a probability space $(X,\mathbf{S},\mu)$. In this case it is clear that $L^\infty(\mu) \subset L^p(\mu)$ for all $p \in (0,+\infty)$, and $\|f\|_p \leq \|f\|_\infty$ for all $f \in \mathcal{L}^0(\mu)$. More generally, if $0 < p < q < +\infty$, we have $L^q(\mu) \subset L^p(\mu)$ and

$$\|f\|_p \leq \|f\|_q, \quad f \in \mathcal{L}^0(\mu).$$

This is obtained by an application of the Hölder inequality (9.3) for the exponent $r = q/p > 1$:

$$\|f\|_p^p = \int_X |f|^p \cdot 1 \, d\mu \leq \left(\int_X (|f|^p)^r \, d\mu \right)^{1/r} \left(\int_X 1^{r'} \, d\mu \right)^{1/r'} = \|f\|_q^p.$$

Example 9.21. In contrast with the previous two examples, the space $L^p(\lambda_d)$, where λ_d is Lebesgue measure on $\mathbb{R}^d$, is not contained in $L^q(\lambda_d)$ if $q \neq p$; see Problem 9H.

Example 9.22. For an arbitrary measure space $(X, \mathbf{S}, \mu)$ and an arbitrary exponent $p \in (0, +\infty]$, the space $L^p(\mu)$ is contained in $L^0(\mu)$ by definition, and the inclusion is continuous. Thus, a sequence $(u_n)_{n=1}^\infty$ in $L^p(\mu)$ which converges to $u \in L^p(\mu)$, that is $\lim_{n \to \infty} \|u_n - u\|_p = 0$, also converges in measure to u. This is seen as follows. Let $f : X \to \mathbb{C}$ be a measurable function, and let $E \in \mathbf{S}$ be a set with $\mu(E) < +\infty$. When $p \in (0, 1]$ we have $\min\{|f|, 1\} \leq |f|^p$ so

$$q_E(f) = \int_E \min\{|f|, 1\} \, d\mu \leq \int_E |f|^p \, d\mu \leq \|f\|_p^p,$$

while for $p \in (1, +\infty]$ we have

$$q_E(f) \leq \int_E |f| \, d\mu \leq \left(\int_E |f|^p \, d\mu \right)^{1/p} \left(\int_X \chi_E^{p'} \, d\mu \right)^{1/p'} \leq \|f\|_p \mu(E)^{1/p'}$$

by the Hölder inequality (9.3).

Corollary 9.23. *Let $(X, \mathbf{S}, \mu)$ be a measure space, let $p \in (0, +\infty]$, and let $f_n : X \to \mathbb{C}$ be measurable functions in $\mathcal{L}^p(\mu)$ for $n \in \mathbb{N}$. If $\lim_{n \to \infty} \|f_n\|_p = 0$ then $(f_n)_{n=1}^\infty$ has a subsequence which converges to zero almost everywhere $[\mu]$.*

Proof. The statement is obvious when $p = +\infty$. When $p \in (0, +\infty)$ the result follows from the preceding example and Proposition 9.14(2) once we show that the measure $\mu|N$ is σ-finite, where $N = \bigcup_n N_{f_n}$. This is seen by noting that

$$N(f_n) = \bigcup_{k \in \mathbb{N}} \{x \in X : f_n(x) \geq 1/k\}$$

and $\mu(\{x \in X : f_n(x) \geq 1/k\}) \leq k^p \|f_n\|_p^p$ for $n, k \in \mathbb{N}$. $\quad\square$

The simple functions play an important role in the study of the spaces $L^p(\mu)$, as they do everywhere in integration theory.

Proposition 9.24. *For every measure space $(X, \mathbf{S}, \mu)$ and every $p \in (0, +\infty]$ the linear manifold of simple functions in $L^p(\mu)$, defined by*

$$\{[f] : f \in \mathcal{L}^p(\mu) \text{ is a simple function}\},$$

is dense in $L^p(\mu)$.

Proof. Fix $p \in (0, +\infty]$. It suffices to show that positive elements in $L^p(\mu)$ can be approximated by simple functions. Let $[f]$ be such an element. Redefining f on a null set, we may assume that $f : X \to [0, +\infty)$. Consider the simple functions $\{s_n\}_{n=1}^{\infty}$ constructed in Proposition 2.36. If $p = +\infty$, f can be assumed to be bounded, in which case $s_n \to f$ uniformly on X; in particular $\lim_{n \to \infty} \|f - s_n\|_{\infty} = 0$. If $p < +\infty$, we have $0 \leq s_n \leq f$, so $\|s_n\|_p < +\infty$. Moreover, $|f - s_n|^p \leq f^p$ and the equality $\lim_{n \to \infty} \|f - s_n\|_p = 0$ follows from the dominated convergence theorem. $\square$

The preceding result allows us in many cases to prove results about functions in $\mathcal{L}^p(\mu)$ by verifying them for characteristic functions and then using linearity and density. In some situations we also need to use other dense linear manifolds in $L^p(\mu)$.

Example 9.25. Fix an integer $d \geq 1$ and a set $A \subset \mathbb{R}^d$ which is measurable $[\lambda_d]$. The functions χ_{A_n}, where $A_n = A \cap \{x \in \mathbb{R}^d : \|x\|_d \leq n\}$, satisfy

$$\lim_{n \to \infty} \|\chi_{A_n} - \chi_A\|_p = 0, \quad p \in (0, +\infty).$$

It follows that equivalence classes of simple functions f for which N_f is bounded form a dense linear manifold in $L^p(\lambda_d)$.

Example 9.26. Let $A \subset \mathbb{R}^d$, where $d \in \mathbb{N}$, be a bounded Borel set, and let ε be a positive real number. The regularity of λ_d (Proposition 7.10) implies the existence of a compact set K and of an open set G such that $K \subset A \subset G$ and $\lambda_d(G \setminus K) < \varepsilon$. Urysohn's lemma implies the existence of a function $f \in \mathcal{C}_c(\mathbb{R}^d)$ with the property that $\chi_K \leq f \leq \chi_G$. We have

$$\|\chi_A - f\|_p \leq \|\chi_{G \setminus K}\|_p < \varepsilon^{1/p}, \quad p \in (0, +\infty),$$

and therefore χ_A can be approximated arbitrarily well in the p-norm by continuous functions with compact support. Thus $\{[f] : f \in \mathcal{C}_c(\mathbb{R}^d)\}$ is a dense linear manifold in $L^p(\lambda_d)$, $p \in (0, +\infty)$. The reader will have no difficulty extending this observation to arbitrary regular Borel measures on locally compact Hausdorff spaces that assign finite measure to compact sets.

The Hölder inequality (9.3) allows us to produce continuous linear functionals on the space $L^p(\mu)$ when $p \in [1, +\infty]$. Given an element $[h] \in L^{p'}(\mu)$, define $\varphi_{[h]} = \varphi_h : L^p(\mu) \to \mathbb{C}$ by setting

$$\varphi_h([f]) = \int_X f h \, d\mu, \quad [f] \in L^p(\mu). \tag{9.6}$$

When $p \in (1, +\infty]$, the equality (9.4) with p in place of p' shows that

$$\|\varphi_h\| = \|h\|_{p'}, \quad [h] \in L^{p'}(\mu). \tag{9.7}$$

Under favorable circumstances, all functionals on $L^p(\mu)$, $p \in [1, +\infty)$ are of the form φ_h for some $[h] \in L^{p'}(\mu)$, thus allowing us to identify the dual $L^p(\mu)^*$ with $L^{p'}(\mu)$.

Proposition 9.27. *Assume that $(X, \mathbf{S}, \mu)$ is a σ-finite measure space and $p \in [1, +\infty)$. Then the map $u \mapsto \varphi_u$ is an isometric linear map from $L^{p'}(\mu)$ onto the dual space $L^p(\mu)^*$.*

Proof. We first verify that (9.7) is also true when $p = 1$. Indeed, assume that $u = [h] \in L^\infty(\mu)$ and $\|u\|_\infty = \|h\|_\infty > \alpha \geq 0$. Since μ is σ-finite, there exists a set $E \in \mathbf{S}$ such that $0 < \mu(E) < +\infty$ such that $|h(x)| > \alpha$ for all $x \in E$. Then the function

$$f(x) = \begin{cases} \dfrac{\overline{h(x)}}{\mu(E)|h(x)|}, & x \in E, \\ 0, & x \in X \setminus E, \end{cases}$$

is integrable $[\mu]$, $\|f\|_1 = 1$, and $\int_X f h \, d\mu = \mu(E)^{-1} \int_E |h| \, d\mu > \alpha$. Thus

$$\|u\|_\infty \leq \sup\{|\varphi_u([f])| : \|f\|_1 \leq 1\} = \|\varphi_u\|.$$

The opposite inequality follows from (9.3). It remains to prove that the map $u \mapsto \varphi_u$ is surjective. We assume first that μ is a finite measure, which implies that $[\chi_E] \in L^p(\mu)$ for every $E \in \mathbf{S}$. Given a functional $\varphi \in L^p(\mu)^*$, define a set function $\nu : \mathbf{S} \to \mathbb{C}$ by setting $\nu(E) = \varphi([\chi_E])$, $E \in \mathbf{S}$. We show that ν is countably additive. Indeed, consider a countable union $E = \bigcup_{n=1}^\infty E_n$ of pairwise disjoint measurable sets, and observe that

$$\nu(E) - \sum_{n=1}^N \nu(E_n) = \varphi([\chi_{\bigcup_{n=N+1}^\infty E_n}])$$

$$\leq \|\varphi\| \|\chi_{\bigcup_{n=N+1}^\infty E_n}\|_p = \|\varphi\| \left[\sum_{n=N+1}^\infty \mu(E_n) \right]^{1/p}.$$

Countable additivity of ν follows by letting $N \to +\infty$. Clearly $\nu(E) = 0$ if $\mu(E) = 0$. The Radon-Nikodým Theorem 6.10 combined with Proposition 6.19 yields a measurable function $h : X \to \mathbb{C}$ such that

$$\varphi([\chi_E]) = \nu(E) = \int_E h \, d\mu = \int_X \chi_E h \, d\mu, \quad E \in \mathbf{S}.$$

It follows that

$$\varphi([s]) = \int_X s h \, d\mu = \varphi_h([s])$$

for every simple function $s \in \mathcal{L}^0(\mu)$. Define

$$E_n = \{x \in X : |h(x)| \leq n\},$$

so $[\chi_{E_n} h] \in L^{p'}(\mu)$ for $n \in \mathbb{N}$. We have then

$$|\varphi_{\chi_{E_n} h}([s])| = \left| \int_X s \chi_{E_n} h \, d\mu \right| = |\varphi([s \chi_{E_n}])| \leq \|\varphi\| \|s\|_p$$

for any simple function s. Proposition 9.24 implies that $\|\varphi_{\chi_{E_n} h}\| \leq \|\varphi\|$ and therefore $\|\chi_{E_n} h\|_{p'} \leq \|\varphi\|$. The monotone convergence theorem (or, in case $p' = +\infty$, the fact that $\bigcup_{n=1}^\infty E_n = X$) implies now that $\|h\|_{p'} < +\infty$, and therefore that the functional φ_h is defined and continuous on $L^p(\mu)$. Since $\varphi([s]) = \varphi_h([s])$ for simple functions s, we conclude by Proposition 9.24 that $\varphi = \varphi_h$.

We consider now the general case of a σ-finite measure μ. There is a partition $X = \bigcup_{n=1}^\infty X_n$ such that $X_n \in \mathbf{S}$ and $\mu(X_n) < +\infty$ for $n \in \mathbb{N}$. The special case just proved can be applied to the subspaces $L^p(\mu)$ consisting of elements of the form $[\chi_{X_n} f]$ with $f \in \mathcal{L}^p(\mu)$, and it yields measurable functions $h_n : X \to \mathbb{C}$ such that $\|h_n\|_{p'} < +\infty$, $h_n = \chi_{X_n} h_n$, and

$$\varphi([\chi_{X_n} f]) = \int_X f h_n \, d\mu, \quad f \in \mathcal{L}^p(\mu).$$

Define now a measurable function $h : X \to \mathbb{C}$ by setting $h(x) = h_n(x)$ for $x \in X_n$. Set $Y_N = \bigcup_{n=1}^N X_n$, and note that

$$\varphi([\chi_{Y_N} f]) = \varphi_{\chi_{Y_N}}([f]), \quad f \in \mathcal{L}^p(\mu), N \in \mathbb{N}.$$

We conclude that $\|\chi_{Y_N} h\|_{p'} \leq \|\varphi\|$ for $N \in \mathbb{N}$, and letting $N \to +\infty$ yields $\|h\|_{p'} \leq \|\varphi\|$. Finally, we observe that $\lim_{N \to \infty} \|f - \chi_{Y_N} f\|_p = 0$, $f \in \mathcal{L}^p(\mu)$, by the dominated convergence theorem, and thus

$$\varphi([f]) = \lim_{N \to \infty} \varphi([\chi_{Y_N} f]) = \lim_{N \to \infty} \varphi_h([\chi_{Y_N} f]) = \varphi_h([f]),$$

which concludes the proof of the proposition. $\qquad\qquad\qquad\qquad\qquad$ $\square$

Note incidentally that the preceding argument provides another proof of the completeness of $L^{p'}(\mu)$ when μ is σ-finite because all duals of normed spaces are complete.

Example 9.28. Let $(X, \mathbf{S}, \mu)$ be an arbitrary measure space. Relation (9.3) shows that the dual space of $L^\infty(\mu)$ contains a linearly isometric copy of $L^1(\mu)$, but this dual is generally much larger. Thus, in general, one cannot identify $L^1(\mu)$ with the dual of $L^\infty(\mu)$. In fact, it turns out that the space $L^1(\lambda_1)$ cannot be identified with the dual of *any* Banach space. On the other hand ℓ^1 can be identified with the dual of another sequence space, namely the space c_0 consisting of those sequences $\alpha = (\alpha_n)_{n \in \mathbb{N}}$ which converge to zero, endowed with the ∞-norm. Under this norm, c_0 is a Banach space which also cannot be identified with the dual of any Banach space. Proofs of the facts stated in this example are sketched in Problems 9D, 9W, and 9X.

For an arbitrary measure space $(X, \mathbf{S}, \mu)$, the space $L^2(\mu)$ occupies a special place because its norm is given by a *scalar product*.

Definition 9.29. Let $\mathcal{V}$ be an arbitrary complex vector space. A *scalar product* (or *inner product*) on $\mathcal{V}$ is a map $\langle \cdot, \cdot \rangle : \mathcal{V} \times \mathcal{V} \to \mathbb{C}$ with the following properties:

(1) The map $x \mapsto \langle x, y \rangle$ is linear for every $y \in \mathcal{V}$,
(2) $\overline{\langle x, y \rangle} = \langle y, x \rangle$, $x, y \in \mathcal{V}$, and
(3) $\langle x, x \rangle > 0$, $x \neq 0$.

A complex Banach space $(\mathcal{V}, \|\cdot\|)$ is called a *complex Hilbert space* if there exists a scalar product $\langle \cdot, \cdot \rangle$ on $\mathcal{V}$ such that $\langle x, x \rangle = \|x\|^2$, $x \in \mathcal{V}$.

Example 9.30. For every $d \in \mathbb{N}$, the Euclidean space $\mathbb{C}^d$ is a (complex) Hilbert space with scalar product given by

$$\langle x, y \rangle = \sum_{j=1}^{d} \xi_j \overline{\eta_j}, \quad x = (\xi_1, \dots, \xi_d), y = (\eta_1, \dots, \eta_d) \in \mathbb{C}^d.$$

For an arbitrary measure space $(X, \mathbf{S}, \mu)$, the Banach space $L^2(\mu)$ is a (complex) Hilbert space under the scalar product

$$\langle [f], [g] \rangle = \langle f, g \rangle = \int_X f \overline{g} \, d\mu, \quad f, g \in \mathcal{L}^2(\mu).$$

The Hölder inequality (9.16) for $p = 2$ takes the form

$$|\langle [h], [g] \rangle| \leq \|[h]\|_2 \|[g]\|_2, \quad [h], [g] \in L^2(\mu),$$

and it is referred to as the *Schwarz inequality*. In particular, ℓ^2 is a Hilbert space. Given an arbitrary complex Hilbert space $\mathcal{V}$, the following easily verified *polarization identity*

$$\langle x, y \rangle = \frac{1}{4} \sum_{k=0}^{3} i^k \|x + i^k y\|^2, \quad x, y \in \mathcal{V}, \tag{9.8}$$

shows that the scalar product on $\mathcal{V}$ is uniquely determined by the norm. The Schwarz inequality is valid in an arbitrary complex Hilbert space $\mathcal{V}$, that is,

$$|\langle x, y \rangle| \leq \|x\| \|y\|, \quad x, y \in \mathcal{V}. \tag{9.9}$$

This is easily seen by observing that $\langle x - \lambda y, x - \lambda y \rangle \geq 0$, $x, y \in \mathcal{V}, \lambda \in \mathbb{C}$, and setting $\lambda = \langle x, y \rangle / \langle y, y \rangle$ if $y \neq 0$.

Given two vectors x, y in a complex Hilbert space $\mathcal{V}$, we say that x is orthogonal (or perpendicular) to y, and we write $x \perp y$, if $\langle x, y \rangle = 0$. Note that

$$\|x + y\|_2^2 = \|x\|_2^2 + \|y\|_2^2$$

whenever $x \perp y$. A collection of vectors $\{e_\gamma\}_{\gamma \in \Gamma} \subset \mathcal{V}$ is said to be *orthonormal* if $\|e_\gamma\|_2 = 1$ for all $\gamma \in \Gamma$ and $\langle e_\gamma, e_{\gamma'} \rangle = 0$ if $\gamma \neq \gamma'$. An orthonormal set which is maximal (relative to inclusion) is called an *orthonormal basis* for $\mathcal{V}$. It is clear (from Zorn's lemma) that every orthonormal set in $\mathcal{V}$ is a subset of an orthonormal basis in $\mathcal{V}$.

Remark 9.31. An orthonormal basis for a Hilbert space $\mathcal{V}$ is not a Hamel basis of $\mathcal{V}$ unless $\mathcal{V}$ is finite dimensional.

Proposition 9.32. *Let $\mathcal{V}$ be a complex Hilbert space and let $\{e_\gamma\}_{\gamma \in \Gamma}$ be an orthonormal set in $\mathcal{V}$. Then:*

(1) *For every vector $x \in \mathcal{V}$ the Bessel inequality*

$$\|x\|^2 \geq \sum_{\gamma \in \Gamma} |\langle x, e_\gamma \rangle|^2 \tag{9.10}$$

holds. If $\{e_\gamma\}_{\gamma \in \Gamma}$ is an orthonormal basis for $\mathcal{V}$, we have equality in (9.10) and, more generally, Parseval's identity holds:

$$\langle x, y \rangle = \sum_{\gamma \in \Gamma} \langle x, e_\gamma \rangle \overline{\langle y, e_\gamma \rangle}, \quad x, y \in \mathcal{V}.$$

(2) *Conversely, given a family $(\alpha_\gamma)_{\gamma \in \Gamma} \in \ell^2(\Gamma)$, the sum $\sum_{\gamma \in \Gamma} \alpha_\gamma e_\gamma$ converges unconditionally to a vector x, that is, for every $\varepsilon > 0$ there exists a finite set $F \subset \Gamma$ such that*

$$\left\| x - \sum_{\gamma \in G} \alpha_\gamma e_\gamma \right\| < \varepsilon$$

for every finite set G such that $F \subset G \subset \Gamma$. The vector x satisfies $\langle x, e_\gamma \rangle = \alpha_\gamma$, $\gamma \in \Gamma$.

Proof. For an arbitrary vector $x \in \mathcal{V}$ and every finite subset $F \subset \Gamma$ we set

$$x_F = \sum_{\gamma \in F} \langle x, e_\gamma \rangle e_\gamma.$$

It is immediate that

$$\langle x_F, e_\gamma \rangle = \begin{cases} \langle x, e_\gamma \rangle, & \gamma \in F, \\ 0, & \gamma \in \Gamma \setminus F, \end{cases}$$

and this implies that $x - x_F \perp x_F$. Thus

$$\|x\|_2^2 = \|x - x_F\|_2^2 + \|x_F\|_2^2 \geq \|x_F\|_2^2 = \sum_{\gamma \in F} |\langle x, e_\gamma \rangle|^2.$$

This implies (9.10) since F is arbitrary.

Assume now that $(\alpha_\gamma)_{\gamma \in \Gamma} \in \ell^2(\Gamma)$, and define for each finite subset $F \subset \Gamma$,

$$x_F = \sum_{\gamma \in F} \alpha_\gamma e_\gamma.$$

We have

$$\|x_{F_1} - x_{F_2}\|_2^2 = \sum_{\gamma \in F_1 \triangle F_2} |\alpha_\gamma|^2,$$

for any two finite subsets F_1 and F_2 of Γ, and the summability of the family $\{|\alpha_\gamma|^2\}_{\gamma \in \Gamma}$ implies that the net $\{x_F\}$ satisfies the Cauchy property. The vector $x = \lim_F x_F$ exists because $L^2(\mu)$ is complete. Moreover, $\langle x_F, e_\gamma \rangle = \alpha_\gamma$ if $\gamma \in F$, and (2) follows by passing to the limit.

We return now to the proof of the remainder of (1). Suppose therefore that $\{e_\gamma\}_{\gamma \in \Gamma}$ is an orthonormal basis for $\mathcal{V}$ and $x \in \mathcal{V}$. The Bessel inequality and (2) imply that the sum $x' = \sum_{\gamma \in \Gamma} \langle x, e_\gamma \rangle e_\gamma$ converges unconditionally and $\langle x - x', e_\gamma \rangle = 0$ for all $\gamma \in \Gamma$. The vector $x - x'$ must then be zero, for otherwise the normalized vector $(x - x')/\|x - x'\|_2$ could be adjoined to the collection $\{e_\gamma\}_{\gamma \in \Gamma}$ to form a larger orthonormal set. Let $y \in \mathcal{V}$ be another vector, and calculate the scalar product

$$\langle x, y \rangle = \left\langle \sum_{\gamma \in \Gamma} \langle x, e_\gamma \rangle e_\gamma, \sum_{\gamma \in \Gamma} \langle y, e_\gamma \rangle e_\gamma \right\rangle$$

$$= \lim_F \left\langle \sum_{\gamma \in F} \langle x, e_\gamma \rangle e_\gamma, \sum_{\gamma \in F} \langle y, e_\gamma \rangle e_\gamma \right\rangle$$

$$= \lim_F \sum_{\gamma \in F} \langle x, e_\gamma \rangle \overline{\langle y, e_\gamma \rangle} = \sum_{\gamma \in \Gamma} \langle x, y_\gamma \rangle \overline{\langle y, e_\gamma \rangle},$$

where the second limit is taken along the net determined by the finite subsets $F \subset \Gamma$. This is precisely Parseval's identity. $\qquad\square$

Part (2) of the theorem shows that, given an orthonormal basis $\{e_\gamma\}_{\gamma \in \Gamma}$ for a complex Hilbert space $\mathcal{V}$, there exists a linear isometric bijection between $\mathcal{V}$ and $\ell^2(\Gamma)$. Note that $\|e_\gamma - e_{\gamma'}\|_2 = \sqrt{2}$, $\gamma \neq \gamma'$, so $\mathcal{V}$ is not separable if Γ is uncountable. On the other hand, the set of linear combinations with rational coefficients of finitely many basis elements of $\mathcal{V}$ is dense in $\mathcal{V}$, and therefore $\mathcal{V}$ is separable if Γ is at most countable. We conclude that if $\mathcal{V}$ is separable and infinite dimensional, then it is linearly isometric to $\ell^2 = \ell^2(\mathbb{N})$.

Example 9.33. Denote by μ the restriction of λ_1 to the interval $(0, 1)$. For each $n \in \mathbb{Z}$, the function $e_n(t) = e^{2\pi i n t}$, $t \in (0, 1)$, belongs to $L^2(\mu)$ and

$$\int_{(0,1)} e_n \, d\mu = \int_0^1 e^{2\pi i n t} \, dt = \begin{cases} 1, & n = 0, \\ 0, & n \neq 0. \end{cases}$$

Observe that $e_n\overline{e_m} = e_{n-m}$, $n, m \in \mathbb{Z}$, so the above calculation shows that $\{e_n\}_{n\in\mathbb{Z}}$ is an orthonormal set in $L^2(\mu)$. The set $\{e_n\}_{n\in\mathbb{Z}}$ is in fact an orthonormal basis in $L^2(\mu)$. To see this, recall from Example 9.26 that $\mathcal{C}_c((0,1))$ provides a dense linear manifold in $L^2(\mu)$, so it suffices to show that every $f \in \mathcal{C}_c((0,1))$ can be approximated in the 2-norm by linear combinations of the functions $\{e_n\}_{n\in\mathbb{Z}}$. The Weierstrass approximation theorem (see Problem 4CC) shows that every such function f can be approximated uniformly on $(0,1)$ by linear combinations of the functions $\{e_n\}_{n\in\mathbb{Z}}$, and uniform convergence implies convergence in the 2-norm because

$$\|u\|_2 \leq \|u\|_\infty, \quad u \in L^2(\mu).$$

We conclude that every function $f \in \mathcal{L}^2((0,1))$ has a *Fourier expansion*

$$f = \sum_{n=-\infty}^{+\infty} c_n e_n, \quad c_n = \int_0^1 e^{-2\pi i n t} f(t)\, dt, \quad n \in \mathbb{Z},$$

in the sense that

$$\lim_{N\to\infty} \left\| f - \sum_{n=-N}^{N} c_n e_n \right\|_2 = 0 = \lim_{F} \left\| f - \sum_{n\in F} c_n e_n \right\|_2,$$

where the limit is taken along the net determined by the finite subsets F of $\mathbb{Z}$.

Example 9.34. Fix an integer $d \in \mathbb{N}$ and denote by μ the restriction of λ_d to the unit cube $[0,1]^d$. Given $n = (n_1, n_2, \ldots, n_d) \in \mathbb{Z}^d$ and $x = (\xi_1, \xi_2, \ldots, \xi_d) \in \mathbb{R}^d$, we denote by

$$n \cdot x = \sum_{j=1}^{d} n_j \xi_j$$

the ordinary scalar product of α and n. The functions $\{e_n\}_{n\in\mathbb{Z}^d}$ defined by

$$e_n(x) = e^{2\pi i n \cdot x}, \quad x \in \mathbb{R}^d,$$

form an orthonormal system in $L^2(\mu)$. The Weierstrass approximation theorem (see Problem 4BB) implies that the set of all linear combinations of the functions e_n is uniformly dense in the set of all continuous functions $f : \mathbb{R}^d \to \mathbb{C}$ with the property that

$$f(x + n) = f(x), \quad x \in \mathbb{R}^d, n \in \mathbb{Z}^d.$$

It follows (see also the following example) that $\{e_n\}_{n\in\mathbb{Z}^d}$ is, in fact, an orthonormal basis in $L^2(\mu)$. In particular, every function $f \in L^2(\mu)$ can be written as a multiple series

$$f = \sum_{n \in \mathbb{Z}^d} \langle f, e_n \rangle e_n \qquad (9.11)$$

which converges unconditionally in the 2-norm. The numbers

$$\widehat{f}(n) = \langle f, e_n \rangle = \int_{[0,1]^d} f(x) e^{-2\pi i n \cdot x} \, d\lambda_d(x), \quad n \in \mathbb{Z}^d,$$

are called the *Fourier coefficients* of the function f, and they are defined even for functions $f \in L^1(\mu)$. The expansion (9.11) and its generalizations are the subject of the following chapter.

Example 9.35. The spaces $L^p(\lambda_d)$ are separable for $p \in (0, +\infty)$. One way to see this is to show that the linear manifold generated by characteristic functions of cells of the form $P = \prod_{i=1}^{d} (r_i, s_i)$ with $r_i, s_i \in \mathbb{Q}$ for $i = 1, 2, \ldots, d$ is dense. It is also easy to see that $C_c(\mathbb{R}^d)$ is a dense linear manifold in $L^p(\lambda_d)$, $p < +\infty$. Indeed, with P as above, χ_P is approximated arbitrarily well in the $L^p(\lambda_d)$ by continuous functions f satisfying $\chi_P \leq f \leq \chi_Q$ where Q is a cell with the same center as P but with slightly larger sides. These arguments fail for $p = \infty$ and, in fact, $L^\infty(\lambda_d)$ is not separable as can be seen by noting that $\|\chi_E - \chi_F\|_\infty = 1$ if $\lambda_d(E \triangle F) \neq 0$.

Proposition 9.36. *Fix an integer $d \in \mathbb{N}$ and an element $[f] \in L^p(\lambda_d)$, where $p \in (0, +\infty)$. Denote by*

$$f_x(y) = f(y - x), \quad y \in \mathbb{R}^d,$$

the translate of f by $x \in \mathbb{R}^d$. Then the map $x \mapsto [f_x]$ is continuous from $\mathbb{R}^d$ to $L^p(\lambda_d)$.

Proof. It is easily seen that $\|f_x\|_p = \|f\|_p$ for all $x \in \mathbb{R}^d$. Suppose that $f \in C_c(\mathbb{R}^d)$, that is, f is continuous on $\mathbb{R}^d$ and has compact support. Then $\lim_{x \to x_0} \|f_x - f_{x_0}\|_\infty = 0$ because f is uniformly continuous on $\mathbb{R}^d$. For a fixed $x_0 \in \mathbb{R}^d$ and for $\|x - x_0\|_{\mathbb{R}^d} < 1$, the functions f_x vanish outside some compact set K independent of x. It follows that

$$\|f_x - f_{x_0}\|_p \leq \lambda_d(K)^{1/p} \|f_x - f_{x_0}\|_\infty$$

for $\|x - x_0\|_{\mathbb{R}^d} < 1$, and the conclusion follows in this case. Proposition 9.6 applied to the linear transformations $f \mapsto f_x$, $x \in \mathbb{R}^d$, implies that the collection consisting of those elements $[f] \in L^p(\lambda_d)$ for which the desired conclusion holds is a closed linear manifold in $L^p(\lambda_d)$. The proposition follows because $\{[f] : f \in C_c(\mathbb{R}^d)\}$ is dense in $L^p(\lambda_d)$. $\qquad \square$

Example 9.37. Let $(X, \mathbf{S}, \mu)$ be an arbitrary measure space. We have seen in Example 8.13 that for any $f \in \mathcal{L}^0(\mu)$ and $p \in (0, +\infty)$ we can calculate the p-norm of f by the formula

$$\|f\|_p^p = \int_0^\infty pt^{p-1}\mu(\{x : |f(x)| > t\})\,dt.$$

Consider the Borel measure ν on $(0, +\infty)$ defined by $d\nu(t) = dt/t$, and define

$$h(t) = t\mu(\{x : |f(x)| > t\})^{1/p}, \quad t > 0.$$

The above equation can then be written as

$$\|f\|_p = p^{1/p}\|h\|_p,$$

where $\|h\|_p$ is calculated relative to the measure ν. This observation leads naturally to a refinement of the spaces L^p. Namely, fix $p \in (0, +\infty)$, define the function h as above, and set

$$\|f\|_{p,q} = q^{1/p}\|h\|_q, \quad q \in (0, +\infty),$$

$$\|f\|_{p,\infty} = \|h\|_\infty.$$

The *Lorentz space* $L^{p,q}(\mu)$ consists of those classes $[f] \in L^0(\mu)$ with the property that $\|f\|_{p,q} < +\infty$. Note that the map $[f] \mapsto \|f\|_{p,q}$ is not generally a norm on $L^{p,q}(\mu)$. Clearly, $\|f\|_p = \|f\|_{p,p}$ so $L^{p,p}(\mu) = L^p(\mu)$. The space $L^{p,\infty}(\mu)$ is usually called *weak* $L^p(\mu)$ because it is a 'slightly' larger space. To see this, we recall the Tchebysheff inequality (Problem 4C):

$$\mu(\{x : |f(x)| > t\}) = \int_{\{x:|f(x)|>t\}} 1\,d\mu$$

$$\leq \int_{\mu(\{x:|f(x)|>t\})} \frac{|f|^p}{t^p}\,d\mu \leq \frac{\|f\|_p^p}{t^p}, \quad t > 0.$$

This yields immediately that $t\mu(\{x : |f(x)| > t\})^{1/p} \leq \|f\|_p$, that is, $\|f\|_{p,\infty} \leq \|f\|_p$. For some values of p and q, a closely related expression does define a norm on $L^{p,q}(\mu)$.

The weak L^1 space appears in many estimates, for instance in the study of maximal functions which we now introduce.

Definition 9.38. Given $d \in \mathbb{N}$ and a function $f \in \mathcal{L}^0(\lambda_d)$, the *maximal function* $Mf : \mathbb{R}^d \to [0, +\infty]$ is defined by

$$(Mf)(x) = \sup_{\{B:x\in B\}} \frac{1}{\lambda_d(B)} \int_B |f|\,d\lambda_d, \quad x \in \mathbb{R}^d,$$

where the supremum is taken over all open balls B in $\mathbb{R}^d$ containing x.

There is an obvious connection between the differentiation Theorem 6.31 and the maximal function since

$$\left|\frac{1}{\lambda_d(B)} \int_B f\,d\lambda_d\right| \leq \frac{1}{\lambda_d(B)} \int_B |f|\,d\lambda_d \leq (Mf)(x)$$

when $x \in B$. More generally, many convergence results can be obtained by showing that the relevant limit is dominated by the maximal function.

Note that $\{x \in \mathbb{R}^d : (Mf)(x) > \alpha\}$ is an open set for every $\alpha > 0$, and therefore Mf is measurable $[\lambda_d]$. It is clear that

$$M(f + g) \leq Mf + Mg, \quad M(tf) = tM(f), \quad f, g \in \mathcal{L}^0(\lambda_d), t \geq 0,$$

that is, M is *subadditive* and *positive homogeneous* on $\mathcal{L}^0(\lambda_d)$. If f is essentially bounded, then Mf is bounded and

$$\|Mf\|_\infty \leq \|f\|_\infty.$$

Example 9.39. Consider the function $f = \chi_{(-1,1)}$ on $\mathbb{R}$. The maximal function of f can be calculated explicitly:

$$(Mf)(x) = \begin{cases} 1, & x \in (-1, 1), \\ 2/(1 + x), & x > 1, \\ 2/(1 - x), & x < -1. \end{cases}$$

This maximal function is not integrable $[\lambda_1]$, but $\|Mf\|_{1,\infty} = 4$.

The following result was first proved by Hardy and Littlewood when $d = 1$.

Theorem 9.40. *For every $d \in \mathbb{N}$ and every $f \in \mathcal{L}^1(\lambda_d)$ we have $[Mf] \in L^{1,\infty}(\lambda_d)$, and*

$$\|Mf\|_{1,\infty} \leq 3^d \|f\|_1.$$

Proof. Fix $t > 0$ and let $K \subset \{x \in \mathbb{R}^d : (Mf)(x) > t\}$ be a compact set. The collection of those open balls B with the property that

$$\frac{1}{\lambda_d(B)} \int_B |f| \, d\lambda_d > t$$

covers K. Let $B_1, B_2, \ldots, B_N$ be the balls constructed in Lemma 6.30. We have

$$\lambda_d(K) \leq 3^n \sum_{j=1}^N \lambda_d(B_j) \leq \frac{3^n}{t} \sum_{j=1}^N \int_{B_j} |f| \, d\lambda_d \leq \frac{3^n}{t} \|f\|_1,$$

and therefore $t\lambda_d(\{x : (Mf)(x) > t\}) \leq 3^n \|f\|_1$. These inequalities for $t > 0$ yield the desired conclusion. $\qquad \square$

The estimate just proved shows that the maximal operator is of weak type $(1, 1)$ in the sense of the following definition.

Definition 9.41. Let $(X, \mathbf{S}, \mu)$ and $(Y, \mathbf{T}, \nu)$ be measure spaces, let $\mathcal{D} \subset \mathcal{L}^0(\mu)$ be a linear manifold, and let $T : \mathcal{D} \to \mathcal{L}^0(\nu)$ be an arbitrary map. We say that T is of *(strong) type* (p, q), $p, q \in (0, +\infty]$, if there exists a constant

$c \geq 0$ such that $\|Tf\|_q \leq c\|f\|_p$ for every $f \in \mathcal{D}$. We say that T is of *weak type* (p,q), $p \in (0, +\infty]$, $q \in (0, +\infty)$, if there exists a constant $c \geq 0$ such that $\|Tf\|_{q,\infty} \leq \|f\|_p$ for every $f \in \mathcal{D}$.

In order to make some statements uniform, it is convenient to agree that weak type (p, ∞) is the same as strong type (p, ∞), $p \in (0, +\infty]$. The Marcinkiewicz interpolation theorem produces strong type inequalities at intermediate indices from weak type estimates at the endpoints of an interval. For the statement, we require the notation

$$f^{(\alpha)}(x) = \begin{cases} f(x), & |f(x)| > \alpha, \\ 0, & \text{otherwise,} \end{cases}$$

and $f_{(\alpha)} = f - f^{(\alpha)}$ for the truncations of a function $f \in \mathcal{L}^0(\mu)$ at some level $\alpha > 0$. These functions satisfy the inequalities

$$\|f^{(\alpha)}\|_{p_0} \leq \alpha^{1 - \frac{p}{p_0}} \|f\|_p^{\frac{p}{p_0}}$$

and

$$\|f_{(\alpha)}\|_{p_1} \leq \alpha^{1 - \frac{p}{p_1}} \|f\|_p^{\frac{p}{p_1}}$$

provided that $0 < p_0 < p < p_1 < +\infty$. These inequalities follow immediately because

$$|f(x)|^{p_0} = |f(x)|^p |f(x)|^{p_0 - p} \leq \alpha^{p_0 - p} |f(x)|^p$$

if $|f(x)| > \alpha$ and

$$|f_{(\alpha)}(x)|^{p_1} \leq \alpha^{p_1 - p} |f(x)|^p$$

if $|f(x)| \leq \alpha$.

Theorem 9.42. (Marcinkiewicz) *Let* $(X, \mathbf{S}, \mu)$ *and* $(Y, \mathbf{T}, \nu)$ *be measure spaces, let* $\mathcal{D} \subset \mathcal{L}^0(\mu)$ *be a linear manifold, and let* $T : \mathcal{D} \to \mathcal{L}^0(\nu)$ *be an arbitrary map. Assume that* $f^{(\alpha)}$ *and* $f_{(\alpha)}$ *belong to* $\mathcal{D}$ *for every* $f \in \mathcal{D}$ *and every* $\alpha > 0$, *and that* T *is subadditive, that is,*

$$|T(f + g)| \leq |T(f)| + |T(g)| \quad \text{almost everywhere } [\mu], \quad f, g \in \mathcal{D}.$$

If T *is of weak type* (p_0, q_0) *and of weak type* (p_1, q_1), *where* $0 \leq p_0 \leq q_0 \leq +\infty$, $0 \leq p_1 \leq q_1 \leq +\infty$ *and* $q_0 \neq q_1$, *then* T *is of strong type* (p_τ, q_τ) *for each* $\tau \in (0,1)$, *where*

$$\frac{1}{p_\tau} = \frac{1-\tau}{p_0} + \frac{\tau}{p_1}, \frac{1}{q_\tau} = \frac{1-\tau}{q_0} + \frac{\tau}{q_1}, \quad \tau \in (0,1).$$

Proof. We give complete details under the additional assumptions that $p_0 \neq p_1$, and $q_0 \neq +\infty \neq q_1$. The special cases when $p_0 = p_1$ or one of the numbers q_0, q_1 is infinite are sketched in the problems. Fix $\tau \in (0,1)$ and write $p = p_\tau, q = q_\tau$. We assume without loss of generality that $p_0 < p_1$ and

$$\|T(f)\|_{q_0,\infty} \le c_0\|f\|_{p_0}, \|T(f)\|_{q_1,\infty} \le c_1\|f\|_{p_1,\infty}, \quad f \in \mathcal{D}.$$

In other words,

$$\nu(\{y : |(Tf)(y)| > t\}) \le c_j^{q_j} \frac{\|f\|_{p_j}^{q_j}}{t^j}, \quad f \in \mathcal{D}, j = 0, 1.$$

We need to prove that there exists $c > 0$ such that

$$\|T(f)\|_q \le c\|f\|_p, \quad f \in \mathcal{D}. \tag{9.12}$$

Start with a function $f \in \mathcal{D} \cap \mathcal{L}^p(\mu)$, and recall that

$$\|T(f)\|_q^q = q \int_0^{+\infty} t^{q-1} \nu(\{y : |(Tf)(y)| > t\}) \, dt. \tag{9.13}$$

Given $s > 0$, we have $|T(f)| \le |T(f^{(s)})| + |T(f_{(s)})|$, so

$$\nu(\{y : |(Tf)(y)| > t\}) \le \nu(\{y : |T(f^{(s)})| > \frac{t}{2}\}) + \nu(\{y : |T(f_{(s)})| > \frac{t}{2}\})$$

$$\le (2c_0)^{q_0} t^{-q_0} \|f^{(s)}\|_{p_0}^{q_0} + (2c_1)^{q_1} t^{-q_1} \|f_{(s)}\|_{p_1}^{q_1}.$$

It is useful to use different cutoff levels for different values of t. More specifically, we use the value

$$s = s(t) = s_1 t^\gamma,$$

where

$$\gamma = \frac{p_1(q - q_1)}{q_1(p - p_1)} = \frac{p_0(q - q_0)}{q_0(p - p_0)} \tag{9.14}$$

and s_1 is a positive constant. The rationale for these choices is made clear later. We have

$$\|T(f)\|_q^q \le (2c_0)^{q_0} q \int_0^{+\infty} t^{q-q_0-1} \|f^{(s(t))}\|_{p_0}^{q_0} \, dt \tag{9.15}$$

$$+ (2c_1)^{q_1} q \int_0^{+\infty} t^{q-q_0-1} \|f_{(s(t))}\|_{p_1}^{q_1} \, dt,$$

and we need to estimate the two integrals in terms of $\|f\|_p$. We consider first the case in which $q_0 < q_1$, when we must have $\gamma > 0$, $s(t)$ is monotone increasing, and $q_0 < q < q_1$. The integral

$$\int_0^{+\infty} t^{q-q_0-1} \|f^{(s(t))}\|_{p_0}^{q_0} \, dt$$

can be written as $\|g_0\|_{r_0}^{r_0} = \int_0^{+\infty} g_0(t)^{r_0} \, d\rho_0(t)$, where

$$g_0(t) = \|f^{(s(t))}\|_{p_0}^{p_0}, \quad d\rho_0(t) = t^{q-q_0-1} \, dt, \quad r_0 = q_0/p_0 > 1.$$

By Lemma 9.16, we have

$$\|g_0\|_{r_0} = \int_0^{+\infty} t^{q-q_0-1} h(t) g_0(t)\, dt$$

for some Borel measurable function $h : (0, +\infty) \to \mathbb{R}$ with the property that

$$\|h\|_{r_0'}^{r_0'} = \int_0^{+\infty} h(t)^{r_0'}\, d\rho_0(t) \leq 1,$$

where $r_0' = q_0/(q_0 - p_0)$ is the conjugate exponent of r_0. Note that

$$s^{-1}(t) = (t/s_1)^{1/\gamma},$$

so the Fubini-Tonelli Theorem 8.4 yields

$$\begin{aligned}
\|g_0\|_{r_0} &= \int_0^{+\infty} t^{q-q_0-1} h(t) \left[\int_{\{x: |f(x)| > s(t)\}} |f(x)|^{p_0}\, d\mu(x) \right] dt \\
&= \int_X |f(x)|^{p_0} \left[\int_0^{+\infty} \chi_{(0, s_1^{-1/\gamma} |f(x)|^{1/\gamma})} h(t)\, d\rho(t) \right] d\mu(x).
\end{aligned}$$

We apply the Hölder inequality applied to the inner integral to obtain

$$\begin{aligned}
\int_0^{+\infty} \chi_{(0, s_1^{-1/\gamma} |f(x)|^{1/\gamma})} h(t)\, d\rho(t) &\leq \|\chi_{(0, s_1^{-1/\gamma} |f(x)|^{1/\gamma})}\|_{r_0} \|h\|_{r_0'} \\
&\leq \frac{s_1^{(q_0 - q)/\gamma r_0} |f(x)|^{(q-q_0)/\gamma r_0}}{(q - q_0)^{1/r_0}},
\end{aligned}$$

and therefore

$$\|g_0\|_{r_0} \leq \frac{s_1^{(q_0-q)/\gamma}}{q - q_0} \int_X |f(x)|^{p_0 + (q-q_0)/\gamma r_0}\, d\mu(x) = \frac{s_1^{(q_0-q)/\gamma}}{q - q_0} \|f\|_p^p$$

by the choice (9.14) of γ. Thus the first term on the right side of (9.15) is

$$(2c_0)^{q_0} q \int_0^{+\infty} t^{q-q_0-1} \|f^{(s(t))}\|_{p_0}^{q_0}\, dt = (2c_0)^{q_0} q \|g\|_{r_0}^{r_0}$$

$$\leq (2c_0)^{q_0} q \frac{s_1^{(q_0-q)/\gamma}}{q - q_0} \|f\|_p^{r_0 p}.$$

Similarly, to estimate the second term we write

$$\int_0^{+\infty} t^{q-q_1-1} \|f_{(s(t))}\|_{p_1}^{q_1}\, dt = \int_0^{+\infty} g_1(t)^{r_1}\, d\rho_1(t) = \|g_1\|_{r_1}^{r_1},$$

where

$$g_1(t) = \|f_{(s(t))}\|_{p_1}^{p_1}, \quad d\rho_1(t) = t^{q-q_1-1}\,dt, \quad r_1 = q_1/p_1 > 1.$$

The quantity $\|g_1\|_{r_1}$ is evaluated the same way, with $\chi_{(s_1^{-1/\gamma}|f(x)|^{1/\gamma},+\infty)}$ in place of $\chi_{(0,s_1^{-1/\gamma}|f(x)|^{1/\gamma})}$, and this yields

$$(2c_0)^{q_0}q\int_0^{+\infty} t^{q-q_1-1}\|f_{(s(t))}\|_{p_1}^{q_1}\,dt = (2c_1)^{q_1}q\|g_1\|_{r_1}^{r_1}$$

$$\le (2c_1)^{q_1}q\frac{s_1^{(q_1-q)/\gamma}}{q_1-q}\|f\|_p^{r_1p},$$

and thus

$$\|T(f)\|_q^q \le (2c_0)^{q_0}q\frac{s_1^{(q_0-q)/\gamma}}{q-q_0}\|f\|_p^{r_0p} + (2c_1)^{q_1}q\frac{s_1^{(q_1-q)/\gamma}}{q_1-q}\|f\|_p^{r_1p}. \qquad (9.16)$$

We can now set $s_1 = \|f\|_p^\eta$ to obtain the desired inequality (9.12) provided that

$$\frac{(q_0-q)\eta}{\gamma} + r_0p = \frac{(q_1-q)\eta}{\gamma} + r_1p = q.$$

A simple calculation shows that

$$\eta = \gamma p\frac{r_1-r_0}{q_1-q_0}$$

satisfies these requirements.

It remains to consider the case in which $q_0 > q_1$. In this case we have $\gamma < 0$ so the function $s(t)$ is decreasing, and thus one must interchange the intervals $(0, s_1^{-1/\gamma}|f(x)|^{1/\gamma})$ and $[s_1^{-1/\gamma}|f(x)|^{1/\gamma}, +\infty)$ in the preceding argument. One obtains the same inequality as (9.16) with $q_0 - q$ and $q - q_1$ in place of $q - q_0$ and $q_1 - q$, respectively. Strong type (p,q) is obtained with the same choice of the constant s_1. $\qquad\square$

Corollary 9.43. *Let d be a positive integer and let $p \in (1,+\infty)$. There exists a constant $c = c(d,p)$ such that*

$$\|M(f)\|_p \le c\|f\|_p, \quad f \in \mathcal{L}^0(\lambda_d),$$

where $M(f)$ denotes, as usual, the maximal function.

Proof. Denote by $\mathcal{D}$ the linear manifold in $\mathcal{L}^0(\lambda_d)$ consisting of measurable functions defined on the entire $\mathbb{R}^d$. Then $M : \mathcal{D} \to \mathcal{D}$ is subadditive, it has weak type $(1,1)$ by Theorem 9.40, and clearly

$$\|M(f)\|_\infty \le \|f\|_\infty, \quad f \in \mathcal{D}.$$

The desired result follows immediately by interpolation. $\qquad\square$

The inequalities $p_0 \leq q_0$ and $p_1 \leq q_1$ were necessary in the proof of Theorem 9.42 because of the use of Hölder's inequality for the exponents $r_0 = q_0/p_0$ and $r_1 = q_1/p_1$. These restrictions can be removed under the assumption that T is linear and of strong type (p_0, q_0) and (p_1, q_1), although we need $p_0, p_1 \geq 1$. We require a lemma from complex analysis which we present without proof (see Problem 9EE).

Lemma 9.44. *Set $\Omega = \{x + iy : 0 < x < 1, y \in \mathbb{R}\}$, and let $u : \overline{\Omega} \to \mathbb{C}$ be a bounded continuous function such that the restriction $u|\Omega$ is analytic. If $|u(iy)| \leq 1$ and $|u(1 + iy)| \leq 1$ for every $y \in \mathbb{R}$ then $|u(z)| \leq 1$ for every $z \in \Omega$.*

Theorem 9.45. (M. Riesz and G. O. Thorin) *Let $(X, \mathbf{S}, \mu)$ and $(Y, \mathbf{T}, \nu)$ be measure spaces, let $\mathcal{D} \subset \mathcal{L}^0(\mu)$ be a linear manifold containing all simple integrable functions on X, and let $T : \mathcal{D} \to \mathcal{L}^0(\nu)$ be a linear map. Assume that $f^{(\alpha)}$ and $f_{(\alpha)}$ belong to $\mathcal{D}$ for every $f \in \mathcal{D}$ and every $\alpha > 0$. If T is of strong type (p_0, q_0) and of strong type (p_1, q_1), where $1 \leq p_0, q_0 \leq +\infty$, $1 \leq p_1, q_1 \leq +\infty$, and $q_0 \neq q_1$, then T is of strong type (p_τ, q_τ) for each $\tau \in (0, 1)$, where*

$$\frac{1}{p_\tau} = \frac{1 - \tau}{p_0} + \frac{\tau}{p_1}, \frac{1}{q_\tau} = \frac{1 - \tau}{q_0} + \frac{\tau}{q_1}, \quad \tau \in (0, 1).$$

More precisely, if

$$\|T(f)\|_{q_j} \leq c_j \|f\|_{p_j}, \quad f \in \mathcal{D}, j = 0, 1, \tag{9.17}$$

then

$$\|T(f)\|_{q_\tau} \leq c_0^{1-\tau} c_1^\tau \|f\|_{p_\tau}, \quad f \in \mathcal{D}, \tau \in (0, 1). \tag{9.18}$$

Proof. We argue first that it suffices to prove the inequality (9.18) when f is a simple function. Given an arbitrary function $f \in \mathcal{D} \cap \mathcal{L}^{p_\tau}$, we find integrable simple functions $(s_n)_{n=1}^\infty$ such that $\lim_{n \to \infty} \|f - s_n\|_{p_\tau} = 0$ and $\lim_{n \to \infty} T(s_n) = T(f)$ almost everywhere $[\nu]$. Then the Fatou Lemma (Theorem 4.29) implies

$$\|T(f)\|_{q_\tau} \leq \limsup_{n \to \infty} \|T(s_n)\|_{q_\tau}$$

and (9.18) follows from the corresponding inequality with s_n in place of f. The construction of s_n is routine and the details are sketched in Problem 9FF. We also assume that the exponents p_0, p_1, q_0, q_1 are all finite and leave to the reader the easy adaptation of the argument to the case in which one or more of these equals $+\infty$.

Consider simple functions

$$f = \sum_{k=1}^M \lambda_k \chi_{E_k} \in \mathcal{L}^0(\mu), \quad g = \sum_{\ell=1}^N \rho_\ell \chi_{F_\ell} \in \mathcal{L}^0(\nu),$$

with nonzero complex coefficients λ_k, ρ_ℓ, pairwise disjoint collections $\{E_k\}_{k=1}^N$ in $\mathbf{S}$, $\{F_\ell\}_{\ell=1}^N$ in $\mathbf{T}$, and

$$\sum_{k=1}^{M} |\lambda_k| \mu(E_k) = \sum_{\ell=1}^{N} |\rho_\ell| \nu(F_\ell) = 1.$$

Define $\alpha_\zeta, \beta_\zeta \in \mathbb{C}$ by

$$\alpha_\zeta = \frac{1-\zeta}{p_o} + \frac{\zeta}{p_1}, \quad \beta_\zeta = \frac{1-\zeta}{q_0'} + \frac{\zeta}{q_1'}, \quad \zeta \in \mathbb{C},$$

and observe that the functions

$$f_\zeta = \sum_{k=1}^{M} \frac{\lambda_k}{|\lambda_k|} |\lambda_k|^{\alpha_\zeta} \chi_{E_k}, \quad g_\zeta = \sum_{\ell=1}^{N} \frac{\rho_\ell}{|\rho_\ell|} |\rho_\ell|^{\beta_\zeta} \chi_{F_\ell}, \quad \zeta \in \mathbb{C},$$

belong to $\mathcal{D}$ by hypothesis. The function $u : \mathbb{C} \to \mathbb{C}$ defined by

$$u(\zeta) = \frac{1}{k_0^{1-\zeta} k_1^\zeta} \int_Y g_\zeta T(f_\zeta) \, d\nu =$$

$$= \frac{1}{k_0^{1-\zeta} k_1^\zeta} \sum_{k=1}^{M} \sum_{\ell=1}^{N} |\lambda_k|^{\alpha_\zeta} |\rho_\ell|^{\beta_\zeta} \int_Y \chi_{F_\ell} T(\chi_{E_k}) \, d\nu, \quad \zeta \in \mathbb{C},$$

is clearly analytic. If $\zeta = \tau + i\sigma$ for some $\tau \in [0,1]$ we have

$$|f_\zeta| = \sum_{k=1}^{M} |\lambda_k|^{\Re\alpha_\zeta} \chi_{E_k} = \sum_{k=1}^{M} |\lambda_k|^{1/p_\tau} \chi_{E_k},$$

and thus $\|f_\zeta\|_{p_\tau} = 1$. Analogously, $\|g_\zeta\|_{q_\tau'} = 1$, $\zeta = \tau + i\sigma$, $\tau \in [0,1]$. The hypothesis implies that $|u(\zeta)| \leq 1$ if $\Re\zeta \in \{0,1\}$, and Lemma 9.44 yields the inequality $|u(\zeta)| \leq 1$ when $\Re\zeta \in (0,1)$. In particular, if $\zeta = \tau \in (0,1)$ we obtain

$$\left| \int_Y g_\tau T(f_\tau) \, d\nu \right| \leq k_0^{1-\tau} k_1^\tau. \tag{9.19}$$

Every simple function $h \in \mathcal{L}^{p_\tau}(\mu)$ with $\|h\|_{p_\tau} = 1$ can be written as $h = f_\tau$ for some simple function f satisfying $\|f\|_1 = 1$, and an analogous observation about $\mathcal{L}^{q_\tau'}(\nu)$ shows that (9.19) implies that

$$\left| \int_Y g T(f) \, d\nu \right| \leq k_0^{1-\tau} k_1^\tau,$$

when $f \in \mathcal{L}^0(\mu)$ and $g \in \mathcal{L}^0(\nu)$ are simple and satisfy $\|f\|_{p_\tau} = \|g\|_{q_\tau'} = 1$. Replacing f by $f/\|f\|_{p_\tau}$ for an arbitrary simple function f,

$$\left| \int_Y g T(f) \, d\nu \right| \le k_0^{1-\tau} k_1^\tau \|f\|_{p_\tau}$$

when $f \in \mathcal{L}^0(\mu)$ and $g \in \mathcal{L}^0(\nu)$ are simple and satisfy $\|g\|_{q'_\tau} = 1$. Lemma 9.16 implies now the desired inequality

$$\|T(f)\|_{q_\tau} \le k_0^{1-\tau} k_1^\tau \|f\|_{p_\tau}$$

for simple functions $f \in \mathcal{L}^0(\mu)$. $\square$

Problems

9A. Let I be an uncountable set, and endow $\mathcal{V} = \mathbb{C}^I$ with the family of seminorms $\{q_i\}_{i \in I}$ defined by $q_i((\lambda_j)_{j \in I}) = |\lambda_i|$ for $i \in I$. Show that $0 \in \mathcal{V}$ does not have a countable neighborhood base in the topology defined by these seminorms. In particular $\mathcal{V}$ is not metrizable. Determine the dual space $\mathcal{V}^*$ of $\mathcal{V}$.

9B. Find a sequence of functions $\{f_n\}_{n=1}^\infty \subset \mathcal{L}^0(\mathbb{R}, \mathbf{B}_\mathbb{R}, \lambda_1)$ which converges in measure to zero, but has the property that the sequence $\{f_n(t)\}_{n=1}^\infty$ does not converge for any $t \in \mathbb{R}$.

9C. Prove an analog of Theorem 9.15 where L^0 is replaced by L^p with $p \in (0, +\infty)$ and the measure μ is not assumed to be σ-finite.

9D. Denote by c_0 the vector subspace of ℓ^∞ consisting of those sequences in ℓ^∞ which converge to zero.

  (i) Show that c_0, endowed with the ∞-norm, is a Banach space.
  (ii) Given $x = (\xi_n)_{n=1}^\infty \in \ell^1$, define a functional $\varphi_x \in (c_0)^*$ by setting

  $$\varphi_x((\zeta_n)_{n=1}^\infty) = \sum_{n=1}^\infty \xi_n \zeta_n, \quad (\zeta_n)_{n=1}^\infty \in c_0.$$

  Show that φ_x is indeed continuous, and $\|\varphi_x\| = \|x\|_1$. Also show that every functional $\varphi \in (c_0)^*$ is of the form $\varphi = \varphi_x$ for some $x \in \ell^1$. Thus ℓ^1 can be identified with the dual of a Banach space. (In fact ℓ^1 can be identified with the dual of many quite distinct Banach spaces.)
  (iii) Fix $p \in (0, 1)$. Given $x = (\xi_n)_{n=1}^\infty \in \ell^\infty$, define a functional $\varphi_x \in (\ell^p)^*$ by setting

  $$\varphi_x((\zeta_n)_{n=1}^\infty) = \sum_{n=1}^\infty \xi_n \zeta_n, \quad (\zeta_n)_{n=1}^\infty \in \ell^p.$$

  Verify that φ_x is continuous and show that every continuous linear functional on ℓ^p is of this form.

9E. (Day) Consider the restriction μ of λ_1 to the interval $[0, 1]$, and fix $p \in (0, 1)$. Show that there is no nonzero continuous linear functional on the F-space $L^p(\mu)$. Prove the same statement for the space $L^0(\mu)$ with the topology of convergence in measure given by the quasinorm $q([f]) = \int_{[0,1]} \min\{1, |f|\} \, d\mu$, $f \in \mathcal{L}^0(\mu)$. (Hint: If φ is a continuous linear functional, then the set $\{u \in L^p(\mu) : |\varphi(u)| < 1\}$ is convex and it must contain $V_\varepsilon = \{u \in L^p(\mu) : \|u\|_p < \varepsilon\}$ for some $\varepsilon > 0$. Every element in $L^p(\mu)$ can be written as $(u_1 + u_2 + \cdots + u_n)/n$ with $u_1, \ldots, u_n \in V_\varepsilon$ for sufficiently large n.)

9F. Let $(X, \mathbf{S}, \mu)$ be a measure space.

(i) Let φ be a continuous linear functional on $L^\infty(\mu)$. Define a function $\nu : \mathbf{S} \to \mathbb{C}$ by setting $\nu(E) = \varphi([\chi_E])$ for $E \in \mathbf{S}$. Show that ν is finitely additive, that $\mu(E) = 0$ implies $\nu(E) = 0$, and that ν has finite total variation. More precisely,

$$\sum_{j=1}^{n} |\nu(E_j)| \leq \|\varphi\|$$

for every finite collection $(E_j)_{j=1}^{n}$ of pairwise disjoint sets in $\mathbf{S}$.

(ii) Conversely, let $\nu : \mathbf{S} \to \mathbb{C}$ be a finitely additive set function of finite variation c such that $\mu(E) = 0$ implies $\nu(E) = 0$. Show that there exists $\varphi \in L^\infty(\mu)^*$ such that $\varphi([\chi_E]) = \nu(E)$ for all $E \in \mathbf{S}$ and $\|\varphi\| = c$.

(iii) Consider the case in which $X = \mathbb{N}$ and ν is the counting measure. Let $\mathcal{F}$ be a family of subsets of $\mathbb{N}$ which is closed under the formation of finite intersections, does not contain any finite set, and is maximal subject to these conditions. Define $\nu : 2^{\mathbb{N}} \to [0, +\infty)$ by setting $\nu(E) = 1$ for $E \in \mathcal{F}$ and $\nu(E) = 0$ for $E \notin \mathcal{F}$. Show that there exists a functional $\varphi \in (\ell^\infty)^*$ of norm one such that $\varphi(\chi_E) = \nu(E)$ for all $E \in 2^{\mathbb{N}}$. This functional φ is not represented by any sequence in ℓ^1. (Hint: Show that for every set $A \subset \mathbb{N}$ that does not belong to $\mathcal{F}$ there exists a set $B \in \mathcal{F}$ such that $A \cap B = \varnothing$.)

9G. Consider two σ-finite measure spaces $(X, \mathbf{S}, \mu)$, $(Y, \mathbf{T}, \nu)$, and let $f : X \times Y \to [0, +\infty)$ be measurable $\mathbf{S} \times \mathbf{T}$. Prove the following extension of Minkowski's inequality:

$$\left[\int_X \left(\int_Y f(x, y) \, d\nu(y) \right)^p d\mu(x) \right]^{1/p} \leq \int_Y \left[\int_X f(x, y)^p \, d\mu(x) \right]^{1/p} d\nu(y).$$

(Hint: Use Lemma 9.16. The usual Minkowski inequality corresponds to the case where ν is the counting measure on a set with two elements.)

9H. Show that $L^p(\lambda_1) \not\subset L^q(\lambda_1)$ if $p, q \in (0, +\infty]$ and $p \neq q$.

9I. Let $(X, \mathbf{S}, \mu)$ and $(Y, \mathbf{T}, \nu)$ be two σ-finite measure spaces. Given functions $f \in \mathcal{L}^0(\mu)$ and $g \in \mathcal{L}^0(\nu)$, define a function $f \otimes g \in \mathcal{L}^0(\mu \times \nu)$ by setting

$$(f \otimes g)(x, y) = f(x)g(y), \quad x \in X, y \in Y.$$

(i) Show that $\|f \otimes g\|_p = \|f\|_p \|g\|_p$ for all $f \in \mathcal{L}^0(\mu)$, $g \in \mathcal{L}^0(\nu)$, and for every $p \in (0, +\infty]$. Here we employ the usual conventions that $0 \cdot (+\infty) = 0$ and $(+\infty) \cdot (+\infty) = +\infty$.

(ii) Show that the linear manifold generated by $\{[f \otimes g] : f \in \mathcal{L}^p(\mu), g \in \mathcal{L}^p(\nu)\}$ is dense in $L^p(\mu \times \nu)$ if $p \in (0, +\infty)$. For $p = +\infty$, this linear manifold is dense in the weak* topology on $L^\infty(\mu \times \nu)$ given by the identification of $L^\infty(\mu \times \nu)$ with $L^1(\mu \times \nu)^*$.

(iii) Let $\{f_\alpha\}_{\alpha \in A}$ and $\{g_\beta\}_{\beta \in B}$ be orthonormal bases in $L^2(\mu)$ and $L^2(\nu)$, respectively. Prove that $\{f_\alpha \otimes g_\beta\}_{(\alpha, \beta) \in A \times B}$ is an orthonormal basis in $L^2(\mu \times \nu)$.

9J. Let $\mathcal{V}$ be a complex Hilbert space, and let $\mathcal{M} \subset \mathcal{V}$ be a closed subspace. Denote by $\mathcal{M}^\perp = \{f \in L^2(\mu) : \langle f, g \rangle = 0, g \in \mathcal{M}\}$ the *orthogonal complement* of $\mathcal{M}$. Show that $\mathcal{M} \cap \mathcal{M}^\perp = \{0\}$ and that $\mathcal{M} + \mathcal{M}^\perp = \mathcal{V}$. (Hint: Choose an orthonormal basis $\{f_\alpha\}_{\alpha \in A}$ for $\mathcal{M}$ and find by maximality an orthonormal set $\{g_\beta\}_{\beta \in B}$ such that $\{f_\alpha\}_{\alpha \in A} \cup \{g_\beta\}_{\beta \in B}$ is an orthonormal basis for $\mathcal{V}$. Then $\{g_\beta\}_{\beta \in B}$ is an orthonormal basis for $\mathcal{M}^\perp$.)

9K. Let $(X, \mathbf{S})$ be a measurable space, and let μ and ν be two equivalent, σ-finite measures on $\mathbf{S}$ (Definition 6.15). Denote by $\varphi = d\nu/d\mu$ the Radon-Nikodým derivative of ν with respect to μ. Show that the map $[f] \mapsto [f\varphi^{1/p}]$ is a linear isometry from $L^p(\nu)$ onto $L^p(\mu)$ for each $p \in (0, +\infty)$. (When $p \in (0, 1)$, the isometry is relative to the metric determined by the quasinorm $f \mapsto \|f\|_p^p$. For $p = +\infty$, we have $L^\infty(\mu) = L^\infty(\nu)$.)

9L. Let $(X, \mathbf{S}, \mu)$ be a measure space, let $\mathbf{T} \subset \mathbf{S}$ be a σ-algebra such that $\mu|\mathbf{T}$ is σ-finite, and fix $p \in [1, +\infty)$.

(i) Let $f, g : X \to \mathbb{C}$ be integrable $[\mu]$ and such that $\|f\|_p < +\infty$, $\|g\|_{p'} < +\infty$, and g is measurable $[\mathbf{T}]$. Show that $\mathbb{E}[gf|\mathbf{T}] = g\mathbb{E}[f|\mathbf{T}]$ almost everywhere $[\mu]$.
(ii) Let $h : X \to \mathbb{C}$ be integrable $[\mu]$. Show that $\|\mathbb{E}[h|\mathbf{T}]\|_p \le \|h\|_p$ for all $p \in [1, +\infty]$. (Hint: Use (i) and Lemma 9.16.)

9M. Let $p, p_1, p_2, \ldots, p_n \in (0, +\infty]$ satisfy the identity

$$\frac{1}{p} = \sum_{j=1}^{n} \frac{1}{p_j},$$

where we use the usual convention $1/+\infty = 0$. Consider a measure space $(X, \mathbf{S}, \mu)$ and functions $f_j \in \mathcal{L}^{p_j}(\mu)$ for $j = 1, 2, \ldots, n$. Show that the product $f = f_1 f_2 \cdots f_n$ belongs to $\mathcal{L}^p(\mu)$ and

$$\|f\|_p \le \|f_1\|_{p_1} \|f_2\|_{p_2} \cdots \|f_n\|_{p_n}.$$

9N. Let $(X, \mathbf{S}, \mu)$ be a σ-finite measure space, and let $p, q \in (0, +\infty]$ satisfy $p < q$.

(i) Show that
$$L^p(\mu) \cap L^q(\mu) \subset L^r(\mu) \subset L^p(\mu) + L^q(\mu)$$
for every $r \in (p, q)$.
(ii) If $f \in \mathcal{L}^p(\mu) \cap \mathcal{L}^q(\mu)$, the function $r \mapsto \|f\|_r$ is continuous on the interval $[p, q]$. (For $q = +\infty$, continuity at q means that $\lim_{r \to +\infty} \|f\|_r = \|f\|_\infty$.)

9O. Given $d \in \mathbb{N}$, a function $f \in \mathcal{L}^0(\lambda_d)$, and a vector $x \in \mathbb{R}^d$ denote, as usual, by f_x the translate of f by x, that is, $f_x(y) = f(y - x)$, $y \in \mathbb{R}^d$. Fix $p \in (0, +\infty)$ and two functions $f, g \in \mathcal{L}^p(\lambda_d)$. Prove that

$$\lim_{\|x\|_{\mathbb{R}^d} \to +\infty} \|f_x + g\|_p^p = \|f\|_p^p + \|g\|_p^p.$$

9P. Define a function $h : \mathbb{R} \to \mathbb{R}$ by setting

$$h(t) = \begin{cases} 1, & 0 \le t < \frac{1}{2}, \\ -1, & \frac{1}{2} \le t < 1, \\ 0, & t \in \mathbb{R} \setminus [0, 1). \end{cases}$$

Define the functions $h_{j,k}(t) = 2^{j/2} h(2^j(t - k))$ for $j, k \in \mathbb{Z}$ and $t \in \mathbb{R}$. Show that the vectors $\{[h_{j,k}]\}_{j,k \in \mathbb{Z}}$ form an orthonormal basis for $L^2(\lambda_1)$. Moreover, those functions $h_{j,k}$ which are supported in $[0, 1]$ provide an orthonormal basis for the space $\{[f] \in L^2(\lambda_1|[0,1]) : \int_0^1 f \, d\lambda_1 = 0\}$.

9Q. Exhibit a finite measure space $(X, \mathbf{S}, \mu)$ and a function $f \in \mathcal{L}^\infty(\mu)$ with the property that $|\int_X fh \, d\mu| < \|f\|_\infty$ for every $h \in \mathcal{L}^1(\mu)$ such that $\|h\|_1 \le 1$.

9R. Consider a measure space $(X, \mathbf{S}, \mu)$ and a sequence $\{[g_n]\}_{n \in \mathbb{N}} \subset L^\infty(\mu)$ which converges to zero in measure.

(i) Suppose in addition that $\sup_{n \in \mathbb{N}} \|g_n\|_\infty < +\infty$. Show that $\lim_{n \to \infty} \int_X f g_n \, d\mu = 0$ for every $f \in \mathcal{L}^1(\mu)$. (In other words, $\{[g_n]\}_n$ tends to zero in the weak* topology of $L^\infty(\mu)$ when this space can be identified with the dual of $L^1(\mu)$.)

(ii) Show that the conclusion of (i) need not hold if the norms $\|g_n\|_\infty$ are not bounded.

9S. Let $\{e_n\}_{n \in \mathbb{N}}$ be an orthonormal sequence in a complex Hilbert space $\mathcal{V}$. Show that 0 belongs to the closure of the set $A = \{e_n + n e_m : n, m \in \mathbb{N}\}$ in the weak topology. Show, moreover, that there is no sequence $\{x_n\}_{n \in \mathbb{N}} \subset A$ which converges to zero weakly. (This proves that the weak topology is not metrizable on $\mathcal{V}$. Hint: Observe that $\{e_n\}$ converges weakly to 0. Use uniform boundedness.)

9T. Consider the space ℓ^1 of absolutely convergent series of complex numbers.

(i) Show that 0 belongs to the closure of the set $\{x \in \ell^1 : \|x\|_1 = 1\}$ in the weak topology.

(ii) (Schur) Let $\{x_n\}_{n=1}^\infty \subset \ell^1$ be a sequence which converges weakly to 0. Show that $\lim_{n \to \infty} \|x_n\|_1 = 0$. (Thus the unit ball of ℓ^1 fails to be metrizable in the weak topology of ℓ^1.)

9U. Let $(X, \mathbf{S}, \mu)$ be a σ-finite measure space, and fix $p \in (0, +\infty)$.

(i) Show that $L^p(\mu)$ is a separable space if and only if the following condition is satisfied: there exists a sequence $\{E_n\}_{n \in \mathbb{N}} \subset \mathbf{S}$ such that for every $F \in \mathbf{S}$ we have $\mu(E \triangle F) = 0$ for some set in the σ-algebra generated by the collection $\{E_n\}_{n \in \mathbb{N}}$. If $L^p(\mu)$ is separable, show that $L^0(\mu)$ is separable when given the topology of convergence in measure.

(ii) Under what conditions is $L^\infty(\mu)$ separable?

9V. Let $\mathcal{V}$ be a real vector space and let $C \subset \mathcal{V}$ be a nonempty convex set. A nonempty convex set $F \subset C$ is called a *face* of C if $C \setminus F$ is convex. A point $x \in C$ is said to be an *extreme point* of C if $\{x\}$ is a face of C.

(i) Let F be a face of C, and let F' be a face of F. Show that F' is a face of C.

(ii) Let $\{F_\gamma\}_{\gamma \in \Gamma}$ be a family of faces of C that is totally ordered by inclusion. Show that $\bigcap_{\gamma \in \Gamma} F_\gamma$ is either empty or a face of C.

(iii) Let $\varphi : \mathcal{V} \to \mathbb{R}$ be a linear functional such that $\varphi | C$ has a maximum value α. Show that the set $\{x \in C : \varphi(x) = \alpha\}$ is a face of C.

9W. (Kreĭn-Mil'man) Let $\mathcal{V}$ be a normed vector space, and let $K \subset \mathcal{V}^*$ be a nonempty convex set that is compact in the weak* topology.

(i) Use Problem 9V(ii) and Zorn's lemma to show that the set of weak*-closed faces of K has minimal elements relative to inclusion.

(ii) Show that a minimal closed face of K consists of exactly one point. Thus K has at least one extreme point. (Hint: Let F be a weak*-closed face of K that contains two points $\varphi \neq \psi$. Choose $x \in \mathcal{V}$ such that $\varphi(x) \neq \psi(x)$ and apply Problem 9V(iii) to the functional $\varphi \mapsto \varphi(x)$ to show that F has a face $F' \subsetneq F$.)

9X. Show that the closed unit balls of c_0 and $L^1(\lambda_1)$ possess no extreme point. Deduce that there is no Banach space $\mathcal{V}$ whose dual is linearly isometric to either c_0 or $L^1(\lambda_1)$.

9Y. (Jordan-von Neumann) Let $(\mathcal{V}, \|\cdot\|)$ be a Banach space. Show that $\mathcal{V}$ is a Hilbert space if and only if it satisfies the *parallelogram identity*

$$\|x+y\|^2 + \|x-y\|^2 = 2\|x\|^2 + 2\|y\|^2, \quad x, y \in \mathcal{V}.$$

(Hint: Use (9.8) to define the scalar product.)

9Z. Let $n \geq 3$ be an integer and let $\omega \in \mathbb{C}$ be a primitive root of order n of unity, that is, $\omega^n = 1 \neq \omega^k$, $k = 1, \ldots, n-1$. Prove the identity

$$\langle x, y \rangle = \frac{1}{n} \sum_{k=0}^{n-1} \omega^k \|x + \omega^k y\|^2$$

for any two vectors x, y in a complex Hilbert space.

9AA. Let $(X, \mathbf{S}, \mu)$ be a measure space. Show that the inequality $\|f+g\|_{1,\infty} \leq \|f\|_{1,\infty} + \|g\|_{1,\infty}$ is not generally true for $f, g \in \mathcal{L}^0(\mu)$, but that the space $L^{1,\infty}(\mu)$ is a topological vector space if a set $G \subset L^{1,\infty}(\mu)$ is declared to be open precisely when, for every $u \in G$ we have $\{v \in L^{1,\infty}(\mu) : \|u-v\|_{1,\infty} < \varepsilon\} \subset G$ for some $\varepsilon > 0$ (depending on u).

9BB. Let $(X, \mathbf{S}, \mu)$ be a measure space, and suppose that $0 < q_1 < q \leq q_2 \leq +\infty$. Show that $L^{q_1,\infty} \cap L^{q_2,\infty} \subset L^q$. Deduce a proof of the case $p_1 = p_2$ of Theorem 9.42. Alternatively, verify that the proof of Theorem 9.42 applies with minor changes when $p_1 = p_2$.

9CC. Maintaining the notation in the proof of Theorem 9.42, suppose that $p_0 \leq p_1$ and either q_0 or q_1 is infinite.

(i) Calculate the (limiting) values of the exponents γ and η.

(ii) Show that for an appropriate value of $b > 0$, and for $a = b\|f\|_p^\eta$, we have $\nu(\{y : |T(f^{(s(t))})(y)| > t/2\}) = 0$ for every t.

(iii) Use (ii) to replace the right-hand side of (9.15) by a single integral. Estimate this integral to prove that the conclusion of Theorem 9.42 holds.

9DD. In the proof of Theorem 9.42, use the constant $a = c\|f\|_p^\eta$ for an appropriate constant $c > 0$ to show that

$$\|T(f)\|_q^q \leq 2^q q \left[\frac{1}{q-q_0} + \frac{1}{q_1-q} \right] c_0^{1-\tau} c_1^\tau \|f\|_p^q, \quad f \in \mathcal{L}^p(\mu), q_0 < q_1.$$

9EE. Let u be a function satisfying the hypothesis of Lemma 9.44. Apply the maximum modulus principle in a large rectangle to the function $u(z)e^{\varepsilon(z^2-1)}$, $\varepsilon > 0$, to obtain the conclusion of the lemma.

9FF. Suppose that a linear map T satisfies the hypotheses of Theorem 9.45, and let $f \in \mathcal{D}$.

a. Suppose that $f \in \mathcal{L}^{p_0}(\mu) \cap \mathcal{L}^{p_1}(\mu)$, and let $(s_n)_{n=1}^\infty$ be a sequence of simple functions such that $|s_n| \leq |s_{n+1}| \leq |f|$ and $\lim_{n\to\infty} \|s_n(x) - f(x)\|_{p_j} = 0$ for $j = 1, 2$. Show that the sequence $(T(s_n))_{n=1}^\infty$ has a subsequence that converges to $T(f)$ almost everywhere $[\nu]$.

b. Suppose that $f \in \mathcal{L}^{p_\tau}(\mu)$ for some $\tau \in (0,1)$. Use the truncations $f^{(1)}$ and $f_{(1)}$ to prove the existence of a sequence $(s_n)_{n=1}^\infty$ of simple functions such that $\lim_{n\to\infty} \|s_n - f\|_{p_\tau} = 0$ and $(T(s_n))_{n=1}^\infty$ converges to $T(f)$ almost everywhere $[\nu]$.

9GG. Prove Theorem 9.45 when one or more of the exponents p_0, p_1, q_0, q_1 is infinite.

9HH. Suppose that $(X, \mathbf{S}, \mu)$ is a measure space, $0 < p_0 < p_1 < +\infty$, $t \in (0,1)$, $p = (1-t)p_0 + tp_1$, and $f \in \mathcal{L}^{p_0} \cap \mathcal{L}^{p_1}$. Theorem 9.45 implies that $\|f\|_p \leq \|f\|_{p_0}^{1-t} \|f\|_{p_1}^t$. Give a direct proof of this inequality using the following outline.

a. Reduce the inequality to the case in which $\|f\|_{p_0} = \|f\|_{p_1} = 1$ via replacing f and μ by af and $b\mu$, respectively, for some positive constants a, b.

b. Suppose that $\|f\|_{p_0} = \|f\|_{p_1} = 1$. Use $f^{(1)}$ and $f_{(1)}$ to show that $\|f\|_p \leq 2$.

c. Apply the argument in (b) to the function $f_n : X^n \to \mathbb{C}$ defined by

$$f_n(x_1, \ldots, x_n) = f(x_1) \cdots f(x_n), \quad x_1, \ldots, x_n \in X,$$

and to the product of n copies of μ to deduce that $\|f\|_p^n \leq 2$. Let $n \to \infty$ to reach the desired conclusion.

9II. Let $(X, \mathbf{S}, \mu)$ be a probability space, and let $\{f_n\}_{n \in \mathbb{N}}$ be a sequence in $L^\infty(\mu)$ with the property that $\sup_n \|f\|_\infty < +\infty$. Show that $\lim_{n \to \infty} \|f_n\|_1 = 0$ if and only if $\lim_{n \to \infty} \|f_n\|_2 = 0$. (Of course, 1 and 2 can be replaced by any two positive numbers.)

Chapter 10
Fourier analysis

Periodic phenomena such as sunrises and seasons have fascinated people since prehistoric times. The astronomers of antiquity tried to describe the various observed periodic motions of the planets in terms of uniform circular motions or sums of such motions (cycles and epicycles). A mathematical form of this idea was studied sporadically in the 17th century when it was realized that some functions $f : \mathbb{R} \to \mathbb{C}$ that satisfy $f(x + 1) = f(x)$ for $x \in \mathbb{R}$ can be written as

$$f(x) = \sum_{n=-\infty}^{\infty} c_n e^{2\pi i n x}, \quad x \in \mathbb{R}, \tag{10.1}$$

where the coefficients are calculated as

$$c_n = \int_0^1 e^{-2\pi i n t} f(t)\, dt, \quad n \in \mathbb{Z}.$$

For instance, the Taylor series for $\log(1 - x)$ yields such a representation for a *discontinuous* function, namely

$$\sum_{n \in \mathbb{Z} \setminus \{0\}} \frac{e^{2\pi i n x}}{2\pi i n} = \sum_{n=1}^{\infty} \frac{\sin(\pi n x)}{\pi n} = \begin{cases} \frac{1}{2}(1 - x), & x \in (0, 1), \\ 0, & x = 0. \end{cases}$$

(One can proceed then to write $\int_0^1 (1 - x)^2\, dx = 1/3$ in terms of the coefficients $1/(\pi n)$, and thereby obtain $\sum_{n=1}^{\infty} n^{-2} = \pi^2/6$. Euler used different techniques to calculate this sum.) Fourier found this kind of development to be crucial in describing the solutions of the heat equation, and audaciously declared that *every* periodic function can be written in the form (10.1). (This unproved claim delayed the publication of Fourier's work.) There were several failed attempts to prove the claim as well as a number of positive results and

© Springer International Publishing Switzerland 2016
H. Bercovici et al., *Measure and Integration*,
DOI 10.1007/978-3-319-29046-1_10

interesting examples. A large part of modern analysis evolved from this study. It eventually became clear that the functions f to be so represented must satisfy some requirements, and that the very meaning of (10.1) may need to be modified in order to allow for sufficient generality. For instance, we have seen in Chapter 9 that functions $f : (0,1) \to \mathbb{C}$ that are square integrable $[\lambda_1]$ have a *Fourier expansion* of the form (10.1) in the sense that

$$\lim_{N \to \infty} \left\| f - \sum_{n=-N}^{N} c_n e_n \right\|_2 = 0,$$

where $e_n(t) = e^{2\pi i n t}$ for $t \in (0,1)$ and $n \in \mathbb{Z}$.

In this chapter we discuss in greater detail the representation of functions as Fourier series and Fourier integrals, focusing on functions of one variable mainly to simplify notation. We present a fairly limited outline of the subject and we only consider very briefly issues of pointwise convergence.

In the discussion of Fourier series it is often natural to replace the interval $(0,1)$ by the unit circle

$$\mathbb{T} = \{\zeta \in \mathbb{C} : |\zeta| = 1\} = \{e^{2\pi i t} : t \in [0,1)\}$$

and to endow $\mathbb{T}$ with the measure m obtained by pushing forward $\lambda_1 | [0,1)$ via the map $t \mapsto e^{2\pi i t}$. This measure satisfies

$$m(\{e^{2\pi i t} : t \in [\alpha, \beta)\}) = \beta - \alpha, \quad 0 \leq \alpha \leq \beta \leq 1,$$

and is called *normalized arclength measure* on $\mathbb{T}$. The measurable sets $[m]$ are the sets of the form $\{e^{2\pi i t} : t \in E\}$, where $E \subset \mathbb{R}$ is measurable $[\lambda_1]$. The translation invariance of λ_1 corresponds to rotation invariance for m. More precisely,

$$m(\omega E) = m(E)$$

for every $\omega \in \mathbb{T}$ and every set $E \subset \mathbb{T}$ that is measurable $[m]$. We also have

$$m(E^{-1}) = m(E)$$

if E is measurable $[m]$, where $E^{-1} = \{1/\zeta : \zeta \in E\}$. This follows from the invariance of λ_1 under the change of variable $t \mapsto -t$.

Fourier coefficients were defined for functions in $L^2(m) = L^2(\mathbb{T}, m)$, but their definition makes sense for functions in $L^1(m)$, and even for complex Borel measures on $\mathbb{T}$; see Problem 10A.

Definition 10.1. Given $f \in L^1(m)$, the *Fourier coefficients* of f are the numbers

$$\widehat{f}(n) = \int_{\mathbb{T}} \zeta^{-n} f(\zeta) \, dm(\zeta), \quad n \in \mathbb{Z}.$$

The series

$$\sum_{n \in \mathbb{Z}} \widehat{f}(n) \zeta^n$$

is called the *Fourier series* of f.

The following result is due to Riemann and Lebesgue.

Lemma 10.2. *For every $f \in L^1(m)$ we have*

$$|\widehat{f}(n)| \leq \|f\|_1, \quad n \in \mathbb{Z},$$

and $\lim_{|n| \to \infty} |\widehat{f}(n)| = 0$.

Proof. The first assertion is immediate because

$$|\widehat{f}(n)| \leq \int_{\mathbb{T}} |\zeta^{-n} f(\zeta)| \, dm(\zeta) = \int_{\mathbb{T}} |f(\zeta)| \, dm(\zeta) = \|f\|_1, \quad n \in \mathbb{Z}.$$

Proposition 9.6, with the maps $f \mapsto \widehat{f}(n)$ in place of T_n, allows us to reduce the proof of the last assertion to elements f in a set that spans a dense linear manifold in $L^1(\mathbb{T})$. Thus it suffices to verify that assertion when $f(\zeta) = \zeta^k$ for some $k \in \mathbb{Z}$, and the assertion is immediate in this case because $\widehat{f}(n) = 0$ for $n \neq k$. □

Corollary 10.3. (The Hausdorff-Young inequalities) *Suppose that $p \in [1, 2]$ and $f \in L^p(m)$. Then we have*

$$\left[\sum_{n \in \mathbb{Z}} |\widehat{f}(n)|^{p'} \right]^{1/p'} \leq \|f\|_p,$$

where $p' = p/(p-1)$ denotes the conjugate exponent of p.

Proof. The Riesz-Thorin interpolation theorem (Theorem 9.45) applied to the map $f \mapsto \{\widehat{f}(n)\}_{n \in \mathbb{Z}}$ with $\mu = m$ and ν the counting measure on $\mathbb{Z}$ shows that it suffices to prove the corollary for $p = 1$ and $p = 2$. These two cases follow from the preceding lemma and from the Bessel inequality (9.10), respectively. □

Example 10.4. Equality is achieved in Corollary 10.3 when $f(\zeta) = \zeta^n$ for some $n \in \mathbb{Z}$.

Unlike the case of functions in $L^2(m)$, the partial sums

$$(S_N f)(\zeta) = \sum_{n=-N}^{N} \widehat{f}(n) \zeta^n, \quad N \in \mathbb{N}, \zeta \in \mathbb{T},$$

do not generally converge to f in $L^1(m)$. To understand this phenomenon it is useful to consider the convolution of functions.

Definition 10.5. Given two functions $f, g : \mathbb{T} \to \mathbb{C}$ that are measurable $[m]$ and a point $\zeta \in \mathbb{T}$, the *convolution* $f * g$ is defined at ζ if

$$\int_{\mathbb{T}} |f(\omega)g(\zeta/\omega)|\, dm(\omega) < +\infty,$$

and we define

$$(f * g)(\zeta) = \int_{\mathbb{T}} f(\omega)g(\zeta/\omega)\, dm(\omega).$$

It is easy to see that the set of points where $f * g$ is defined is measurable $[m]$ and $f * g$ is measurable $[m]$ as well. A simple change of variables shows that $f * g = g * f$. Indeed, the measure m is invariant under the transformation $\omega \mapsto \zeta/\omega$, so setting $w = \zeta/\omega$ yields

$$\int_{\mathbb{T}} |f(\omega)g(\zeta/\omega)|\, dm(\omega) = \int_{\mathbb{T}} |f(\zeta/w)g(w)|\, dm(w).$$

The same calculation, without the absolute values, yields $(g * f)(\zeta) = (f * g)(\zeta)$ at points ζ where $(f * g)(\zeta)$ is defined. In many cases of interest, $f * g$ is defined everywhere or almost everywhere $[m]$.

Example 10.6. The sequence of functions

$$D_N(\zeta) = \sum_{n=-N}^{N} \zeta^n, \quad N \in \mathbb{N}, \zeta \in \mathbb{T},$$

is called the *Dirichlet kernel*. For any $f \in L^1(m)$ and $N \in \mathbb{N}$ we have

$$(S_N f)(\zeta) = \int_{\mathbb{T}} f(\omega) \sum_{n=-N}^{N} (\zeta/\omega)^n\, dm(\omega) = (f * D_N)(\zeta), \quad \zeta \in \mathbb{T}.$$

Example 10.7. Fix $p \in [1, +\infty]$, denote by p' the conjugate exponent, and suppose that $f \in L^p(m)$ and $g \in L^{p'}(m)$. Then $f * g$ is defined and continuous on $\mathbb{T}$. Indeed, the invariance of m under the transformation $\omega \mapsto \zeta/\omega$ and Hölder's inequality (9.3) yield

$$\int_{\mathbb{T}} |f(\omega)g(\zeta/\omega)|\, dm(\omega) \le \left[\int_{\mathbb{T}} |f|^p\, dm \right]^{1/p} \left[\int_{\mathbb{T}} |g(\zeta/\omega)|^{p'}\, dm(\omega) \right]^{1/p'}$$
$$= \|f\|_p \|g\|_{p'}, \quad \zeta \in \mathbb{T}.$$

Thus $f * g$ is defined everywhere on $\mathbb{T}$ and

$$\|f * g\|_\infty \le \|f\|_p \|g\|_{p'}.$$

The continuity of $f * g$ follows when $p' < +\infty$ from the fact that the map $\omega \mapsto g_\omega \in L^{p'}(\mathbb{T})$, where

$$g_\omega(\zeta) = g(\zeta/\omega), \quad \zeta, \omega \in \mathbb{T},$$

is continuous from $\mathbb{T}$ to $L^{p'}(m)$. This fact is verified as in Proposition 9.36, noting first that it is obvious when g is a continuous function.

Example 10.8. Suppose now that $f, g \in L^1(m)$. The Fubini-Tonelli theorem (Theorem 8.4) yields

$$\int_{\mathbb{T}} \int_{\mathbb{T}} |f(\omega)g(\zeta/\omega)| \, dm(\omega) \, dm(\zeta) = \int_{\mathbb{T}} |f(\omega)| \left[\int_{\mathbb{T}} |g(\zeta/\omega)| \, dm(\zeta) \right] dm(\omega)$$

$$= \int_{\mathbb{T}} |f(\omega)| \|g\|_1 \, dm(\omega) = \|f\|_1 \|g\|_1,$$

where we also used the rotation invariance of m to evaluate

$$\int_{\mathbb{T}} |g(\zeta/\omega)| \, dm(\zeta) = \int_{\mathbb{T}} |g(\zeta)| \, dm(\zeta) = \|g\|_1, \quad \omega \in \mathbb{T}.$$

We conclude that $\int_{\mathbb{T}} |f(\omega)g(\zeta/\omega)| \, dm(\omega) < +\infty$ for $[m]$-almost every ζ. Therefore $f * g$ is defined almost everywhere $[m]$ and

$$\int_{\mathbb{T}} |(f * g)(\zeta)| \, dm(\zeta) \le \int_{\mathbb{T}} \int_{\mathbb{T}} |f(\omega)g(\zeta/\omega)| \, dm(\omega) \, dm(\zeta).$$

We conclude that $f * g \in L^1(m)$ and

$$\|f * g\|_1 \le \|f\|_1 \|g\|_1.$$

We note for further reference the identity

$$\widehat{f * g}(n) = \widehat{f}(n)\widehat{g}(n), \quad f, g \in L^1(\mathbb{T}), n \in \mathbb{Z}.$$

This is verified using Theorem 8.4 and the rotation invariance of m:

$$\widehat{f * g}(n) = \int_{\mathbb{T}} \zeta^{-n} \int_{\mathbb{T}} f(\omega)g(\zeta/\omega) \, dm(\omega) dm(\zeta)$$

$$= \int_{\mathbb{T}} \omega^{-n} f(\omega) \left[\int_{\mathbb{T}} g(\zeta/\omega)(\zeta/\omega)^{-n} \, dm(\zeta) \right] dm(\omega)$$

$$= \int_{\mathbb{T}} \omega^{-n} f(\omega)\widehat{g}(n) \, dm(\omega) = \widehat{f}(n)\widehat{g}(n).$$

One can interpolate between the observations in the last two examples to obtain an inequality due to Young.

Proposition 10.9. *Suppose that the numbers* $p, q, r \in [1, +\infty]$ *satisfy*

$$\frac{1}{p} + \frac{1}{q} = \frac{1}{r} + 1.$$

For every $f \in L^p(m)$ and $g \in L^q(m)$ we have $f * g \in L^r(m)$ and

$$\|f * g\|_r \leq \|f\|_p \|g\|_q.$$

Proof. Fix first $f \in L^1(m)$ and consider the linear map $T : L^1(m) \to L^1(m)$ defined by $Tg = f * g$, $g \in L^1(m)$. We have the inequalities

$$\|Tg\|_1 \leq \|f\|_1 \|g\|_1, \quad g \in L^1(m),$$

and

$$\|Tg\|_\infty \leq \|f\|_1 \|g\|_\infty, \quad g \in L^\infty(m),$$

and Theorem 9.45 yields

$$\|Tg\|_q \leq \|f\|_1 \|g\|_q, \quad g \in L^q(m), q \in [1, +\infty).$$

To conclude the proof, we fix now $g \in L^q(m)$ and apply Theorem 9.45 to the inequalities

$$\|f * g\|_\infty \leq \|f\|_{p'} \|g\|_q, \quad f \in L^{q'}(m),$$

and

$$\|f * g\|_q \leq \|f\|_1 \|g\|_q, \quad f \in L^1(m),$$

to obtain the desired result. $\qquad\square$

Example 10.10. If p, q, and r are as in Proposition 10.9 and $f(\zeta) = g(\zeta) = \zeta^n$ for some $n \in \mathbb{Z}$, then $\|f * g\|_r = \|f\|_p \|g\|_q$. The result is also sharp when $p = q = r = 1$ in the following sense.

Lemma 10.11. *For any $g \in L^1(\mu)$ we have*

$$\sup\{\|f * g\|_1 : f \in L^1(m), \|f\|_1 \leq 1\} = \|g\|_1.$$

Proof. Set $I_n = \{e^{2\pi i t} : t \in (0, 1/n)\}$ and $f_n = n\chi_{I_n}$ so $\|f_n\|_1 = 1$ for $n \in \mathbb{N}$. It suffices to show that $\lim_{n \to +\infty} \|g - f_n * g\|_1 = 0$. Indeed, by Theorem 8.4,

$$\|g - f_n * g\|_1 = \int_{\mathbb{T}} \left| g(\zeta) - \int_{I_n} ng(\zeta/\omega)\, dm(\omega) \right| dm(\zeta)$$

$$= \int_{\mathbb{T}} \left| n \int_{I_n} [g(\zeta) - g(\zeta/\omega)]\, dm(\omega) \right| dm(\zeta)$$

$$\leq \int_{I_n} n\|g - g_\omega\|_1\, dm(\omega) \leq \sup_{\omega \in I_n} \|g - g_\omega\|_1,$$

where $g_\omega(\zeta) = g(\zeta/\omega)$. The result follows from the continuity, noted earlier, of the map $\omega \mapsto g_\omega$. $\qquad\square$

Example 10.12. The preceding lemma yields, in the special case $g = S_N$,

$$\sup\{\|S_N f\|_1 : f \in L^1(m), \|f\|_1 \leq 1\} = \|D_N\|_1, \quad N \in \mathbb{N}.$$

The numbers

$$L_N = \|D_N\|_1, \quad N \in \mathbb{N},$$

are called the *Lebesgue constants* and can be approximated fairly precisely. Writing $\zeta^{1/2} = e^{\pi i t}$ for $t \in (-1/2, 1/2)$, we have

$$D_N(\zeta) = \zeta^{-N} \sum_{n=0}^{2N} \zeta^n = \frac{\zeta^{-N} - \zeta^{N+1}}{1 - \zeta}$$

$$= \frac{\zeta^{-(N+\frac{1}{2})} - \zeta^{N+\frac{1}{2}}}{\zeta^{-\frac{1}{2}} - \zeta^{\frac{1}{2}}} = \frac{\sin((2N+1)\pi t)}{\sin(\pi t)},$$

so

$$L_N = 2 \int_0^{1/2} \frac{|\sin((2N+1)\pi t)|}{\sin(\pi t)} \, dt.$$

We have

$$\lim_{t \downarrow 0} \frac{1}{t} \left[\frac{1}{\sin(\pi t)} - \frac{1}{\pi t} \right] = \frac{\pi}{3!},$$

so the difference

$$\frac{1}{\sin(\pi t)} - \frac{1}{\pi t}$$

is bounded for $t \in [0, 1/2)$. Therefore the difference between L_N and

$$L_N' = \frac{2}{\pi} \int_0^{1/2} \frac{|\sin((2N+1)\pi t)|}{t} \, dt$$

is bounded. A change of variable yields

$$L_N' = \frac{2}{\pi} \int_0^{(2N+1)\pi} \frac{|\sin s|}{s} \, ds = \frac{2}{\pi} \int_0^\pi \sin s \left[\sum_{n=0}^{2N} \frac{1}{t + n\pi} \right] ds.$$

We observe next that

$$0 \le \sum_{n=1}^{2N} \left[\frac{1}{n\pi} - \frac{1}{s + n\pi} \right] = \sum_{n=1}^{2N} \frac{t}{n\pi(s + n\pi)} \le \sum_{n=1}^\infty \frac{1}{\pi n^2} < +\infty.$$

Approximating $\sum_{n=1}^{2N} 1/n$ by $\int_1^{2N} (1/t) \, dt = \log(2N)$, we see that $\sum_{n=1}^{2N} 1/(s + n\pi)$ differs by a bounded quantity from $\pi^{-1} \log N$, and this shows that L_N differs by a bounded quantity from

$$\frac{2}{\pi^2} \log N \int_0^\pi \sin s \, ds = \frac{2}{\pi^2} \log N.$$

In particular, $\lim_{N \to \infty} L_N = +\infty$.

Corollary 10.13. *There exists a function $f \in L^1(m)$ such that*

$$\sup\{\|S_N f\|_1 : N \in \mathbb{N}\} = +\infty.$$

For such functions f, the sequence $\{S_N f\}_{N \in \mathbb{N}}$ does not have a limit in $L^1(m)$.

Proof. The result follows immediately from Theorem 9.5 applied to the sequence of linear operators $S_N : L^1(\mathbb{T}) \to L^1(\mathbb{T})$. Since

$$\sup\{\|S_N\| : N \in \mathbb{N}\} = \sup\{L_N : N \in \mathbb{N}\} = +\infty,$$

the result follows immediately from Theorem 9.5 applied to the sequence of linear operators $S_N : L^1(\mathbb{T}) \to L^1(\mathbb{T})$. $\square$

We know, of course, that $\lim_{N \to \infty} \|f - S_N f\|_2 = 0$ if $f \in L^2(m)$, and thus $\lim_{N \to \infty} \|f - S_N f\|_1 = 0$ as well for such functions. For general functions $f \in L^1(m)$, the behavior of the sequence $\{S_N\}_{N \in \mathbb{N}}$ can be improved by averaging it in various ways.

Definition 10.14. Let $\{a_n\}_{n \in \mathbb{Z}}$ be a bounded family of complex numbers. The series $\sum_{n=-\infty}^{\infty} a_n$ is said to be *Abel summable* to the complex number s if

$$\lim_{r \uparrow 1} \sum_{n=-\infty}^{\infty} r^{|n|} a_n = s.$$

The boundedness assumption implies that the above series converges absolutely for $r \in [0, 1)$.

Proposition 10.15. *Let $\{a_n\}_{n \in \mathbb{Z}}$ be a bounded family of complex numbers such that the limit $s = \lim_{N \to \infty} \sum_{n=-N}^{N} a_n$ exists. Then $\sum_{n=-\infty}^{\infty} a_n$ is Abel summable to s.*

Proof. Replacing a_0 by $a_0 - s$ allows us to assume that $s = 0$. Setting $s_N = \sum_{n=-N}^{N} a_n$ for $n \in \mathbb{N}$ and $s_{-1} = 0$, we have

$$\sum_{n=-\infty}^{\infty} r^{|n|} a_n = \sum_{n=0}^{\infty} r^n (s_n - s_{n-1}) = \sum_{n=0}^{\infty} (r^n - r^{n+1}) s_n.$$

For fixed $N \in \mathbb{N}$ and $r \in [0, 1)$, we have the estimate

$$\left| \sum_{n=0}^{\infty} (r^n - r^{n+1}) s_n \right| \leq \left| \sum_{n=0}^{N} (r^n - r^{n+1}) s_n \right| + \sup_{n > N} |s_n| \sum_{n=N+1}^{\infty} (r^n - r^{n+1}).$$

The last sum above equals $r^{N+1} < 1$. We conclude that

$$\limsup_{r \uparrow 1} \left| \sum_{n=-\infty}^{\infty} r^{|n|} a_n \right| \leq \sup_{n > N} |s_n|,$$

and the proposition follows by letting $N \to \infty$ in this inequality. $\square$

Example 10.16. Let us examine the Abel summability of the Fourier series of a function $f \in L^1(m)$. The series

$$f_r(\zeta) = \sum_{n=-\infty}^{\infty} r^{|n|} \widehat{f}(n) \zeta^n, \quad \zeta \in \mathbb{T},$$

converges uniformly for any $r \in [0, 1)$, and f_r is therefore a continuous function of ζ. Corollary 4.25 allows us to interchange summation and integration to obtain

$$f_r(\zeta) = \int_{\mathbb{T}} f(\omega) \left[\sum_{n=-\infty}^{\infty} r^{|n|} (\zeta/\omega)^n \right] dm(\omega) = (P_r * f)(\zeta),$$

where

$$P_r(\zeta) = \sum_{n=-\infty}^{\infty} r^{|n|} \zeta^n = \frac{1}{1 - r\zeta} + \frac{1}{1 - r\zeta^{-1}} - 1 = \frac{1 - r^2}{|r - \zeta|^2}, \quad r \in [0, 1), \zeta \in \mathbb{T}.$$

The family of functions $\{P_r\}_{r \in [0,1)}$ is called the *Poisson kernel*.

Lemma 10.17. *The Poisson kernel has the following properties:*

(1) $P_r(\zeta) \geq 0$ and $P_r(\zeta) = P_r(1/\zeta)$ for $r \in [0, 1)$ and $\zeta \in \mathbb{T}$.
(2) $\int_{\mathbb{T}} P_r(\zeta)\, dm(\zeta) = 1$ for $r \in [0, 1)$.
(3) *The function* $t \mapsto P_r(e^{2\pi i t})$ *is decreasing for* $t \in [0, 1/2)$.
(4) *For every* $\delta \in (0, 1/2]$,

$$\lim_{r \uparrow 1} \max\{P_r(e^{2\pi i t}) : t \in [\delta, 1/2]\} = 0.$$

Proof. Property (1) is immediate from the closed form of P_r, and (2) is obtained if we integrate the series defining P_r. Property (3) is seen by observing that

$$|r - e^{2\pi i t}|^2 = 1 + r^2 - 2r \cos(2\pi t)$$

and that $\cos(2\pi t)$ is decreasing on $[0, 1/2)$. Finally, the maximum in (4) is

$$\frac{1 - r^2}{1 + r^2 - 2r \cos(2\pi \delta)},$$

and the denominator has limit $2 - 2 \cos(2\pi \delta) \neq 0$ as $r \uparrow 1$. □

The following result is due to H. A. Schwarz.

Proposition 10.18. *For every* $f \in \mathcal{C}(\mathbb{T})$ *we have*

$$\lim_{r \uparrow 1} \|f - P_r * f\|_\infty = 0.$$

Proof. Lemma 10.17(2) allows us to write

$$f(\zeta) - (P_r * f)(\zeta) = \int_{\mathbb{T}} [f(\zeta) - f_\omega(\zeta)] P_r(\omega)\, dm(\omega), \quad \zeta \in \mathbb{T}, r \in [0,1),$$

where $f_\omega(z) = f(\zeta/\omega)$. For fixed $\delta \in (0, 1/2]$ we have

$$|f(\zeta) - (P_r * f)(\zeta)| \le \int_{|\omega-1|<\delta} |f(\zeta) - f_\omega(\zeta)| P_r(\omega)\, dm(\omega)$$

$$+ \int_{|\omega-1|\ge\delta} |f(\zeta) - f_\omega(\zeta)| P_r(\omega)\, dm(\omega)$$

$$\le \sup_{|\omega-1|<\delta} \|f - f_\omega\|_\infty + 2\|f\|_\infty \sup_{|\omega-1|\ge\delta} P_r(\omega),$$

so

$$\limsup_{r\uparrow 1} \|f - P_r * f\|_\infty \le \sup_{|\omega-1|<\delta} \|f - f_\omega\|_\infty$$

by Lemma 10.17(4). The conclusion is obtained by letting $\delta \to 0$ and using the fact that f is uniformly continuous on $\mathbb{T}$. $\qquad\square$

Corollary 10.19. *Fix $p \in [1, +\infty)$ and $f \in L^p(m)$. We have $P_r * f \in L^p(m)$ for every $r \in [0,1)$, and $\lim_{r\uparrow 1} \|f - P_r * f\|_p = 0$.*

Proof. We have $\|P_r * f\|_p \le \|f\|_p$ by Proposition 10.9. Proposition 9.6 shows that it suffices to prove the last assertion for f in a dense linear manifold in $L^p(\mathbb{T})$, and the conclusion follows from the Proposition 10.18 because $\mathcal{C}(\mathbb{T})$ is such a linear manifold and $\|f - P_r * f\|_p \le \|f - P_r * f\|_\infty$. $\qquad\square$

The preceding result shows, in particular, that a function in $L^1(m)$ is uniquely determined by its Fourier coefficients. The easily verified vector version of Proposition 10.15 implies the following result.

Corollary 10.20. *Suppose $f \in L^1(m)$ is such that the sequence $\{S_N f\}_{N \in \mathbb{N}}$ converges in $L^1(m)$. Then $\lim_{N\to\infty} \|f - S_N f\|_1 = 0$.*

The hypotheses of the last result are satisfied, for instance, when f is sufficiently smooth.

Example 10.21. Suppose that $f : \mathbb{T} \to \mathbb{C}$ is a continuously differentiable function, in the sense that the map $t \mapsto f(e^{2\pi it})$ is continuously differentiable on $\mathbb{R}$. Denote by $f' : \mathbb{T} \to \mathbb{C}$ the derivative, that is, $f'(e^{2\pi it}) = (d/dt)f(e^{2\pi it})$. The Fourier coefficients of f' are easily related to those of f via integration by parts:

$$\widehat{f'}(0) = \int_0^1 \frac{d}{dt} f(e^{2\pi it})\, dt = 0,$$

and

$$\widehat{f'}(n) = \int_0^1 e^{-2\pi int} \frac{d}{dt} f(e^{2\pi it})\, dt = -\int_0^1 f(e^{2\pi int}) \frac{d}{dt} e^{-2\pi int}\, dt$$

$$= 2\pi in \widehat{f}(n), \quad n \ne 0.$$

Lemma 10.2 implies that $\lim_{|n|\to\infty} n\widehat{f}(n) = 0$. If f is twice continuously differentiable, this argument shows that $\lim_{n\to\infty} n^2\widehat{f}(n) = 0$. In particular, the series $\sum_{n=-\infty}^{\infty} \widehat{f}(n)\zeta^n$ converges uniformly for $\zeta \in \mathbb{T}$. As noted above, the sum of this series is necessarily $f(\zeta)$.

We examine next the pointwise behavior of the functions $P_r * f$ for $f \in L^1(m)$. This is best understood by comparison with the behavior of functions of the form $g * f$, where g is the characteristic function of an arc. More precisely, for each $s \in (0, 1/2]$, set $J_s = \{e^{2\pi it} : t \in (-s, s)\}$ and $h_s = (1/2s)\chi_{J_s}$ so $\int_{\mathbb{T}} h_s \, dm = 1$. Then $(h_s * f)(\zeta)$ is the average of f over an arc of (normalized) length $2s$ centered at ζ. Thus Theorem 6.22 implies that $\lim_{s\downarrow 0} h_s * f = f$ almost everywhere $[m]$. We relate P_r and h_s using the following result.

Lemma 10.22. *Let $g : \mathbb{T} \to [0, +\infty)$ be a continuous function such that*

(1) $g(\zeta) = g(\zeta^{-1})$ *for $\zeta \in \mathbb{T}$, and*
(2) $t \mapsto g(e^{2\pi it})$ *is a nonincreasing function on $[0, 1/2]$.*

Then there exists a finite, positive Borel measure μ on $(0, 1/2]$ such that

$$g(\zeta) = \int_{(0,1/2]} h_s(\zeta) \, d\mu(s), \quad \zeta \in \mathbb{T},$$

and $\mu((0, 1/2]) = \int_{\mathbb{T}} g \, dm$.

Proof. When $g = c$ is constant, the conclusion is satisfied with $\mu = c\delta_{1/2}$. Subtracting the constant $g(-1)$ from g we restrict ourselves to the case in which $g(-1) = 0$. It follows from Example 5.21 that there exists a positive Borel measure ν on $(0, 1/2)$ such that

$$\nu([t, 1/2)) = g(e^{2\pi it}) - g(-1) = g(e^{2\pi it}), \quad t \in (0, 1/2).$$

Alternatively, this can be written as

$$g(\zeta) = \int_{(0,1/2)} \chi_{J_s}(\zeta) \, d\nu(s), \quad \zeta \in \mathbb{T},$$

and the conclusion of the lemma is satisfied by the measure μ defined by $d\mu(s) = 2s \, d\nu(s)$. Theorem 8.4 easily gives $\mu((0, 1/2]) = \int_{\mathbb{T}} g \, dm$. $\qquad\square$

Example 10.23. The functions P_r, $r \in [0, 1]$, satisfy the hypotheses of Lemma 10.22. The measure μ_r on $(0, 1/2]$ satisfying

$$P_r(\zeta) = \int_{(0,1/2]} h_s(\zeta) \, d\mu_r(s), \quad \zeta \in \mathbb{T},$$

is such that the difference $\nu_r = \mu_r - P_r(-1)\delta_{1/2}$ ν_r is absolutely continuous $[\lambda_1]$ and

$$\frac{d\nu_r}{d\lambda_1}(s) = -2s\frac{d}{ds}P_r(e^{2\pi is}), \quad s \in (0,1/2).$$

We have $\mu_r((0,1/2]) = 1$, and the fact that $\lim_{r\uparrow1} P_r(z) = 0$ uniformly outside any neighborhood of 1 implies that

$$\lim_{r\uparrow1}\mu_r((\eta,1/2]) = 0$$

for every $\eta \in (0,1/2]$. This is also easily verified by explicit calculation.

The above observation allows us to compare the pointwise behavior of the convolutions $(P_r * f)(\zeta)$ with the averages

$$(h_s * f)(\zeta) = \frac{1}{2s}\int_{I_s} f(\zeta/\omega)\, dm(\omega).$$

Corollary 10.24. *Given a function $f \in L^1(m)$ and a point $\zeta \in \mathbb{T}$ such that $\lim_{s\downarrow0}(h_s * f)(\zeta)$ exists, we also have $\lim_{r\uparrow1}(P_r * f)(\zeta) = \lim_{s\downarrow0}(h_s * f)(\zeta)$.*

Proof. Subtracting a constant from f we may, and do, assume that $\lim_{s\downarrow0}(h_s * f)(\zeta) = 0$. Fix $\eta \in (0,1/2]$ and use Theorem 8.4 to see that

$$|(P_r * f)(\zeta)| = \left|\int_{[0,1/2)} (h_s * f)(\zeta)\, d\mu_r(s)\right|$$
$$\leq \sup_{s\in(0,\eta)} |(h_s * f)(\zeta)| + \mu_r(\eta,1/2]) \sup_{s\in(0,1/2]} |(h_s * f)(\zeta)|.$$

Thus

$$\limsup_{r\uparrow1} |(P_r * f)(\zeta)| \leq \sup_{s\in(0,\eta)} |(h_s * f)(\zeta)|,$$

and the desired conclusion is obtained by letting $\eta \to 0$.   □

It is natural to ask whether functions defined on $\mathbb{R}$ can also be represented in an appropriate sense by Fourier series.

Example 10.25. Let $f : \mathbb{R} \to \mathbb{C}$ be twice continuously differentiable and supported in some bounded interval. Choose $k \in \mathbb{N}$ such that f vanishes outside the interval $[-k/2, k/2]$, and define $g : \mathbb{T} \to \mathbb{C}$ by

$$g(e^{2\pi it}) = f(kt), \quad t \in [-1/2, 1/2].$$

As seen in Example 10.21,

$$g(\zeta) = \sum_{n=-\infty}^{\infty} \widehat{g}(n)\zeta^n, \tag{10.2}$$

and the series converges uniformly for $\zeta \in \mathbb{T}$. A change of variable shows that

$$\widehat{g}(n) = \int_{-1/2}^{1/2} e^{-2\pi i n t} f(kt) \, dt = \frac{1}{k} \int_{-k/2}^{k/2} e^{-2\pi i (t/k)} f(t) \, dt$$

$$= \frac{1}{k} \int_{\mathbb{R}} e^{-2\pi i (n/k) t} f(t) \, dt, \quad n \in \mathbb{Z}.$$

Equation (10.2) amounts to

$$f(kx) = \sum_{n=-\infty}^{\infty} \left[\frac{1}{k} \int_{\mathbb{R}} e^{-2\pi i (n/k) t} f(t) \, dt \right] e^{2\pi i n x}, \quad x \in [-1/2, 1/2),$$

or, equivalently,

$$f(x) = \sum_{n=-\infty}^{\infty} \left[\frac{1}{k} \int_{\mathbb{R}} e^{-2\pi i (n/k) t} f(t) \, dt \right] e^{2\pi i (n/k) x}, \quad x \in [-k/2, k/2).$$

Of course, the above series defines a function with period $2\pi k$ on $\mathbb{R}$, so the equality does not persist outside $[-k/2, k/2)$. The series is also reminiscent of a Riemann integral and suggests a new concept.

Definition 10.26. Given a function $f \in L^1(\lambda_1)$, the *Fourier transform* $\widehat{f}$: $\mathbb{R} \to \mathbb{C}$ is defined by

$$\widehat{f}(t) = \int_{\mathbb{R}} e^{-2\pi i t x} f(x) \, d\lambda_1(x), \quad t \in \mathbb{R}.$$

Example 10.25 suggests that the analog of the Fourier series representation is

$$f(x) = \int_{\mathbb{R}} e^{2\pi i t x} \widehat{f}(t) \, d\lambda_1(t), \quad x \in \mathbb{R}, \tag{10.3}$$

known as the *Fourier inversion formula,* which we now explore further.

Example 10.27. Fix $\varepsilon > 0$, and set $f_\varepsilon(x) = e^{-2\pi\varepsilon|x|}$ for $x \in \mathbb{R}$. The Fourier transform $\widehat{f}_\varepsilon$ is easily calculated by evaluating two absolutely convergent integrals:

$$\widehat{f}_\varepsilon(t) = \int_0^\infty e^{-2\pi(it+\varepsilon)x} \, d\lambda_1(x) + \int_{-\infty}^0 e^{-2\pi(it-\varepsilon)x} \, d\lambda_1(x)$$

$$= \frac{1}{2\pi(\varepsilon + it)} + \frac{1}{2\pi(\varepsilon - it)} = \frac{1}{\pi} \frac{\varepsilon}{t^2 + \varepsilon^2}, \quad t \in \mathbb{R}.$$

The function $\widehat{f}_\varepsilon$ is integrable $[\lambda_1]$, and the Fourier inversion formula can be verified directly with some difficulty. The functions $p_\varepsilon = \widehat{f}_\varepsilon$, $\varepsilon > 0$, form the *Poisson kernel* for $\mathbb{R}$.

The Fourier transform of a function in $L^1(\lambda_1)$ however does not usually belong to $L^1(\lambda_1)$, as this next example shows.

Example 10.28. Suppose that $a, b \in \mathbb{R}$ satisfy $a < b$, and set $f = \chi_{(a,b)}$. Then the Fourier transform

$$\widehat{f}(t) = \int_a^b e^{-2\pi itx} dx = \frac{e^{-2\pi itb} - e^{-2\pi ita}}{-2\pi it}, \quad t \in \mathbb{R} \setminus \{0\},$$

is not integrable over $\mathbb{R}$.

There is however an analog of the Riemann-Lebesgue lemma (Lemma 10.2).

Lemma 10.29. *For every $f \in L^1(\lambda_1)$, the Fourier transform $\widehat{f}$ is continuous on $\mathbb{R}$,*

$$|\widehat{f}(t)| \leq \|f\|_1, \quad t \in \mathbb{R},$$

and $\lim_{|t| \to \infty} \widehat{f}(t) = 0$.

Proof. The inequality $|\widehat{f}(t)| \leq \|f\|_1$ is immediate. The other assertions are also easily verified when $f = \chi_{(a,b)}$ because

$$\lim_{t \to 0} \frac{e^{-2\pi itb} - e^{-2\pi ita}}{-2\pi it} = b - a = \widehat{f}(0).$$

The lemma therefore follows for functions in the linear manifold generated by $\{\chi_{(a,b)} : a < b\}$ and thus for functions in the closure of this manifold in $L^1(\lambda_1)$, which is $L^1(\lambda_1)$. Indeed, $\mathcal{C}_0(\mathbb{R})$ is closed under the formation of uniform limits. $\qquad\square$

The analogs for Fourier integrals of the partial sums $S_N(f)$ used in the context of Fourier series are the integrals $\int_{-T}^{T} e^{2\pi it} \widehat{f}(t) \, d\lambda_1(t)$. Certain averages of these integrals do converge to the original function f. In order to write these averages, we extend the concept of convolution to functions on $\mathbb{R}$.

Definition 10.30. Given two functions $f, g : \mathbb{R} \to \mathbb{C}$ that are measurable $[\lambda_1]$, and a point $t \in \mathbb{R}$, the *convolution* $f * g$ is defined at t if

$$\int_{\mathbb{R}} |f(s)g(t - s)| \, d\lambda_1(s) < +\infty,$$

and, in this case, we define

$$(f * g)(t) = \int_{\mathbb{R}} f(s)g(t - s) \, d\lambda_1(s).$$

All the results and observations described in Examples 10.7, 10.8, Proposition 10.9, and Lemma 10.11 remain true for convolutions on $\mathbb{R}$, and the proofs apply with minor notational changes. The translation invariance of λ_1 implies the identity

$$\widehat{f * g}(t) = \widehat{f}(t)\widehat{g}(t), \quad f, g \in L^1(\lambda_1), t \in \mathbb{R}.$$

Indeed, by Theorem 8.4,

$$
\begin{aligned}
\widehat{f * g}(t) &= \int_{\mathbb{R}} e^{-2\pi i x t} \int_{\mathbb{R}} f(y) g(x - y) \, d\lambda_1(y) d\lambda_1(x) \\
&= \int_{\mathbb{R}} e^{-2\pi i y t} f(y) \left[\int_{\mathbb{R}} e^{-2\pi i (x-y)t} g(x - y) \, d\lambda_1(x) \right] d\lambda_1(y) \\
&= \int_{\mathbb{R}} e^{-2\pi i y t} f(y) \widehat{g}(t) \, d\lambda_1(y) = \widehat{f}(t)\widehat{g}(t).
\end{aligned}
$$

Example 10.31. Let us examine the Fourier inversion formula as an improper integral. Given $f \in L^1(\lambda_1)$ and $T > 0$, we have, for each $x \in \mathbb{R}$,

$$
\begin{aligned}
\int_{-T}^{T} e^{2\pi i x t} \widehat{f}(t) \, d\lambda_1(t) &= \int_{-T}^{T} e^{2\pi i x t} \left[\int_{\mathbb{R}} e^{-2\pi i y t} f(y) \, d\lambda_1(y) \right] d\lambda_1(t) \\
&= \int_{\mathbb{R}} f(y) \left[\int_{-T}^{T} e^{2\pi i (x-y)t} \, d\lambda_1(t) \right] d\lambda_1(y),
\end{aligned}
$$

where the change in the order of integration is justified by Theorem 8.4 because the integrand is absolutely integrable. We conclude that

$$\int_{-T}^{T} e^{2\pi i x t} \widehat{f}(t) \, d\lambda_1(t) = (d_T * f)(x), \quad x \in \mathbb{R},$$

where

$$d_T(x) = \int_{-T}^{T} e^{2\pi i x t} \, d\lambda_1(t) = \frac{\sin(2\pi T x)}{\pi x}$$

is the analog of the Dirichlet kernel. The function d_T is bounded but not integrable over $\mathbb{R}$. It follows easily that there exist functions $f \in L^1(\lambda_1)$ for which the functions $d_T * f$ do not even belong to $L^1(\lambda_1)$ and, in particular, cannot approximate f in $L^1(\lambda_1)$.

Lemma 10.32. *Given* $f, g \in L^1(\lambda_1)$, *we have* $\int_{\mathbb{R}} f\widehat{g} \, d\lambda_1 = \int_{\mathbb{R}} \widehat{f} g \, d\lambda_1$.

Proof. This follows immediately from Theorem 8.4 applied to the integrable function of two variables $e^{-2\pi i x t} f(x) g(t)$. $\qquad\square$

Example 10.33. An application of the preceding lemma with $g(x) = e^{-2\pi |x|}$ yields

$$\int_{\mathbb{R}} \widehat{f}(t) e^{-2\pi |t|} \, d\lambda_1(t) = \int_{\mathbb{R}} f(x) \frac{\varepsilon}{\pi(\varepsilon^2 + x^2)} \, d\lambda_1(x) \qquad (10.4)$$

$$= \int_{\mathbb{R}} f(x) p_\varepsilon(x) \, d\lambda_1(x).$$

This observation is even more useful when applied to translates of the function f.

Lemma 10.34. *Fix $f \in L^1(\lambda_1)$, $a \in \mathbb{R}$, and $b \in \mathbb{R} \setminus \{0\}$. Denote by $g(x) = f(x - a)$ the translate of f by a and by $h(x) = f(x/b)$ the dilation of f by $1/b$. Then $\widehat{g}(t) = e^{-2\pi i a t} \widehat{f}(t)$ and $\widehat{h}(t) = |b| \widehat{f}(bt)$ for all $t \in \mathbb{R}$.*

Proof. Both identities follow by simple changes of variable. Indeed,

$$\widehat{g}(t) = \int_{\mathbb{R}} e^{-2\pi i t x} f(x - a) \, d\lambda_1(x) = \int_{\mathbb{R}} e^{-2\pi i t (y + a)} f(y) \, d\lambda_1(y)$$

follows from the change of variable $x = y + a$ which leaves λ_1 invariant, while

$$\widehat{h}(t) = \int_{\mathbb{R}} e^{-2\pi i t x} f(x/b) \, d\lambda_1(x) = \int_{\mathbb{R}} e^{-2\pi i t b y} f(y) |b| \, d\lambda_1(y)$$

follows by using the transformation $x = by$. $\square$

Example 10.35. Substitute $f(a - x)$ for $f(x)$ in (10.4) to obtain

$$\int_{\mathbb{R}} e^{2\pi i a t} \widehat{f}(t) e^{-2\pi \varepsilon |t|} \, d\lambda_1(t) = (f * p_\varepsilon)(a), \quad a \in \mathbb{R}, \varepsilon > 0, \tag{10.5}$$

where $p_\varepsilon(x) = \varepsilon/(\pi(\varepsilon^2 + t^2))$ is the Poisson kernel of Example 10.27. The functions p_ε are even, decreasing on $[0, +\infty)$, and $\int_{\mathbb{R}} p_\varepsilon \, d\lambda_1 = 1$. It follows as in Lemma 10.22 that there exist Borel probability measures μ_ε on $(0, +\infty)$ such that

$$p_\varepsilon(x) = \int_{(0,+\infty)} \frac{1}{2s} \chi_{(-s,s)}(x) \, d\mu_\varepsilon(s), \quad x \in \mathbb{R}, \varepsilon > 0.$$

Indeed, it is easy to calculate μ_ε explicitly:

$$d\mu_\varepsilon(s) = 2s \left[-\frac{dp_\varepsilon(s)}{ds} \right] ds = \frac{4\varepsilon s^2}{\varepsilon^2 + t^2} \, ds.$$

This formula, or the equality $\mu_\varepsilon((\eta, +\infty)) = \mu_1((\eta/\varepsilon, +\infty))$, shows that

$$\lim_{\varepsilon \downarrow 0} \mu_\varepsilon((\eta, +\infty)) = 0, \quad \eta > 0.$$

The proofs of Proposition 10.18, Corollary 10.19, and Corollary 10.24, with minor changes, yield the following result.

Theorem 10.36. *Let $f : \mathbb{R} \to \mathbb{C}$ be a measurable function. Then:*

(1) *If f is bounded and uniformly continuous on $\mathbb{R}$, then*

$$\lim_{\varepsilon \downarrow 0} \| f - p_\varepsilon * f \|_\infty = 0.$$

(2) If $f \in L^p(\lambda_1)$ for some $p \in [1, +\infty)$, then $\lim_{\varepsilon \downarrow 0} \|f - p_\varepsilon * f\|_p = 0$.
(3) If $f \in L^1(\lambda_1)$, then $\lim_{\varepsilon \downarrow 0} p_\varepsilon * f = f$ almost everywhere $[\lambda_1]$.

Corollary 10.37. An element $f \in L^1(\lambda_1)$ is uniquely determined by its Fourier transform. If $\widehat{f} \in L^1(\lambda_1)$ then f equals almost everywhere $[\lambda_1]$ a function in $\mathcal{C}_0(\mathbb{R})$, and the Fourier inversion formula (10.3) holds almost everywhere $[\lambda_1]$.

Proof. Relation (10.5) shows that $p_\varepsilon * f$ is uniquely determined by $\widehat{f}$ and therefore so is f. When $\widehat{f}$ is continuous, the left-hand side of (10.5) converges to $\int_{\mathbb{R}} e^{2\pi i a t} \widehat{f}(t) \, dt$ by the dominated convergence theorem. Therefore this function, which is the Fourier transform of $\widehat{f}$ evaluated at $-a$, belongs to $\mathcal{C}_0(\mathbb{R})$ and equals $f(a)$ almost everywhere $[\lambda_1]$. □

Example 10.38. The Poisson kernel p_ε was defined in Example 10.27 as the Fourier transform of the function $e^{-2\pi\varepsilon|x|}$, $x \in \mathbb{R}$. Since p_ε is integrable $[\lambda_1]$, the Fourier inversion formula yields $\widehat{p_\varepsilon}(t) = e^{-2\pi\varepsilon|t|}$, $t \in \mathbb{R}$.

Example 10.39. Suppose that $f \in \mathcal{C}_0(\mathbb{R})$ is continuously differentiable on $\mathbb{R}$ and both f and f' are integrable $[\lambda_1]$. Then $\widehat{f'}(t) = 2\pi i t \widehat{f}(t)$, $t \in \mathbb{R}$. Indeed, integration by parts yields

$$\int_{-R}^{R} e^{-2\pi i t x} f'(x) \, d\lambda_1(x) = e^{-2\pi i t R} f(R) - e^{2\pi i t R} f(-R)$$

$$+ 2\pi i t \int_{-R}^{R} e^{-2\pi i t x} f(x) \, d\lambda_1(x), \quad R > 0,$$

and the desired conclusion is obtained as $R \to +\infty$. If, in addition, $f' \in \mathcal{C}_0(\mathbb{R})$ is continuously differentiable and f'' is integrable, we can iterate this argument to obtain $\widehat{f''}(t) = -4\pi^2 t^2 \widehat{f}(t)$. It follows in this case that $\widehat{f}$ is integrable and therefore the Fourier inversion formula holds. Observe also that in this situation $\widehat{f}$ belongs to $L^p(\lambda_1)$ for all $p \in (1/2, +\infty]$.

The analog of Young's inequality is not as straightforward for Fourier transforms. Indeed, neither of the spaces $L^1(\lambda_1)$ or $L^2(\lambda_1)$ is contained in the other. A version of Parseval's identity is however true for elements in $L^1(\lambda_1) \cap L^2(\lambda_1)$.

Lemma 10.40. For every $f \in L^1(\lambda_1) \cap L^2(\lambda_1)$, we have $\widehat{f} \in L^2(\lambda_1)$ and $\|\widehat{f}\|_2 = \|f\|_2$.

Proof. Suppose first that $\widehat{f} \in L^1(\lambda_1)$. Set $g(t) = \overline{\widehat{f}(t)}$, $t \in \mathbb{R}$, and observe that

$$\widehat{g}(x) = \int_{\mathbb{R}} e^{-2\pi i x t} \overline{\widehat{f}(t)} \, d\lambda_1(t) = \left[\int_{\mathbb{R}} e^{2\pi i x t} \widehat{f}(t) \, d\lambda_1(t) \right]^{-} = \overline{f(x)}$$

by the Fourier inversion formula. Therefore

$$\|f\|_2^2 = \int_{\mathbb{R}} f\overline{f}\, d\lambda_1 = \int_{\mathbb{R}} f\overline{g}\, d\lambda_1 = \int_{\mathbb{R}} \widehat{f}\overline{g}\, d\lambda_1 = \|\widehat{f}\|_2^2.$$

In case $f \in L^1(\lambda_1) \cap L^2(\lambda_1)$, we approximate f in both $L^1(\lambda_1)$ and in $L^2(\lambda_1)$ by the functions $f_\varepsilon = p_\varepsilon * f$ for which $\widehat{f_\varepsilon}(t) = e^{-2\pi\varepsilon|t|}\widehat{f}(t)$ is integrable $[\lambda_1]$. We have

$$\|f\|_2 = \lim_{\varepsilon\downarrow 0} \|f_\varepsilon\|_2 = \lim_{\varepsilon\downarrow 0} \|\widehat{f_\varepsilon}\|_2 = \|\widehat{f}\|_2,$$

where the last equality follows from the dominated convergence theorem. $\square$

The preceding result allows us to define the Fourier transform of a function in $L^1(\lambda_1) + L^2(\lambda_1)$. If $f \in L^2(\lambda_1)$ and $N \in \mathbb{N}$, we have $f_N = \chi_{(-N,N)}f \in L^1(\lambda_1) \cap L^2(\lambda_1)$ and $\lim_{N\to\infty} \|f - f_N\|_2 = 0$ by the dominated convergence theorem. Lemma 10.40 implies that the sequence $\{\widehat{f_N}\}_{N\in\mathbb{N}}$ is Cauchy in $L^2(\lambda_1)$, and therefore it has a limit in that space.

Definition 10.41. Given a function $f \in L^2(\lambda_1)$, we define the Fourier transform $\widehat{f}$ to be the limit in $L^2(\lambda_1)$ of the sequence $\{\widehat{f_N}\}_{N\in\mathbb{N}}$, where $f_N = \chi_{(-N,N)}f$, $N \in \mathbb{N}$.

When $f \in L^1(\lambda_1) \cap L^2(\lambda_1)$, the new definition of $\widehat{f}$ agrees with the original one. Indeed, we also have $\lim_{N\to\infty} \|f - f_N\|_1 = 0$, so the sequence $\{\widehat{f_N}\}_{N\in\mathbb{N}}$ converges uniformly to the function $\widehat{f}$ of Definition 10.26. Since sequences which converge in $L^2(\lambda_1)$ have subsequences which converge almost everywhere $[\lambda_1]$, we conclude that Definitions 10.26 and 10.41 yield the same element of $L^2(\lambda_1)$. It follows that setting

$$\widehat{f+g} = \widehat{f} + \widehat{g}, \quad f \in L^1(\lambda_1), g \in L^2(\lambda_1),$$

gives a consistent definition of the Fourier transform on $L^1(\lambda_1) + L^2(\lambda_1)$. Of course, we have $L^p(\lambda_1) \subset L^1(\lambda_1) + L^2(\lambda_1)$ for $p \in (1,2)$, and the Young inequalities follow by interpolation.

Proposition 10.42. *For every $p \in [1,2]$ and for every $f \in L^p(\lambda_1)$, we have $\widehat{f} \in L^{p'}(\lambda_1)$ and $\|\widehat{f}\|_{p'} \le \|f\|_p$, where p' denotes the conjugate exponent of p.*

Unlike in the case of $\mathbb{T}$, the Young and Hausdorff-Young inequalities are generally not sharp on $L^p(\lambda_1)$. For instance, it is true that $\|\widehat{f}\|_{p'} \le c_p\|f\|_p$, where

$$c_p = \sqrt{p^{1/p}/p'^{1/p'}} < 1$$

if $p \in (1,2)$. The proof of this result of Babenko and Beckner is beyond the scope of this text.

Problems

10A. Let μ be a complex Borel measure on $\mathbb{T}$. Define the Fourier coefficients of μ by the formula

$$\widehat{\mu}(n) = \int_{\mathbb{T}} \zeta^{-n} \, d\mu(\zeta), \quad n \in \mathbb{Z}.$$

Show that $|\widehat{\mu}(n)| \leq |\mu|(\mathbb{T})$ for all $n \in \mathbb{Z}$. Prove that the equality

$$|\widehat{\mu}(n)| = |\mu|(\mathbb{T})$$

holds only when there exist a constant $c \in \mathbb{C}$ and a finite positive Borel measure ν on $\mathbb{T}$ such that $d\mu(\zeta) = c\zeta^n d\nu(\zeta)$.

10B. Given $w \in \mathbb{T}\backslash\{1\}$, consider the measure $\mu = \delta_1 + i\delta_w$ with total variation $|\mu|(\mathbb{T}) = 2$. Determine the values w for which

$$\sup_{n \in \mathbb{Z}} |\widehat{\mu}(n)| = 2.$$

(Hint: Problem 4EE is useful when $w = e^{2\pi i t}$ and $t \notin \mathbb{Q}$.)

10C. Let $f, g : \mathbb{T} \to \mathbb{C}$ be measurable $[m]$. Verify that the domain of the function $f * g$ is measurable $[m]$ and that $f * g$ is measurable $[m]$ and equal to $g * f$ on that domain.

10D. Extend Lemma 10.11 to exponents $q > 1$, that is, show that

$$\sup\{\|f * g\|_q : f \in L^1(m), \|f\|_1 \leq 1\} = \|g\|_q$$

for every $g \in L^q(m)$.

10E. Given $f \in L^1(m)$ and $\zeta \in \mathbb{T}$, define the rotation f_ζ of f by $f_\zeta(z) = f(z/\zeta)$, $z \in \mathbb{T}$. Show that $\widehat{f_\zeta}(n) = \zeta^{-n} \widehat{f}(n)$ for $n \in \mathbb{Z}$.

10F. Let $f : \mathbb{T} \to \mathbb{C}$ be a function of *finite variation* in the sense that the function $t \mapsto f(e^{2\pi i t})$ has finite variation on $[0, 1)$. The total variation v of f is defined as the least upper bound of all sums of the form

$$|f(\zeta_1) - f(\zeta_2)| + \cdots + |f(\zeta_{n-1}) - f(\zeta_n)| + |f(\zeta_n) - f(\zeta_0)|,$$

where n is a natural number and $\zeta_j = e^{2\pi i t_j}$ with

$$0 \leq t_1 \leq \cdots \leq t_n \leq 1.$$

(i) Show that $|\widehat{f}(n)| \leq v/n$ for $n \in \mathbb{Z} \setminus \{0\}$.

(ii) Set $\zeta_n = e^{2\pi i/n}$ and show that

$$n\|f - f_{\zeta_n}\|_p^p = \sum_{k=0}^{n-1} \|f_{\zeta_n^k} - f_{\zeta_n^{k+1}}\|_p^p \leq v\|f - f_{\zeta_n}\|_\infty^{p-1}$$

for all $p \in (1, +\infty)$.

(iii) Show that f is continuous on $\mathbb{T}$ if and only if

$$\lim_{n \to \infty} n\|f - f_{\zeta_n}\|_2^2 = 0.$$

(iv) Show that the condition in (iii) is equivalent to

$$\lim_{n \to \infty} n \sum_{k \in \mathbb{Z}} |1 - \zeta_n^k|^2 |\widehat{f}(k)|^2 = 0.$$

(v) Use the inequality

$$\frac{k^2}{n^2} \le |1 - \zeta_n^k|^2 = 4\sin^2\left(\frac{k\pi}{2n}\right) \le \frac{\pi^2 k^2}{n^2}, \quad 1 \le k \le \frac{n}{2},$$

to show that the condition in (iv) is equivalent to

$$\lim_{n\to\infty} \frac{\sum_{k=-n}^{n} k^2 |\widehat{f}(k)|^2}{n} = 0.$$

(vi) Use (i) and Problem 9II to further simplify this condition to

$$\lim_{n\to\infty} \frac{\sum_{k=-n}^{n} k|\widehat{f}(k)|}{n} = 0.$$

(This condition is satisfied if $\lim_{|k|\to\infty} k|\widehat{f}(k)| = 0$.)

10G. Let μ be a complex Borel measure on $\mathbb{T}$ such that $\mu(\mathbb{T}) = 0$, and define a function $f : \mathbb{T} \to \mathbb{C}$ by

$$f(e^{2\pi i t}) = \mu(\{e^{2\pi i s} : s \in [0, t)\}), \quad t \in [0, 1).$$

(i) Show that f has total variation equal to $|\mu|(\mathbb{T})$.
(ii) Verify that $\widehat{f}(n) = \widehat{\mu}(n)/(2\pi i n)$ for $n \in \mathbb{Z} \setminus \{0\}$.
(iii) (Wiener) Show that μ has no point masses if and only if

$$\lim_{n\to\infty} \frac{\sum_{k=-n}^{n} \widehat{\mu}(k)}{n} = 0.$$

(iv) Use a measure of the form $\mu - cm$ to show that (iii) holds without the assumption that $\mu(\mathbb{T}) = 0$.

10H. Use the fact that $\sup\{L_N : N \in \mathbb{N}\} = +\infty$ and the formula $(S_N f)(1) = \int_{\mathbb{T}} f D_N \, dm$ to show that there exist functions $f \in \mathcal{C}(\mathbb{T})$ for which the sequence $\{(S_N f)(1)\}_{N\in\mathbb{N}}$ does not converge.

10I. Use Proposition 10.18 to provide another proof of the Weierstrass approximation theorem for trigonometric polynomials.

10J. Prove a version of Lemma 10.22 without the assumption that the function g is continuous. Show that the measure μ is uniquely determined by g.

10K. Verify that the function

$$x \mapsto \begin{cases} \frac{\sin x}{x}, & x \ne 0 \\ 1, & x = 0, \end{cases}$$

does not belong to $L^1(\lambda_1)$. Deduce that there are functions $f \in L^1(\lambda_1)$ such that $d_T * f \notin L^1(\lambda_1)$ for any positive integer T. (The function d_T is defined in Example 10.31.)

10L. Characterize those functions $f \in L^1(\lambda_1)$ that satisfy

$$\|f\|_1 = \sup\{|\widehat{f}(t)| : t \in \mathbb{R}\}.$$

10M. Let μ be a complex Borel measure on $\mathbb{R}$. Define the Fourier transform $\widehat{\mu} : \mathbb{R} \to \mathbb{C}$ of μ by $\widehat{\mu}(t) = \int_{\mathbb{R}} e^{-2\pi i x t} \, d\mu(x)$. Show that $|\widehat{\mu}|$ is bounded by the total variation $|\mu|(\mathbb{R})$ and that $\widehat{\mu}$ is a uniformly continuous function on $\mathbb{R}$.

10N. Suppose that $f \in L^1(\mathbb{R})$ and $\int_{\mathbb{R}} |xf(x)|\, d\lambda_1(x) < +\infty$. Show that $\widehat{f}$ is differentiable on $\mathbb{R}$ and $\widehat{f}' = \widehat{g}$, where $g(x) = -2\pi i x f(x)$, $x \in \mathbb{R}$.

10O. Define $f : \mathbb{R} \to \mathbb{R}$ by $f(x) = e^{-\pi x^2}$, $x \in \mathbb{R}$.

   a. Show that $f'(x) = 2\pi x f(x)$ and $\widehat{f}'(t) = 2\pi t \widehat{f}(t)$, $x, t \in \mathbb{R}$.

   b. Show that $\widehat{f} = f$.

   c. More generally, suppose that p is a complex polynomial of degree n. Show that the Fourier transform of the function $p(x)f(x)$ is of the form $q(t)f(t)$, where q is also a complex polynomial of degree n. (Hint: calculate the Fourier transforms of higher order derivatives $d^n f(x)/dx^n$.)

10P. Suppose that $f, g \in L^{4/3}(\lambda_1)$, so $f * g \in L^2(\lambda_1)$.

   a. Show that $\widehat{f * g} = \widehat{f}\widehat{g}$ almost everywhere $[\lambda_1]$. (The function $\widehat{f}$ is defined because $L^{4/3}(\lambda_1) \subset L^1(\lambda_1) + L^2(\lambda_1)$.)

   b. Prove the inequality $\|f * g\|_2^2 \leq \|f * f\|_2 \|g * g\|_2$.

10Q. Let $\{s_n\}_{n \geq 0}$ be a sequence of complex numbers, and set

$$\sigma_n = \frac{s_0 + s_1 + \cdots + s_{n-1}}{n}, \quad n \in \mathbb{N}.$$

Assume that the sequence $\{s_n\}_{n \geq 0}$ converges to $s \in \mathbb{C}$ and show that $\lim_{n \to \infty} \sigma_n = s$. (When $s_n = \sum_{k=-n}^{n} a_k$ and $\lim_{n \to \infty} \sigma_n = s$, one says that $\sum_{n=-\infty}^{\infty} a_n$ is *Cesàro summable to* s.)

10R. Let $f \in L^1(m)$, and let $\sum_{n=-\infty}^{\infty} a_n \zeta^n$ be its Fourier series. Set $s_n(\zeta) = \sum_{k=-n}^{n} a_k \zeta^k$, $n = 0, 1, \ldots$, and

$$\sigma_n(\zeta) = \frac{s_0(\zeta) + s_1(\zeta) + \cdots + s_{n-1}(\zeta)}{n}, \quad n \in \mathbb{N}, \zeta \in \mathbb{T}.$$

   a. Show that $\sigma_n = f * K_n$, where

$$K_n(e^{2\pi i t}) = \frac{\sin^2(n\pi t)}{n \sin^2(\pi t)}, \quad t \in \mathbb{R}, n \in \mathbb{N},$$

   is the *Fejér kernel*.

   b. Prove an analog of Lemma 10.17 for the sequence $\{K_n\}_{n \in \mathbb{N}}$.

   c. Show that $\lim_{n \to \infty} \sigma_n(\zeta) = f(\zeta)$ for every point $\zeta \in \mathbb{C}$ at which f is continuous, and the limit is uniform on any compact set of continuity for f.

10S. The function $K_n(e^{2\pi i t})$ is not monotone decreasing for $t \in [0, 1/2]$. Show however that there exist probability measures μ_n, $n \in \mathbb{N}$, on $[0, 1/2]$ and a constant $c > 0$ such that

$$K_n(\zeta) \leq c \int_{[0,1/2]} h_s(\zeta)\, d\mu(s), \quad \zeta \in \mathbb{T}, n \in \mathbb{N}.$$

Deduce that, for every $f \in L^1(m)$, we have $\lim_{n \to \infty} f * K_n = f$ almost everywhere $[m]$.

10T. Fix $p \in (1, 2)$ and let p' be the conjugate exponent of p. Let $g \in L^p(\lambda_1)$ be such that $\|g\|_p \neq 0$, let $n \in \mathbb{N}$, and define

$$f(t) = \sum_{k=1}^{n} e^{2\pi i a_k t} f(t - b_k),$$

where $a_1, \ldots, a_n, b_1, \ldots, b_n \in \mathbb{R}$.

    (i) Show that $\|f\|_p^p$ is arbitrarily close to $n\|g\|_p^p$ if the numbers $|b_k - b_\ell|$, $k \neq \ell$, are sufficiently large. (Hint: Use Problem 9O.)

   (ii) Show that $\|\widehat{f}\|_{p'}^{p'}$ is arbitrarily close to $n\|\widehat{g}\|_{p'}^{p'}$ if the numbers $|a_k - a_\ell|$ are sufficiently large.

  (iii) Deduce that the ratio $\|\widehat{f}\|_{p'} / \|f\|_p$ can be made arbitrarily small for appropriate choices of n and $a_1, \ldots, a_n, b_1, \ldots, b_n$.

  (iv) Calculate $\|\widehat{f}\|_{p'} / \|f\|_p$ when $f(t) = e^{-\pi t^2}$, $t \in \mathbb{R}$.

10U. Suppose that $f \in L^2(\lambda_1)$ is supported in an interval of length L. Show that $\widehat{f}$ is infinitely differentiable and its nth derivative g_n satisfies $\|g_n\|_2 \leq (\pi L)^n \|f\|_2$, $n \in \mathbb{N}$.

10V. Suppose that $f : \mathbb{R} \to \mathbb{C}$ is integrable $[m]$.

    (i) Show that the series

$$g(e^{2\pi i t}) = \sum_{k=-\infty}^{\infty} f(t+k), \quad t \in \mathbb{R},$$

converges almost everywhere $[\lambda_1]$ and the function g is integrable $[m]$ over $\mathbb{T}$.

   (ii) Show that the Fourier coefficients of g are $\widehat{f}(n)$, $n \in \mathbb{Z}$.

  (iii) Suppose, in addition, that g is everywhere defined and continuous on $\mathbb{T}$, and that $\sum_{n=-\infty}^{\infty} |\widehat{f}(n)| < +\infty$. Prove the *Poisson summation formula*

$$\sum_{n=-\infty}^{\infty} \widehat{f}(n) = \sum_{n=-\infty}^{\infty} f(n).$$

  (iv) Define $\psi : (0, +\infty) \to \mathbb{R}$ by $\psi(x) = \sum_{n=1}^{\infty} e^{-\pi n^2 x}$, $x > 0$. Prove the following identity due to Jacobi:

$$\frac{1 + 2\psi(x)}{1 + 2\psi(1/x)} = \frac{1}{\sqrt{x}}, \quad x > 0.$$

    (Hint: Apply the Poisson summation formula to $f(t) = e^{-\pi x t^2}$.)

10W. Fix an integer $n \geq 2$, and let $G = \mathbb{Z}/n\mathbb{Z}$ denote the additive group of integers modulo n. We denote by $g_k \in G$ the equivalence class of an integer $k \in \mathbb{Z}$, so $G = \{g_1, \ldots, g_n\}$ and $g_0 = g_n$ is the zero element. Denote by μ the counting measure on G.

    (i) Set $\omega = e^{2\pi i/n}$ and define functions $e_k : G \to \mathbb{C}$ by

$$e_k(g_j) = \omega^{kj}, \quad j, k = 1, \ldots, n.$$

Show that $\{n^{-1/2} e_k\}_{k=1}^n$ is an orthonormal basis in $L^2(\mu)$.

   (ii) Given an arbitrary function $f : G \to \mathbb{C}$, define the Fourier transform $\widehat{f} : G \to \mathbb{C}$ by the formula

$$\widehat{f}(g_k) = \langle f, e_k \rangle = \sum_{j=1}^{n} f(g_j) \overline{e_k(g_j)}, \quad k = 1, \ldots, n.$$

Show that $\|\widehat{f}\|_\infty \leq \|f\|_1$ and $\|\widehat{f}\|_2 = n^{1/2} \|f\|_2$ for every $f : G \to \mathbb{C}$. Deduce an analog of the Hausdorff-Young inequalities.

(iii) Verify the following version of the Fourier inversion formula:

$$f(g_k) = \frac{1}{n} \sum_{j=1}^{n} \widehat{f}(g_j) e_k(g_j), \quad k = 1, \ldots, n,$$

for every function $f : G \to \mathbb{C}$.

(iv) Define an analog of convolution for functions in $L^1(\mu)$ and verify the identity $\widehat{f * g} = \widehat{f}\widehat{g}$.

(v) Prove that the Young inequalities (Proposition 10.9) hold for functions defined on G.

Remark 10.43. The functions e_k in the preceding problem are precisely the *characters* of the group G, that is, the homomorphisms φ from the (additive) group G to the multiplicative group $\mathbb{T}$. Similarly, the functions $\zeta \to \zeta^n$, $n \in \mathbb{Z}$, are the continuous characters of the group $\mathbb{T}$, and the maps $x \mapsto e^{2\pi i t x}$, $t \in \mathbb{R}$, are the continuous characters of the additive group $\mathbb{R}$. Many of the results in this chapter extend to functions defined on an arbitrary Hausdorff, locally compact, Abelian group. The analogs of characters for noncommutative groups are irreducible unitary representations.

Chapter 11
Standard measure spaces

There are many examples of measure spaces that present various pathologies
and on which some of the deeper theorems of measure theory fail. Elements
of the class of standard measure spaces, to be defined shortly, do not display
any of these pathologies and, in addition, can be classified up to a natural
notion of isomorphism. The results we present here use little in addition to
the observation that a separable metric space can be written as a countable
union of closed subsets of arbitrarily small diameter (for instance, the closed
balls of a fixed radius centered at the points of a countable dense set). They
were developed initially in order to resolve questions about the measurability
of the projection onto a one-dimensional subspace of a Borel subset of $\mathbb{R}^2$.
We develop enough of the theory of standard spaces to display some of the
desirable features of measure theory on such spaces. We start with the idea
of isomorphism between measurable spaces.

Definition 11.1. Two measurable spaces $(X, \mathbf{S})$ and $(Y, \mathbf{T})$ are said to be
isomorphic if there is a bijection $f : X \to Y$ such that, for $E \subset X$, we have
$E \in \mathbf{S}$ if and only if $f(E) \in \mathbf{T}$. We say that $(X, \mathbf{S})$ *embeds* in $(Y, \mathbf{S})$ if there
exists $F \in \mathbf{T}$ such that $(X, \mathbf{S})$ is isomorphic to $(F, \mathbf{T}_F)$.

Another way to state the first condition on the function f above is to say
that it is measurable $[\mathbf{S}, \mathbf{T}]$ and the inverse map f^{-1} is measurable $[\mathbf{T}, \mathbf{S}]$. The
following result is an analog of the Cantor-Bernstein theorem of set theory
(see $[\mathbf{I}, \text{Theorem } 4.1]$).

Theorem 11.2. *Assume that the measurable spaces $(X, \mathbf{S})$ and $(Y, \mathbf{T})$ are
such that $(X, \mathbf{S})$ embeds in $(Y, \mathbf{T})$ and $(Y, \mathbf{T})$ embeds in $(X, \mathbf{S})$. Then $(X, \mathbf{S})$
and $(Y, \mathbf{T})$ are isomorphic.*

Proof. Let $E \in \mathbf{S}$ and $F \in \mathbf{T}$ be such that $(X, \mathbf{S})$ and $(Y, \mathbf{T})$ are isomorphic
to $(F, \mathbf{T}_F)$ and $(E, \mathbf{S}_E)$, respectively, and let $f : X \to F$ and $g : Y \to E$
be bijections realizing these isomorphisms. It suffices to find sets $A \in \mathbf{S}$ and

© Springer International Publishing Switzerland 2016
H. Bercovici et al., *Measure and Integration*,
DOI 10.1007/978-3-319-29046-1_11

$B \in \mathbf{T}$ such that $f(A) = Y \setminus B$ and $g(B) = X \setminus A$. Indeed, once such sets are found, an isomorphism between $(X, \mathbf{S})$ and $(Y, \mathbf{T})$ is given by the bijection $h : X \to Y$ defined by

$$h(x) = \begin{cases} f(x), & x \in A, \\ g^{-1}(x), & x \in X \setminus A. \end{cases}$$

To find the set A, consider the map $u : \mathbf{S} \to \mathbf{S}$ defined by

$$u(D) = X \setminus g(Y \setminus f(D)), \quad D \in \mathbf{S}.$$

The map u is monotone increasing (that is, $D_1 \subset D_2$ implies $u(D_1) \subset u(D_2)$), and it carries countable unions to countable unions. It follows that the set $A \in \mathbf{S}$ defined by

$$A = \bigcup_{n=1}^{\infty} \underbrace{u(u(\cdots u(\varnothing)\cdots))}_{n \text{ times}}$$

satisfies the equation $A = u(A)$. To conclude the proof, set $B = Y \setminus f(A)$ and verify that these sets satisfy the desired conditions. $\square$

Definition 11.3. A measurable space $(X, \mathbf{S})$ is said to be *standard* if there exists a complete, separable metric space Y such that $(X, \mathbf{S})$ is isomorphic (as a measurable space) to $(Y, \mathbf{B}_Y)$. A measure space $(X, \mathbf{S}, \mu)$ is said to be *standard* if $(X, \mathbf{S})$ is a standard measurable space.

Another way to state this definition is to say that $(X, \mathbf{S})$ is standard if there is a complete metric d on X such (X, d) is separable and $\mathbf{S} = \mathbf{B}_X$. Observe that the definition of the Borel σ-algebra $\mathbf{B}_X$ only requires the topology of the metric space X, not its metric. There are usually many different topologies on X which generate the same Borel σ-algebra. See, for instance, Theorem 11.9 below.

Example 11.4. The space $((0, 1), \mathbf{B}_{(0,1)})$, where $(0, 1)$ is given its usual topology as a subset of $\mathbb{R}$, is standard. Indeed, $(0, 1)$ is homeomorphic to $\mathbb{R}$ so $((0, 1), \mathbf{B}_{(0,1)})$ is isomorphic to $(\mathbb{R}, \mathbf{B}_{\mathbb{R}})$. Similarly, $[0, 1)$ is homeomorphic to $[0, +\infty)$, and therefore $([0, 1), \mathbf{B}_{[0,1)})$ is standard.

Definition 11.5. A topological space (X, τ) is called a *Polish space* if it is homeomorphic to a complete, separable metric space. The topology τ is said to be *Polish* if (X, τ) is a Polish space.

Remark 11.6. Example 11.4 shows that a subspace of a separable, complete metric space (X, d) may be Polish even though it is not complete in the metric d. Such subspaces are precisely the G_δ sets in X. See Problem D.

Example 11.7. If $\{(X_n, \tau_n)\}_{n=1}^{\infty}$ is a sequence of Polish spaces, then the space $X = \prod_{n=1}^{\infty} X_n$ endowed with the product topology is also a Polish space. Indeed, it is easy to see that X is separable. Moreover, if τ_n is defined by a complete metric d_n on X_n, the metric

$$d((x_n)_{n=1}^\infty, (y_n)_{n=1}^\infty) = \sum_{n=1}^\infty 2^{-n} \min\{1, d_n(x_n, y_n)\}, \quad (x_n)_{n=1}^\infty, (y_n)_{n=1}^\infty \in X,$$

is complete and this metric topology is the product topology. Two important particular cases are the space $\mathcal{S} = \mathbb{N}^\mathbb{N}$ of all sequences of positive integers and the *Cantor space* $\mathcal{C} = \{0,1\}^\mathbb{N}$. Both of these are Polish spaces when endowed with the product topology, where every factor is given the discrete topology. The reason for this terminology is that $\mathcal{C}$ is homeomorphic to the standard ternary Cantor set $C \subset [0,1]$ via the continuous bijection $f : \mathcal{C} \to C$ defined by

$$f((x_n)_{n=1}^\infty) = \sum_{n=1}^\infty \frac{2x_n}{3^n}, \quad (x_n)_{n=1}^\infty \in \mathcal{C}.$$

(Recall that the inverse of a continuous bijection between compact Hausdorff spaces is necessarily continuous.) The space $\mathcal{S}$ is also homeomorphic to a set of real numbers, namely the set $[0,1] \setminus \mathbb{Q}$. The identification $g : \mathcal{S} \to [0,1] \setminus \mathbb{Q}$ is obtained by use of continued fractions:

$$g((x_n)_{n=1}^\infty) = \cfrac{1}{x_1 + \cfrac{1}{x_2 + \cfrac{1}{\ddots}}}, \quad (x_n)_{n=1}^\infty \in \mathcal{S}.$$

Of course, $\mathcal{C}$ is a compact subset of $\mathcal{S}$. The space $\mathcal{S}$ is also homeomorphic to a subspace of $\mathcal{C}$ via the homeomorphism

$$h((x_n)_{n=1}^\infty) = (\underbrace{0, \ldots, 0}_{x_1 \text{ times}}, 1, \underbrace{0, \ldots, 0}_{x_2 \text{ times}}, 1, \underbrace{0, \ldots, 0}_{x_3 \text{ times}}, 1, \ldots).$$

Thus, via Theorem 11.2, $\mathcal{C}$ and $\mathcal{S}$ are isomorphic as measurable spaces.

Another useful Polish space is the *Hilbert cube* $\mathcal{Q} = [0,1]^\mathbb{N}$ with the product topology, where $[0,1]$ is given its usual topology as a subset of $\mathbb{R}$.

Theorem 11.8. *Let (X, τ) be a Polish space. Then*

(1) *X is homeomorphic to a subspace of $\mathcal{Q}$.*
(2) *If X is uncountable, then X contains a subspace homeomorphic to $\mathcal{C}$.*
(3) *There exists a continuous surjective map $h : \mathcal{S} \to X$.*

In particular, if X is not countable then it has cardinality $\mathfrak{c} = 2^{\aleph_0}$.

Proof. Fix a complete metric d defining the topology τ and a dense sequence $\{x_n\}_{n=1}^\infty$ in X. The map $f : X \to \mathcal{Q}$ defined by

$$f(x) = (\max\{1, d(x, x_n)\})_{n=1}^\infty$$

is easily seen to be a homeomorphism of X onto $f(X)$. To prove (2), assume that X is uncountable. We claim that there exist closed, disjoint, uncountable subsets $X_0, X_1 \subset X$ each of which has diameter at most 1. Indeed, assume that such sets cannot be found, and let ε be a positive number. Denote by $B_{n,\varepsilon}$ the closed ball of radius ε centered at x_n. Then any two uncountable sets of the form $B_{n,\varepsilon}$ must intersect. Therefore the union of all uncountable sets $B_{n,\varepsilon}$ is contained in a closed ball of the form $B_{m_\varepsilon,3\varepsilon}$. In other words, $X \setminus B_{m_\varepsilon,3\varepsilon}$ is countable for some m. This argument, applied to $\varepsilon_k = 4^{-k}$, yields a ball B_k of radius $3\varepsilon_k$ such that $X \setminus B_k$ is at most countable, so $X \setminus [\bigcap_{k\in\mathbb{N}} B_k]$ is at most countable as well. The intersection $\bigcap_{k\in\mathbb{N}} B_k$ consists of at most one point, thus leading to the conclusion that X itself is countable, contrary to the hypothesis. We conclude that there exist closed, disjoint uncountable sets $X_0, X_1 \subset X$ of diameter at most 1. We can now construct, by induction, for each $N \in \mathbb{N}$, closed, uncountable, pairwise disjoint sets $X_{n_1,n_2,\ldots,n_N}$ for $(n_1,\ldots,n_N) \in \{0,1\}^N$ such that $X_{n_1,n_2,\ldots,n_N} \subset X_{n_1,n_2,\ldots,n_{N-1}}$ and the diameter of $X_{n_1,n_2,\ldots,n_N}$ is less than 2^{-N}. Indeed, we just apply the preceding observation to the uncountable Polish space $X_{n_1,n_2,\ldots,n_{N-1}}$. Once this is done, a function $g : \mathcal{C} \to X$ such that

$$\bigcap_{N=1}^{\infty} X_{n_1,n_2,\ldots,n_N} = \{g(t)\}, \quad t = (n_j)_{j=1}^{\infty} \in \mathcal{C},$$

exists because d is a complete metric. The map g is a homeomorphism from $\mathcal{C}$ to $g(\mathcal{C})$.

The proof of (3) is similar to that of (2). We construct, for every $k \in \mathbb{N}$ and every $(n_1,\ldots,n_k) \in \mathbb{N}^k$, a closed subset $A_{n_1,\ldots,n_k} \neq \varnothing$ with diameter at most 2^{-k} such that $X = \bigcup_{n_1 \in \mathbb{N}} A_{n_1}$ and

$$A_{n_1,\ldots,n_k} = \bigcup_{n\in\mathbb{N}} A_{n_1,\ldots,n_k,n}, \quad k, n_1,\ldots,n_k \in \mathbb{N}.$$

The function $h : \mathcal{S} \to X$ is defined by the requirement that

$$\{h(\mathbf{n})\} = \bigcap_{k=1}^{\infty} A_{n_1,\ldots,n_k}$$

for every sequence $\mathbf{n} = (n_1, n_2, \ldots) \in \mathcal{S}$. $\square$

It is convenient in what follows to write $\mathbf{B}_\tau$ for the Borel σ-algebra of a topological space (X,τ). Thus $(X,\mathbf{B}_\tau)$ is a standard measurable space if (X,τ) is a Polish space. There are usually different Polish topologies τ, τ' on a set X such that $\mathbf{B}_\tau = \mathbf{B}_{\tau'}$, and which one is used is immaterial for questions of measure theory.

Theorem 11.9. *Let $(X,\mathbf{S})$ be a standard measurable space, and fix a set $F \in \mathbf{S}$. There exists a Polish topology τ on X such that $\mathbf{S} = \mathbf{B}_\tau$ and F is both open and closed in (X,τ).*

Proof. Let d be a metric on X that defines a Polish topology σ on X such that $\mathbf{S} = \mathbf{B}_\sigma$. Denote by $\mathbf{T}$ the collection of those sets $E \in \mathbf{S}$ for which there exists a Polish topology τ on X finer than σ such that $\mathbf{S} = \mathbf{B}_\tau$ and E is both open and closed in τ. We prove the theorem by showing that $\mathbf{T} = \mathbf{S}$. To do this it suffices to show that $\mathbf{T}$ contains all the open sets in σ and that it is closed under the formation of finite intersections and countable unions.

Assume first that E is an arbitrary open set in σ. The subspace $X \setminus E \subset X$, endowed with the restriction of d, is again a complete metric space. We now show that E is also a Polish space in the relative topology on E induced by σ. A complete metric d' on E is obtained by selecting a dense sequence $\{x_n\}_{n=1}^\infty$ in $X \setminus E$ and setting

$$d'(x, y) = d(x, y) + \sum_{n=1}^\infty 2^{-n} |h_n(x) - h_n(y)|, \quad x, y \in E,$$

where

$$h_n(x) = \min\left\{1, \frac{1}{d(x, x_n)}\right\}, \quad n \in \mathbb{N}.$$

We define a metric d_E on X by setting

$$d_E(x, y) = \begin{cases} d(x, y), & x, y \in X \setminus E, \\ d'(x, y), & x, y \in E, \\ \max\{1, d(x, y)\}, & x \in E, y \in X \setminus E \text{ or } x \in X \setminus E, y \in E. \end{cases}$$

The topology τ_E on X defined by this metric is Polish, $\mathbf{S} = \mathbf{B}_{\tau_E}$, and E is both open and closed in τ_E. (See Problem 11C.) This argument shows that $\mathbf{T}$ contains all open sets in σ.

Suppose next that $\{E_n\}_{n=1}^\infty \subset \mathbf{T}$, and d_n is a metric on X defining a Polish topology τ_n finer than σ such that E_n and $X \setminus E_n$ are open in τ_n and $\mathbf{B}_{\tau_n} = \mathbf{S}$ for all $n \in \mathbb{N}$. Define a new metric $\widetilde{d}$ on X by setting

$$\widetilde{d}(x, y) = \sum_{n=1}^\infty 2^{-n} \min\{1, d_n(x, y)\}, \quad x, y \in X.$$

The inequality $\widetilde{d} \geq 2^{-n} \min\{1, d_n\}$ shows that the topology $\widetilde{\tau}$ generated by $\widetilde{d}$ is finer than τ_n for every n. In particular, E_n and $X \setminus E_n$ are open in $\widetilde{\tau}$ for all $n \in \mathbb{N}$. A sequence $\{x_k\}_{k=1}^\infty$ is Cauchy in the metric $\widetilde{d}$ if and only if it is Cauchy in each d_n. Since each d_n is complete, there exists $a_n \in X$ such that $\lim_{k \to \infty} d_n(x_k, a_n) = 0$. Moreover, since τ_n is finer than σ, $\lim_{k \to \infty} d(x_k, a_n) = 0$ as well. We conclude that there exists $a \in X$ such that $a_n = a$ for all $n \in \mathbb{N}$ and $\lim_{k \to \infty} \widetilde{d}(x_k, a) = 0$. Thus $\widetilde{d}$ is a complete metric.

We show next that the space $(X, \widetilde{\tau})$ is second countable. Assume that $\{B_{n,j}\}_{j=1}^\infty$ is a base of open sets for τ_n and note that each $B_{n,j} \in \mathbf{S}$. We show that the finite intersections of sets in $\{B_{n,j}\}_{n,j=1}^{\infty}$ form a base for $\widetilde{\tau}$. Let $G \subset X$ be open in $\widetilde{\tau}$. For every $x \in G$, we have

$$G \supset \{y \in X : \widetilde{d}(x,y) < \varepsilon\}$$

for some $\varepsilon > 0$, and

$$\{y \in X : \widetilde{d}(x,y) < \varepsilon\} \supset \bigcap_{n=1}^{N} \{y \in X : d_n(x,y) < \varepsilon/2\}$$

provided that $2^{-N} < \varepsilon/2$. It follows that $\{y \in X : \widetilde{d}(x,y) < \varepsilon\}$ contains a finite intersection W_x of sets in $\{B_{n,j}\}_{n,j=1}^{\infty}$ such that $x \in W_x$ and thus $G = \bigcup_{x \in G} W_x$. Since $B_{n,j} \in \mathbf{S}$ for all $n, j \in \mathbb{N}$, we conclude that $\mathbf{B}_{\widetilde{\tau}} = \mathbf{S}$. Moreover, E_n is both open and closed in $\widetilde{\tau}$ for $n \in \mathbb{N}$. We see immediately that $E_1 \cap E_2$ is both open and closed in $\widetilde{\tau}$, and this establishes the fact that $\mathbf{T}$ is closed under the formation of finite intersections. It also follows that the union $E = \bigcup_{n \in \mathbb{N}} E_n$ is open in $\widetilde{\tau}$. The first part of the proof provides a Polish topology τ on X finer than $\widetilde{\tau}$ such that $\mathbf{B}_\tau = \mathbf{B}_{\widetilde{\tau}} = \mathbf{S}$ and E is closed as well as open in τ. This shows that $\mathbf{T}$ is closed under the formation of countable unions and thus concludes the proof of the theorem. $\square$

The last part of the proof of Theorem 11.9 yields the following strengthening of the theorem.

Corollary 11.10. *Let $(X, \mathbf{S})$ be a standard measurable space, and fix a sequence $\{F_n\}_{n \in \mathbb{N}} \subset \mathbf{S}$. There exists a Polish topology τ on X such that $\mathbf{S} = \mathbf{B}_\tau$ and each F_n is both open and closed in (X, τ), $n \in \mathbb{N}$.*

Corollary 11.11. *Let $(X, \mathbf{S})$ be a standard measurable space, and let $E \in \mathbf{S}$. Then the measurable space $(E, \mathbf{S}_E)$ is standard.*

Proof. Let τ be a Polish topology on X such that $\mathbf{S} = \mathbf{B}_\tau$. By Theorem 11.9 we may assume that E is closed in X, and therefore is a Polish space in the relative topology. The conclusion follows now because $\mathbf{S}_E$ is the Borel σ-algebra of the topology induced by τ on E. $\square$

Corollary 11.12. *Let $(X, \mathbf{S})$ be a standard measurable space. There exists an injective map $f : X \to \mathcal{C}$ which is measurable $[\mathbf{S}, \mathbf{B}_{\mathcal{C}}]$.*

Proof. Let τ be a Polish topology on X such that $\mathbf{B}_\tau = \mathbf{S}$, and let $\{G_n\}_{n=1}^{\infty}$ be a base of open sets for τ, so $\mathbf{S}$ is the σ-algebra generated by $\{G_n\}_{n=1}^{\infty}$. By Corollary 11.10 there is a Polish topology τ' on X finer than τ such that $\mathbf{B}_{\tau'} = \mathbf{S}$ and such that the sets G_n are also closed in τ'. If $x, y \in X$ and $x \neq y$, there exists $n \in \mathbb{N}$ such that $x \in G_n$ and $y \notin G_n$. Thus the map $f : X \to \mathcal{C}$ defined by

$$f(x) = (\chi_{G_n}(x))_{n=1}^{\infty}$$

is injective and continuous on (X, τ'), and hence measurable $[\mathbf{S}, \mathbf{B}_{\mathcal{C}}]$. $\square$

We have seen in Theorem 11.8 that the Cantor space $\mathcal{C}$ is homeomorphic to a closed subset of any uncountable Polish space. This implies that $(\mathcal{C}, \mathbf{B}_\mathcal{C})$ embeds into any standard space $(X, \mathbf{S})$ such that X is uncountable.

Proposition 11.13. *Let $(X, \mathbf{S})$ and $(Y, \mathbf{T})$ be standard measurable spaces and let $f : X \to Y$ be measurable $[\mathbf{S}, \mathbf{T}]$. Then there exist Polish topologies σ and τ on X and Y, respectively, such that $\mathbf{S} = \mathbf{B}_\sigma$, $\mathbf{T} = \mathbf{B}_\tau$, and $f : (X, \sigma) \to (Y, \tau)$ is continuous.*

Proof. Let τ be a Polish topology on Y such that $\mathbf{T} = \mathbf{B}_\tau$, and let $\{B_n\}_{n \in \mathbb{N}}$ be a base of open sets for this topology. By Corollary 11.10 there exists a Polish topology σ on X such that $\mathbf{S} = \mathbf{B}_\sigma$ and $f^{-1}(B_n)$ is open for every $n \in \mathbb{N}$. The map $f : (X, \sigma) \to (Y, \tau)$ is obviously continuous. $\square$

Theorem 11.14. *Let $(X, \mathbf{S})$ and $(Y, \mathbf{T})$ be two standard measurable spaces, and let $f : X \to Y$ be a measurable function. Assume that $A, B \in \mathbf{S}$ and satisfy $f(A) \cap f(B) = \varnothing$. Then there exist disjoint sets $C, D \in \mathbf{T}$ such that $f(A) \subset C$ and $f(B) \subset D$, that is, $f(A)$ and $f(B)$ are separated by measurable sets.*

Proof. Let d_1 and d_2 be complete metrics on X and Y that define Polish topologies σ and τ on X and Y, respectively, such that $\mathbf{B}_\sigma = \mathbf{S}$, $\mathbf{B}_\tau = \mathbf{T}$, and $f : (X, \sigma) \to (Y, \tau)$ is continuous. This is possible by Proposition 11.13. Suppose, to get a contradiction, that $f(A)$ and $f(B)$ cannot be separated by measurable sets. We claim that there exist decreasing sequences

$$A = A^{(0)} \supset A^{(1)} \supset \cdots \supset A^{(n)} \supset \cdots$$

and

$$B = B^{(0)} \supset B^{(1)} \supset \cdots \supset B^{(n)} \supset \cdots$$

of closed sets such that for each $n \in \mathbb{N}$, $A^{(n)}$ and $B^{(n)}$ have diameter less than $1/n$ and $f(A^{(n)})$ cannot be separated from $f(B^{(n)})$ by measurable sets. These sequences of sets are constructed inductively. Suppose that $A^{(k)}$ and $B^{(k)}$ have been constructed for $k = 0, \ldots, n-1$, and write $A^{(n-1)} = \bigcup_{j \in \mathbb{N}} A_j$ and $B^{(n-1)} = \bigcup_{j \in \mathbb{N}} B_j$ as unions of closed measurable sets A_j and B_j with diameter less than $1/n$. If, for all $j, k \in \mathbb{N}$, $f(A_j)$ and $f(B_k)$ can be separated by measurable sets $C_{j,k} \supset f(A_j)$ and $D_{j,k} \supset f(B_k)$, then the sets

$$C = \bigcup_{j=1}^{\infty} \left[\bigcap_{k=1}^{\infty} C_{j,k} \right], \quad D = Y \setminus C,$$

separate $f(A^{(n-1)})$ and $f(B^{(n-1)})$, contrary to the inductive hypothesis. Thus there exist $j, k \in \mathbb{N}$ such that $f(A_j)$ and $f(B_k)$ cannot be separated by measurable sets. We complete the inductive process by setting $A^{(n)} = A_j$ and $B^{(n)} = B_k$.

272 11 Standard measure spaces

Since d_1 is a complete metric, there exist two unique points

$$x_1 \in \bigcap_{n=1}^{\infty} A^{(n)}, \; x_2 \in \bigcap_{n=1}^{\infty} B^{(n)}.$$

We have $f(x_1) \neq f(x_2)$ because $x_1 \in A$, $x_2 \in B$, and $f(A) \cap f(B) = \varnothing$. Therefore $f(x_1)$ and $f(x_2)$ have disjoint neighborhoods. In other words, there exist $p, q \in \mathbb{N}$ so $G_p \cap G_q = \varnothing$, $f(x_1) \in G_p$, and $f(x_2) \in G_q$. Using the continuity of f at x_1 and x_2, we see that $f(A^{(n)}) \subset G_p$ and $f(B^{(n)}) \subset G_q$ for sufficiently large n, showing that $f(A^{(n)})$ and $f(B^{(n)})$ are separated by measurable sets. This contradiction concludes the proof. $\square$

Repeated application of Theorem 11.14 shows that it is possible to separate more than two sets.

Corollary 11.15. *Let* $(X, \mathbf{S})$ *and* $(Y, \mathbf{T})$ *be two standard measurable spaces, and let* $f : X \to Y$ *be a measurable function. Assume that the sets* $\{E_n\}_{n=1}^{\infty} \subset \mathbf{S}$ *are such that the sets* $\{f(E_n)\}_{n=1}^{\infty}$ *are pairwise disjoint. Then there exist pairwise disjoint sets* $\{F_n\}_{n=1}^{\infty} \subset \mathbf{T}$ *such that* $f(E_n) \subset F_n$ *for all* $n \in \mathbb{N}$.

Proof. By Theorem 11.14, we can select sets $C_n \in \mathbf{T}$ such that $f(E_n) \subset C_n$ and $f(\bigcup_{m \neq n} E_m) \subset Y \setminus C_n$, for $n \in \mathbb{N}$, and set

$$F_n = C_n \cap \bigcap_{m \neq n} (Y \setminus C_m).$$

The sequence $\{F_n\}_{n=1}^{\infty}$ has the two desired properties. $\square$

Theorem 11.16. *Let* $(X, \mathbf{S})$ *and* $(Y, \mathbf{T})$ *be two standard measurable spaces, and let* $f : X \to Y$ *be an injective measurable function. Then for every* $E \in \mathbf{S}$ *we have* $f(E) \in \mathbf{T}$.

Proof. For every $E \in \mathbf{S}$, the space $(E, \mathbf{S}_E)$ is standard, and thus we may, and do, assume with no loss of generality that $E = X$. By Proposition 11.13, there exist Polish topologies σ and τ on X and Y such that $\mathbf{B}_\sigma = \mathbf{S}$, $\mathbf{B}_\tau = \mathbf{T}$ and f is continuous from (X, σ) to (Y, τ). Assume that σ is defined by the complete metric d on X. Construct a partition $X = \bigcup_{n=1}^{\infty} A_n$ into Borel sets of diameter less than $1/2$. Moreover, for each $n_1 \in \mathbb{N}$, A_{n_1} can be partitioned as $A_{n_1} = \bigcup_{m=1}^{\infty} A_{n_1,m}$ where each $A_{n_1,m}$ is a Borel set of diameter less than $1/4$, and continuing by induction, we construct, for each finite sequence $n_1, n_2, \ldots, n_k$ of natural numbers, a (possibly empty) Borel set $A_{n_1,n_2,\ldots,n_k}$ of diameter less than 2^{-k} such that the nonempty sets in the collection $\{A_{n_1,n_2,\ldots,n_{k-1},n} : n \in \mathbb{N}\}$ form a partition of $A_{n_1,n_2,\ldots,n_{k-1}}$. The sets $\{f(A_{n_1,n_2,\ldots,n_k}) : n_1, n_2, \ldots, n_k \in \mathbb{N}\}$ are pairwise disjoint, and Corollary 11.15 yields pairwise disjoint sets $\{B_{n_1,n_2,\ldots,n_k} : n_1, n_2, \ldots, n_k \in \mathbb{N}\} \subset \mathbf{T}$ such that

$$A_{n_1,n_2,\ldots,n_k} \subset B_{n_1,n_2,\ldots,n_k}, \quad n_1, n_2, \ldots, n_k \in \mathbb{N}.$$

Replacing $B_{n_1,\ldots,n_k}$ by $B_{n_1} \cap B_{n_1,n_2} \cap \cdots \cap B_{n_1,\ldots,n_k}$ we may also assume that $B_{n_1,\ldots,n_k} \subset B_{n_1,\ldots,n_{k-1}}$ for all $n_1, n_2, \ldots, n_k \in \mathbb{N}$ and for all $k \in \mathbb{N} \setminus \{1\}$. The theorem will follow from the equality

$$f(X) = \bigcap_{k=1}^{\infty} \left[\bigcup_{n_1,n_2,\ldots,n_k \in \mathbb{N}} (B_{n_1,n_2,\ldots,n_k} \cap f(A_{n_1,n_2,\ldots,n_k})^-) \right],$$

because the right-hand side of the equation is clearly a Borel set. Here we denote by $f(A_{n_1,n_2,\ldots,n_k})^-$ the closure of $f(A_{n_1,n_2,\ldots,n_k})$ in the topology τ. To prove this equality, it suffices to show that each point y in the intersection above is in $f(X)$. Indeed, given such a point y, there exists (see Problem 11I) a sequence of integers $\{n_k\}_{k=1}^{\infty}$ such that

$$y \in B_{n_1,n_2,\ldots,n_k} \cap \overline{f(A_{n_1,n_2,\ldots,n_k})}, \quad k \in \mathbb{N}.$$

In particular, the sets $A_{n_1,n_2,\ldots,n_k}$ corresponding to these indices are not empty. In addition the diameter of $A_{n_1,n_2,\ldots,n_k}$ tends to zero as $k \to \infty$. Since d is a complete metric, there is a unique point $x \in \bigcap_{k=1}^{\infty} \overline{A_{n_1,n_2,\ldots,n_k}}$. Continuity of f at x implies that $f(x) \in \bigcap_{k=1}^{\infty} \overline{f(A_{n_1,n_2,\ldots,n_k})}$ and also that the diameters of the sets in this intersection tend to zero. We conclude that $f(x) = y$. $\qquad \square$

A consequence of the above results is that there are very few equivalence classes of standard measurable spaces with respect to the equivalence relation of isomorphism.

Theorem 11.17. *Two standard measurable spaces are isomorphic if and only if they have the same cardinality. All uncountable standard measurable spaces are isomorphic to $(\mathcal{C}, \mathbf{B}_{\mathcal{C}})$.*

Proof. It suffices to verify the last statement. Let $(X, \mathbf{S})$ be an uncountable standard measurable space. By Theorem 11.8, $(\mathcal{C}, \mathbf{B}_{\mathcal{C}})$ embeds into $(X, \mathbf{S})$, while Theorems 11.12 and 11.16 show that $(X, \mathbf{S})$ embeds into $(\mathcal{C}, \mathbf{B}_{\mathcal{C}})$. Theorem 11.2 then applies to yield the desired conclusion. $\qquad \square$

It was shown in Problem 1Z that analytic sets in a complete metric space X (see Definition 1.24) are precisely the images of continuous functions defined on some other complete metric space Y. (As noted in that problem, the space Y can be taken to be the space $\mathcal{S}$.) The following result allows us to define the concept of an analytic set in an arbitrary standard measurable space.

Proposition 11.18. *Let $(X, \mathbf{S})$ be a standard measurable space and let $A \neq \varnothing$ be a subset of X. The following assertions are equivalent.*

(1) *There exists a Polish topology τ on X such that $\mathbf{S} = \mathbf{B}_{\tau}$ and A is analytic in (X, τ).*

(2) *There exist a standard measurable space $(Y, \mathbf{T})$ and a measurable function $f : Y \to X$ such that $f(Y) = A$.*

(3) *There exists a measurable function $f : \mathcal{S} \to X$ such that $f(\mathcal{S}) = A$.*
(4) *There exist sets $\{F_{n_1,\ldots,n_k} : k, n_1, \ldots, n_k\}$ in $\mathbf{S}$ such that*

$$A = \bigcup_{(n_1, n_2, \ldots) \in \mathbb{N}^\mathbb{N}} (F_{n_1} \cap F_{n_1, n_2} \cap \cdots \cap F_{n_1, \ldots, n_k} \cap \cdots).$$

Sets A that satisfy these equivalent conditions are said to be analytic *in* $(X, \mathbf{S})$.

Proof. The equivalence of (1) and (4) follows immediately from Definition 1.24 and Corollary 11.10. Indeed the sets $\{F_{n_1,\ldots,n_k} : k, n_1, \ldots, n_k\}$ from condition (4) are closed in some Polish topology τ such that $\mathbf{S} = \mathbf{B}_\tau$. Similarly, the equivalence between (2) and (3) follows from Problem 1Z and Proposition 11.13 because the map f in (2) is continuous when X and Y are endowed with appropriate topologies. Finally, the equivalence of (2) and (3) follows from Theorem 11.8(3).   □

For our next result about standard spaces it is useful to consider the *lexicographical order* on the space $\mathcal{S} = \mathbb{N}^\mathbb{N}$. Given $\mathbf{m} = (m_1, m_2, \ldots)$ and $\mathbf{n} = (n_1, n_2, \ldots)$ in $\mathcal{S}$, we write $\mathbf{m} < \mathbf{n}$ if $\mathbf{m} \neq \mathbf{n}$ and for the first index j such that $m_j \neq n_j$ we have $m_j < n_j$. We also write $\mathbf{m} \leq \mathbf{n}$ if either $\mathbf{m} = \mathbf{n}$ or $\mathbf{m} < \mathbf{n}$. The set $\mathcal{S}$ is totally ordered by this relation, but it is not well ordered as illustrated by the decreasing sequence $\{\mathbf{n}_k\}_{k \in \mathbb{N}}$, where

$$\mathbf{n}_k = (\underbrace{0, \ldots, 0}_{k \text{ times}}, 1, 1, \ldots), \quad k \in \mathbb{N}.$$

Proposition 11.19. *Every closed subset $A \neq \varnothing$ of $\mathcal{S}$ contains a least element.*

Proof. Define $m_1 \in \mathbb{N}$ to be the smallest integer with the property that there is some sequence in A with first entry equal to m_1. Then define inductively $m_k \in \mathbb{N}$ for $k \in \mathbb{N} \setminus \{1\}$ to be the smallest integer with the property that there exists a sequence in A starting with $m_1, \ldots, m_{k-1}, m_k$. We show that $\mathbf{m} = (m_1, m_2, \ldots)$ is an element of A. Indeed, the definition of this sequence ensures the existence of elements $\mathbf{n}_k \in A$ starting with $(m_1, \ldots, m_k)$. Clearly $\mathbf{m} = \lim_{k \to \infty} \mathbf{n}_k$. The inequality $\mathbf{m} \leq \mathbf{n}$, $\mathbf{n} \in A$, follows immediately from the definition of $\mathbf{m}$.   □

Theorem 11.16 implies that a measurable bijection between standard measure spaces is automatically an isomorphism. The following result is a substitute of this fact for noninjective maps.

Theorem 11.20. *Let $(X, \mathbf{S})$ and $(Y, \mathbf{T})$ be standard measurable spaces, and let $f : X \to Y$ be a measurable map. There exists a function $g : f(X) \to Y$ such that $f(g(y)) = y$ for all $y \in f(X)$ and, for every $A \in \mathbf{T}$, $g^{-1}(A)$ belongs to the σ-algebra generated by the analytic sets in Y.*

Proof. We may, and do, assume that $X \neq \varnothing$. Let σ and τ be Polish topologies on X and Y, respectively, such that $\mathbf{S} = \mathbf{B}_\sigma$, $\mathbf{T} = \mathbf{B}_\tau$, and $f : (X, \sigma) \to (Y, \tau)$ is continuous. We show first that it suffices to prove the theorem in the special case in which X equals the sequence space $\mathcal{S}$. To do this, choose a surjective continuous function $h : \mathcal{S} \to X$. The existence of such a function is guaranteed by Theorem 11.8(3). Then $f \circ h : \mathcal{S} \to Y$ is surjective. If the theorem is true for $X = \mathcal{S}$, then there exists a map $u : f(X) \to \mathcal{S}$, measurable when $f(X)$ is endowed with the σ-algebra generated by the analytic sets, such that $f(h(u(y))) = y$ for every $y \in f(X)$. Therefore the map $g = h \circ u$ satisfies the conditions in the statement of the theorem.

For the remainder of the proof, we assume that $X = \mathcal{S}$. For each $y \in f(X)$, the set $f^{-1}(\{y\})$ is closed in $\mathcal{S}$ and therefore it has a smallest element $g(y)$. To conclude the proof, it suffices to show that $g^{-1}(F)$ is analytic in Y for F in a collection of sets that generates the σ-algebra $\mathbf{B}_\mathcal{S}$. Such a collection is provided by the open sets

$$F_\mathbf{m} = \{\mathbf{n} \in \mathcal{S} : \mathbf{n} < \mathbf{m}\}, \quad \mathbf{m} \in \mathcal{S}.$$

We show that $g^{-1}(F_\mathbf{m}) = f(F_\mathbf{m})$, and these sets are analytic by Theorem 11.18. The inclusion $g^{-1}(F_\mathbf{m}) \subset f(F_\mathbf{m})$ follows from the equality $f(g(y)) = y$, $y \in g^{-1}(F_\mathbf{m})$. On the other hand, if $y = f(\mathbf{n})$ for some $\mathbf{n} \in F_\mathbf{m}$, then $g(y) \leq \mathbf{n} < \mathbf{m}$ so $g(y) \in F_\mathbf{m}$ The fact that the sets $F_\mathbf{m}$ generate $\mathbf{B}_\mathcal{S}$ is left as an exercise (see Problem 11O). $\qquad \square$

Standard measure spaces are also easy to classify. The appropriate concept of isomorphism is as follows.

Definition 11.21. Two measure spaces $(X, \mathbf{S}, \mu)$ and $(Y, \mathbf{T}, \nu)$ are said to be *isomorphic* if there exist sets $X_0 \in \mathbf{S}$, $Y_0 \in \mathbf{T}$, and a bijection $f : X_0 \to Y_0$ such that:

(1) $\mu(X \setminus X_0) = \nu(Y \setminus Y_0) = 0$,
(2) f and f^{-1} are measurable relative to the σ-algebras $\mathbf{S}_{X_0}$ and $\mathbf{T}_{Y_0}$, and
(3) $\mu(f^{-1}(E)) = \nu(E)$ for every $E \in \mathbf{T}_{Y_0}$.

A measurable map $f : X \to Y$ (injective or not) satisfying condition (3) is said to be *measure preserving*.

Example 11.22. Endow the Cantor space $(\mathcal{C}, \mathbf{B}_\mathcal{C})$, where $\mathcal{C} = \{0, 1\}^\mathbb{N}$, with the product measure $\mu = \rho \times \rho \times \cdots$, where $\rho(\{0\}) = \rho(\{1\}) = 1/2$. The map $f : \mathcal{C} \to [0, 1]$ defined by

$$f((t_n)_{n=1}^\infty) = \sum_{n=1}^\infty \frac{t_n}{2^n}, \quad (t_n)_{n=1}^\infty \in \mathcal{C},$$

is a bijection with range $[0, 1]$ when restricted to the complement of a (countable) set of measure zero $[\mu]$. (The countable set in question consists of those sequences $(t_n)_{n=1}^{\infty}$ such that $t_n = 1$ for all sufficiently large n.) This map realizes an isomorphism between $(\mathcal{C}, \mathbf{B}_{\mathcal{C}}, \mu)$ and $([0,1], \mathbf{B}_{[0,1]}, \lambda_1 | \mathbf{B}_{[0,1]})$.

Example 11.23. The map $h : [0,1) \to [0,1)$ defined by

$$h(t) = \begin{cases} 2t, & 0 \le t < \frac{1}{2}, \\ 2t - 1, & \frac{1}{2} \le t < 1, \end{cases}$$

is measure preserving on $([0,1), \mathbf{B}_{[0,1)}, \lambda_1 | \mathbf{B}_{[0,1)})$, but it is not an isomorphism because it is two-to-one.

Proposition 11.24. *Let $(X, \mathbf{S}, \mu)$ be a standard measure space and let $A \in \mathbf{S}$ be an atom for μ. Then there exists $x \in E$ such that $\mu(\{x\}) = \mu(A)$.*

Proof. Fix a Polish topology τ on X such that $\mathbf{S} = \mathbf{B}_\tau$, and assume that τ is defined by a complete metric d. We show that for each $n \in \mathbb{N}$ we can find a set $A_n \in \mathbf{S}$ such that $A_n \subset A_{n-1}$, $\mu(A_n) = \mu(A)$, and the diameter of A_n is less than $1/n$. This follows by induction because, given $n \in \mathbb{N}$ and supposing that A_{n-1} with the desired properties has been constructed, A_{n-1} can be written as a union of countably many pairwise disjoint Borel sets of diameter less than $1/n$. Only one of these sets can have positive measure (equal to $\mu(A)$) and we define A_n to be that set. Then $B = \bigcup_{n=1}^{\infty}(A \setminus A_n)$ satisfies $\mu(B) = 0$ and $A \setminus B = \bigcap_{n=1}^{\infty} A_n$ has diameter zero, hence it consists of a single point x with $\mu(\{x\}) = \mu(A)$. $\square$

Theorem 11.25. *Let $(X, \mathbf{S}, \mu)$ be a σ-finite standard measure space of positive measure such that $\mu(\{x\}) = 0$ for every $x \in X$ (that is, such that μ is atom-free). Then $(X, \mathbf{S}, \mu)$ is isomorphic to*

$$((0, \mu(X)), \mathbf{B}_{(0,\mu(X))}, \lambda_1 | \mathbf{B}_{(0,\mu(X))}).$$

Proof. Since a σ-finite measure space is isomorphic to a countable direct sum of finite measure spaces, it suffices to prove the theorem for finite measure spaces. Assume for simplicity that $\mu(X) = 1$. The space X cannot be countable, and by Theorem 11.17 we can assume that $X = \mathbb{R}$ and $\mathbf{S} = \mathbf{B}_{\mathbb{R}}$. The function $f : \mathbb{R} \to [0,1]$ defined by $f(t) = \mu((-\infty, t))$ is easily seen to be monotone increasing and continuous, $\lim_{t \to -\infty} f(t) = 0$, and $\lim_{t \to +\infty} f(t) = 1$. Thus the range of f contains $(0,1)$. Denote by $G \subset \mathbb{R}$ the open set consisting of those points $t \in \mathbb{R}$ for which $\mu((t - \varepsilon, t + \varepsilon)) = 0$ for some $\varepsilon > 0$. We can then write G as the disjoint union of its components

$$G = \bigcup_n (\alpha_n, \beta_n),$$

each of which satisfies $\mu((\alpha_n, \beta_n]) = 0$. Setting

$$X_0 = \mathbb{R} \setminus \bigcup_n (\alpha_n, \beta_n],$$

we see that $\mu(X \setminus X_0) = 0$ and f is a bijection from $X \setminus X_0$ to $(0, 1)$. Clearly, $f|(X \setminus X_0)$ and its inverse are Borel functions, and the equality $\mu(f^{-1}(E)) = \lambda_1(E)$ is easily verified for intervals $E \subset [0, 1]$. The theorem follows from the regularity of λ_1 (see Proposition 7.10). $\qquad\square$

Isomorphic models for arbitrary standard σ-finite measures spaces are easily deduced from the preceding result. Indeed, if $(X, \mathbf{S}, \mu)$ is such a space, the set $A = \{x : \mu(\{x\}) > 0\}$ is countable, and the space $(X \setminus A, \mathbf{S}_{X \setminus A}, \mu|(X \setminus A))$ is a standard atom free space, so Theorem 11.25 can be applied. See Problem 11N for the precise statement.

We turn now to the study of conditional expectations in the context of standard spaces. Assume that $(X, \mathbf{S}, \mu)$ is a probability space and $\mathbf{T} \subset \mathbf{S}$ is a σ-algebra. Fix, for every $E \in \mathbf{S}$, a conditional expectation f_E for the function χ_E relative to $\mathbf{T}$, that is, f_E is measurable $\mathbf{T}$ and

$$\int_F f_E \, d\mu = \mu(F \cap E), \quad F \in \mathbf{T}.$$

Given pairwise disjoint sets $E, F \in \mathbf{S}$, the function $f_E + f_F$ is a conditional expectation for $\chi_{E \cup F}$ relative to $\mathbf{T}$, and therefore the equality

$$f_{E \cup F}(x) = f_E(x) + f_F(x)$$

holds for $[\mu]$-almost every $x \in X$. This equality may not hold for every $x \in X$, so in general we cannot conclude that the map $E \mapsto f_E(x)$ is finitely additive for any given $x \in X$. This issue can be overcome for standard probability spaces.

Theorem 11.26. *Let $(X, \mathbf{S}, \mu)$ be a standard probability space, and let $\mathbf{T} \subset \mathbf{S}$ be a σ-algebra. There exists a map $\nu : X \times \mathbf{S} \to [0, 1]$ with the following properties:*

(1) *For each $x \in X$, the map $\nu_x(E) = \nu(x, E)$, $E \in \mathbf{S}$, is a probability measure on $\mathbf{S}$.*

(2) *For each $E \in \mathbf{S}$, the map $x \mapsto \nu(x, E)$ is a conditional expectation of χ_E relative to $\mathbf{T}$.*

If $\nu' : X \times \mathbf{S} \to [0, 1]$ is another map with these properties, then $\nu'_x = \nu_x$ almost everywhere $[\mu]$.

Proof. The result is easily verified when X is countable, so we consider only the uncountable case, and we assume without loss of generality that $X = \mathbb{R}$ and $\mathbf{S} = \mathbf{B}_{\mathbb{R}}$. For every $r \in \mathbb{Q}$, fix a Borel function $f_r : \mathbb{R} \to [0, 1]$ which is a conditional expectation of $\chi_{(-\infty,r]}$ relative to $\mathbf{T}$, that is, f_r is measurable $[\mathbf{T}]$ and

$$\int_E f_r \, d\mu = \mu(E \cap (-\infty,r]), \quad E \in \mathbf{T}.$$

For two rational numbers $r \leq s$, the inequality $f_r(x) \leq f_s(x)$ must hold for x outside some $[\mu]$-null set $E_{r,s} \in \mathbf{T}$. The functions

$$f_{-\infty}(x) = \inf_{r \in \mathbb{Q}} f_r(x), \quad f_{+\infty}(x) = \sup_{r \in \mathbb{Q}} f_r(x)$$

are conditional expectations for the constant functions 0 and 1, and therefore $f_{-\infty}(x) = 1 - f_{+\infty}(x) = 0$ for x outside some $[\mu]$-null set $F \in \mathbf{T}$. Redefine the functions f_r on the $[\mu]$-null set $F \cup \left[\bigcup_{r \leq s} E_{r,s}\right]$ by setting, for example,

$$f_r(x) = \begin{cases} 0, & r < 0, \\ 1, & r \geq 0, \end{cases}$$

and set

$$g_t(x) = \inf\{f_r(x) : r \in \mathbb{Q} \cap (t,+\infty)\}, \quad t, x \in \mathbb{R}.$$

Then g_t is a conditional expectation of $\chi_{(-\infty,t]}$ relative to $\mathbf{T}$ for every $t \in \mathbb{R}$. In addition, for fixed x, the map $t \mapsto g_t(x)$ is monotone increasing, right continuous, and has limits 0 and 1 at $-\infty$ and $+\infty$, respectively. Therefore there exists a Borel probability measure ν_x on $\mathbb{R}$ satisfying

$$\nu_x((-\infty,t]) = g_t(x), \quad t, x \in \mathbb{R}.$$

The function $\nu(x,E) = \nu_x(E)$ obviously satisfies condition (1) in the statement of the theorem. We show now that it also satisfies condition (2). Indeed, the collection $\mathbf{C}$ of those sets $E \in \mathbf{S}$ which satisfy (2) is a σ-algebra containing all intervals of the form $(-\infty,t]$, and therefore $\mathbf{C} = \mathbf{S}$.

Assume finally that ν' is another map satisfying properties (1) and (2) (with ν' in place of ν). Given a rational number r, we must have $\nu'_x((-\infty,r]) = f_r(x)$ for x outside a $[\mu]$-null set E_r. It follows that $\nu'_x = \nu_x$ for x outside the $[\mu]$-null set $\bigcup_{r \in \mathbb{Q}} E_r$.   □

The function ν constructed above is called a *system of conditional measures* or a *disintegration* of the measure μ relative to $\mathbf{T}$.

Example 11.27. Consider standard measurable spaces $(X, \mathbf{S})$ and $(Y, \mathbf{T})$, and let μ be a probability measure on the product space $(X \times Y, \mathbf{S} \times \mathbf{T})$. The collection

$$\mathbf{C} = \{E \times Y : E \in \mathbf{S}\}$$

is a σ-algebra contained in $\mathbf{S} \times \mathbf{T}$, and therefore Theorem 11.26 produces numbers $\nu_{x,y}(G) \in [0,1]$, $x \in X$, $y \in Y$, $G \in \mathbf{S} \times \mathbf{T}$. Since the map $(x,y) \mapsto \nu_{x,y}(G)$ is measurable $[\mathbf{C}]$, it follows that $\nu_{x,y}(G)$ is independent of $y \in Y$. We can thus define $\nu_x(G) = \nu_{x,y}(G)$. The fact that $x \mapsto \nu_x(G)$ is a conditional expectation relative to $\mathbf{C}$ means that

$$\mu(G \cap (E \times Y)) = \int_{E \times Y} \nu_x(G)\,d\mu(x,y), \quad E \times Y \in \mathbf{C}, G \in \mathbf{S} \times \mathbf{T}.$$

The integral above can be written in terms of the *marginal* ρ on $(X, \mathbf{S})$ defined by

$$\rho(E) = \mu(E \times Y), \quad E \in \mathbf{S}.$$

Thus,

$$\mu(G \cap (E \times Y)) = \int_E \nu_x(G)\,d\rho(x), \quad E \in \mathbf{S}, G \in \mathbf{S} \times \mathbf{T}.$$

Consider now the marginal measures $\sigma_x : \mathbf{T} \to [0,1]$ defined by

$$\sigma_x(F) = \nu_x(X \times F), \quad x \in X, F \in \mathbf{T}.$$

Given $E \in \mathbf{S}$ and $F \in \mathbf{T}$, we have

$$\mu(E \times F) = \mu((X \times F) \cap (E \times Y)) = \int_E \nu_x(X \times F)\,d\rho(x) = \int_E \sigma_x(F)\,d\rho(x).$$

This extends easily to

$$\mu(G) = \int_X \sigma_x(G_x)\,d\rho(x), \quad G \in \mathbf{S} \times \mathbf{T},$$

so μ can be viewed as a generalized product measure.

The following result is used in the construction of probability measures, other than product measures, on infinite product spaces. The result, due to Kolmogorov, does not hold for arbitrary measure spaces.

Theorem 11.28. *Let* $(X, \mathbf{S}) = \prod_{n=1}^{\infty}(X_n, \mathbf{S}_n)$ *be a product of standard measurable spaces, and let* $\mathbf{T}_N \subset \mathbf{S}$, $N \in \mathbb{N}$, *be the* σ-*algebra consisting of sets of the form*

$$E \times \prod_{n=N+1}^{\infty} X_n, \quad E \in \prod_{n=1}^{N} \mathbf{S}_n.$$

Assume that $\mu : \mathbf{T} = \bigcup_{N \in \mathbb{N}} \mathbf{T}_N \to [0,1]$ *is a set function such that* $\mu|\mathbf{T}_N$ *is countably additive for all* $N \in \mathbb{N}$. *Then* μ *is countably additive on* $\mathbf{T}$.

Proof. We may assume that each X_n is a complete metric space and $\mathbf{S}_n = \mathbf{B}_{X_n}$. Suppose to the contrary that there exist a set $E \in \mathbf{S}$ and a sequence $\{E_n\}_{n=1}^{\infty}$ of pairwise disjoint sets in $\mathbf{S}$ such that $E = \bigcup_{n=1}^{\infty} E_n$ but

$$\alpha = \mu(E) - \sum_{n=1}^{\infty} \mu(E_n) > 0.$$

The sets $A_n = \bigcup_{m=n}^{\infty} E_m = E \setminus \bigcup_{m=1}^{n-1} E_m$ belong to $\mathbf{T}$, $A_n \supset A_{n+1}$ for $n \in \mathbb{N}$, and $\mu(A_n) \geq \alpha$. We obtain a contradiction by showing that these conditions imply that $\bigcap_{n=1}^{\infty} A_n \neq \varnothing$. Indeed, Theorem 7.3 applied to the measure $\mu | \mathbf{T}_N$ implies the existence of compact sets $K_n \subset A_n$ such that $\mu(A_n \setminus K_n) < 2^{-n}\alpha$ for all $n \in \mathbb{N}$. Note that

$$A_n \setminus (K_1 \cap \cdots \cap K_n) \subset (A_1 \setminus K_1) \cup \cdots \cup (A_n \setminus K_n),$$

and thus

$$\mu(A_n \setminus (K_1 \cap \cdots \cap K_n)) \leq \sum_{k=1}^{n} \frac{\alpha}{2^n} < \alpha.$$

It follows that the intersections $K_1 \cap \cdots \cap K_n$ are not empty and therefore $\bigcap_{n=1}^{\infty} A_n \supset \bigcap_{n=1}^{\infty} K_n \neq \varnothing$ since the K_j have the finite intersection property. $\square$

Theorem 11.20 has a more convenient form in the context of standard measure spaces.

Theorem 11.29. *Let $(X, \mathbf{S})$ be a standard measurable space, let $(Y, \mathbf{T}, \nu)$ be a σ-finite standard measure space, and let $f : X \to Y$ be a surjective measurable map. There exist a set $E \in \mathbf{T}$ and a measurable function $h : Y \setminus E \to X$ such that $\nu(E) = 0$ and $f(h(y)) = y$ for every $y \in Y \setminus E$.*

Proof. There is a partition $Y = \bigcup_{n \in \mathbb{N}} Y_n$ such that for each $n \in \mathbb{N}$, $Y_n \in \mathbf{T}$, $\nu(Y_n) < +\infty$, and Y_n is either a finite atom or $\nu | Y_n$ is atom free. It suffices to prove the theorem for each of the maps $f_n = f | f^{-1}(Y_n)$. Equivalently, we can restrict ourselves to the case in which $\nu(Y) < +\infty$ and either Y is a finite atom or ν_Y is atom free. If Y is a finite atom, simply pick $y \in Y$ such that $\nu(\{y\}) = \nu(Y)$, set $E = Y \setminus \{y\}$, and define $h(y)$ to be any point in $f^{-1}(y)$. Suppose therefore that $0 < \nu(Y) < +\infty$ and ν is atom free. By Theorem 11.25, we may assume that $Y = (0, \nu(Y)]$, $\mathbf{T} = \mathbf{B}_Y$, and $\nu = \lambda_1 | \mathbf{B}_Y$. Let $g : Y \to X$ be a function that satisfies the conditions in Theorem 11.20, let $\{A_n\}_{n \in \mathbb{N}} \subset \mathbf{S}$ be a sequence of sets which generates the σ-algebra $\mathbf{S}$ and set $B_n = f(A_n)$, $n \in \mathbb{N}$. It follows from Proposition 5.33 that each set B_n is measurable $[\lambda_1]$, so there exist Borel sets C_n and D_n such that $C_n \subset B_n \subset D_n \subset Y$ and $\nu(D_n \setminus C_n) = 0$. To conclude the proof, we set $E = \bigcup_{n \in \mathbb{N}}(D_n \setminus C_n)$ and $h = g | (Y \setminus E)$. The measurability of h follows because

$$h^{-1}(A_n) = B_n \cap (Y \setminus E) = C_n \cap (Y \setminus E)$$

is a Borel set for every $n \in \mathbb{N}$. $\square$

Problems

11A. Show that the metric d defined on a product space in Example 11.7 is complete. Also verify that the functions f, g, h of that example are homeomorphisms onto their ranges.

11B. Show that the map g defined in the proof of Theorem 11.8 is indeed a homeomorphism.

11C. Verify that the metric d' defined on the open set E in the first part of the proof of Theorem 11.9 is complete, and show that it induces the same topology as the original metric d.

11D. Let (X_1, d_1) and (X_2, d_2) be metric spaces with X_2 complete, let $A \subset X_1$ be an arbitrary set, and let $f : X_1 \to X_2$ be a continuous map. Denote by A_1 the set of those points $x \in A^-$ with the following property: for every $\varepsilon > 0$ there exists $\delta > 0$ such that $d_2(f(y), f(z)) < \varepsilon$ for any points y, z in X_1 satisfying $d_1(y, x) < \delta$ and $d_1(z, x) < \delta$.

 (i) Show that A_1 is a G_δ set in X_1.
 (ii) Show that f extends to a continuous function $f_1 : A_1 \to X_2$.
 (iii) Suppose that A is a Polish space with the induced topology. Show that A is a G_δ set in X_1. (Hint: consider the identity map on A.)
 (iv) Conversely, show that every G_δ set in a Polish space is a Polish space in the induced topology.

11E. Recall that $\mathcal{S} = \mathbb{N}^{\mathbb{N}}$ is endowed with a Polish topology which is the product topology obtained when each factor $\mathbb{N}$ is given the discrete topology.

 (i) Show that every compact subset $K \subset \mathcal{S}$ has empty interior in $\mathcal{S}$.
 (ii) Let $X \neq \varnothing$ be a Polish space such that every compact subset $K \subset X$ has empty interior in X. Assume, in addition that X is *totally disconnected*, that is, the topology of X has a base consisting of sets which are closed as well as open. Show that X is homeomorphic to $\mathcal{S}$.
 (iii) Let $X \neq \varnothing$ be a compact, totally disconnected Polish space without isolated points. Show that X is homeomorphic to $\mathcal{C}$.

11F. (Cantor-Bendixson) Given a Polish space X, denote by $A \subset X$ the set consisting of those points $x \in X$ which have a countable open neighborhood. Show that A is open and at most countable. Show that the set $X \setminus A$ has no isolated points (and hence is a perfect set), and it is equal to the intersection $\bigcap_{\mathrm{card}(\alpha) \leq \aleph_0} X^\alpha$, where X^α is defined inductively for countable ordinals α as follows: $X^0 = X$, and if X^α has been defined, then $X^{\alpha+1}$ is the set of those $x \in X^\alpha$ which are not isolated in X^α. Moreover, if α is a limit ordinal, then X^α is defined as $\bigcap_{\beta < \alpha} X^\beta$.

11G. Verify that the maps f and g defined in the proof of Theorem 11.8 are indeed homeomorphisms.

11H. In the proof of Theorem 11.14, verify that the sets C and D separate $f(A)$ and $f(B)$. Similarly, verify that the sets F_n in Corollary 11.15 are pairwise disjoint and $f(E_n) \subset F_n$, $n \in \mathbb{N}$.

11.I With the notation as in the proof of Theorem 11.16, consider a point

$$ y \in \bigcap_{k=1}^{\infty} \left[\bigcup_{n_1, n_2, \ldots, n_k \in \mathbb{N}} (B_{n_1, n_2, \ldots, n_k} \cap f(A_{n_1, n_2, \ldots, n_k})^-) \right], $$

so for every $k \in \mathbb{N}$, $y \in B_{n_1(k), n_2(k), \ldots, n_k(k)} \cap f(A_{n_1, n_2, \ldots, n_k})^-$ for some integers $n_1(k), \ldots, n_k(k) \in \mathbb{N}$. Show that $n_j(k) = n_j(k+1)$ when $j \leq k$.

11J. Verify that the function g_t defined in the proof of Theorem 11.26 is indeed a conditional expectation of the function $\chi_{(-\infty,t]}$ relative to $\mathbf{T}$.

11K. Let $(X, \mathbf{S})$ and $(Y, \mathbf{T})$ be standard measurable spaces, and let $f : X \to Y$ be a measurable bijection of X onto Y. Show that f^{-1} is also measurable, and thus that the two measurable spaces are isomorphic.

11L. Let τ and τ' be two Polish topologies on a set X such that $\mathbf{B}_\tau \subset \mathbf{B}_{\tau'}$. Show that $\mathbf{B}_\tau = \mathbf{B}_{\tau'}$.

11M. Let $(X.\mathbf{S})$ be a standard measurable space, and let $E \in \mathbf{S}$ be an uncountable set. Show that the cardinality of E is $\mathfrak{c} = 2^{\aleph_0}$.

11N. Suppose that $(X, \mathbf{S}, \mu)$ is a σ-finite standard measure space. Show that there exist measure spaces $(X_1, \mathbf{S}_1, \mu_1)$ and $(X_2, \mathbf{S}_2, \mu_2)$ such that

   (i) $(X, \mathbf{S}, \mu)$ is isomorphic to the direct sum of $(X_1, \mathbf{S}_1, \mu_1)$ and $(X_2, \mathbf{S}_2, \mu_2)$,
   (ii) μ_1 is atom free and μ_2 is purely atomic,
   (iii) X_1 is a (possibly empty) open interval $(0, \alpha)$ in $\mathbb{R}$, $\mathbf{S}_1 = \mathbf{B}_{X_1}$, and $\mu_1 = \lambda_1 | \mathbf{S}_1$,
   (iv) X_2 is an initial segment of $\mathbb{N}$ (possibly empty) and $\mu_2(\{n\}) > 0$ for every $n \in X_2$.

11O. Given $k, a_1, \ldots, a_k \in \mathbb{N}$, show that the set $V_{a_1,\ldots,a_k} = \{\mathbf{n} = (n_1, \ldots) \in \mathcal{S} : n_1 = a_1, \ldots, n_k = a_k\}$ belongs to the algebra generated by the sets $F_{\mathbf{m}}$ used in the proof of Theorem 11.20, with $\mathbf{m}$ of the form $(m_1, \ldots, m_p, 1, 1, \ldots)$ for some $p, m_1, \ldots, m_p \in \mathbb{N}$. (The sets $V_{a_1,\ldots,a_k}$ form a base for the topology of $\mathcal{S}$.)

11P. Let $(X, \mathbf{S})$ be a standard measurable space, and let $A \subset X$ be an analytic set such that $X \setminus A$ is also analytic. Show that $A \in \mathbf{S}$.

11Q. Let X be a separable metric space and let $\{U_n\}_{n \in \mathbb{N}}$ be a base of open sets in X.

   (i) Show that the set
   $F = \{(x, \mathbf{n}) \in X \times \mathcal{S} : x \in \bigcap_{k=1}^\infty (X \setminus U_k), \mathbf{n} = (n_1, n_2 \ldots)\}$
   is closed in $X \times \mathcal{S}$ and that every closed subset of X is equal to some $\mathcal{S}$-section $F^{\mathbf{n}}$ of F. (The $\mathcal{S}$-sections are defined before the statement of Theorem 8.3.)
   (ii) Let $F \subset \mathcal{S} \times \mathcal{S} \times \mathcal{S}$ be a closed subset such that the collection $\{F^{\mathbf{n}} : \mathbf{n} \in \mathcal{S}\}$ contains all the closed subsets of $\mathcal{S} \times \mathcal{S}$. Define a set $A \subset \mathcal{S} \times \mathcal{S}$ by
   $A = \{(\mathbf{p}, \mathbf{n}) : \text{there exists } \mathbf{p} \in \mathcal{S} \text{ such that } (\mathbf{p}, \mathbf{q}, \mathbf{n}) \in F\}$.
   Show that A is analytic in $\mathcal{S} \times \mathcal{S}$ and that every analytic subset of $\mathcal{S}$ is equal to some $\mathcal{S}$-section $A^{\mathbf{n}}$ of A.

11R. Suppose that $(X, \mathbf{S})$ is an uncountable standard measure space. Show that there exists an analytic set $A \in \mathbf{S} \times \mathbf{S}$ such that every analytic subset of X is equal to some section A^y of A.

   (i) Deduce that the set

$$B = \{x \in X : (x, x) \in A\}$$

   is analytic in X but its complement $X \setminus B$ is not analytic. (Hint: By Problem 11P, $B \notin \mathbf{S}$.)
   (ii) Show that the cardinality of the collection of analytic sets in X equals $\mathfrak{c} = 2^{\aleph_0}$.

11S. Let $(X, \mathbf{S})$ and $(Y, \mathbf{T})$ be standard measurable spaces, let A be an arbitrary subset of X, and let $f : A \to Y$ be measurable $[\mathbf{S}_A, \mathbf{T}]$. Show that there exists a measurable function $g : X \to Y$ such that $g | A = f$. (Hint: Consider first the case of a function with countable range. Assume that Y is a separable complete metric space and approximate f uniformly by countably valued functions f_n. The set where the corresponding functions g_n have a limit as $n \to \infty$ is measurable.)

11T. Let $(X, \mathbf{S})$ and $(Y, \mathbf{T})$ be standard measurable spaces, let A be an arbitrary subset of X, let B be an arbitrary subset of Y, and let $f : A \to B$ be an isomorphism of $(A, \mathbf{S}_A)$ to $(B, \mathbf{T}_B)$. Show that f can be extended to an isomorphism from $(A_1, \mathbf{S}_{A_1})$ to $(B_1, \mathbf{T}_{B_1})$, where $A \subset A_1 \in \mathbf{S}$ and $B \subset B_1 \in \mathbf{T}$. (Hint: Apply Problem 11S to f and f^{-1} to obtain extensions g_1, g_2 and consider the set $\{x \in X : g_1(g_2(x)) = x\}$.)

11U. Suppose that a function $f : [0, 1] \times [0, 1] \to \mathbb{R}$ is such that the map $x \mapsto f(x, y)$ is Borel measurable for every $y \in [0, 1]$ and the map $y \mapsto f(x, y)$ is continuous for every $x \in [0, 1]$. Show that f is Borel measurable. (Hint: $f(x, y) = \lim_{n \to \infty} f(x, 2^{-n}[2^n y])$, where $[2^n y]$ denotes the integer part of $2^n y$. Here we use $[x]$ to denote the integer part of x, that is, $[x]$ is an integer satisfying $[x] \le x < [x] + 1$.)

11V. Denote by X the space $[0, 1]^{[0,1]}$ of all functions $g : [0, 1] \to [0, 1]$ endowed with the topology of pointwise convergence, so X is a compact Hausdorff space. Define $f_n : [0, 1] \to X$, $n \in \mathbb{N}$, by

$$(f_n(x))(y) = \max\{0, 1 - n|x - y|\}, \quad x, y \in [0, 1].$$

Show that each f_n is a continuous function but the pointwise limit f of $\{f_n\}_{n \in \mathbb{N}}$ is not Borel measurable.

11W. Let $(X, \mathbf{S})$ be a measurable space.

   (i) Suppose that there exists a sequence $\{A_n\}_{n \in \mathbb{N}} \subset \mathbf{S}$ that generates $\mathbf{S}$ and separates the points of X, that is, for every $x, y \in X$ such that $x \ne y$ there is $n \in \mathbb{N}$ such that $x \in A_n$ and $y \notin A_n$. Show that $(X, \mathbf{S})$ embeds into $([0, 1], \mathbf{B}_{[0,1]})$. (Hint: $f = \sum_{n \in \mathbb{N}} 3^{-n} \chi_{A_n}$.)

   (ii) Suppose that $(X, \mathbf{S})$ is standard and $\{A_n\}_{n \in \mathbb{N}} \subset \mathbf{S}$ is a sequence that separates the points of X. Show that $\{A_n\}_{n \in \mathbb{N}}$ generates $\mathbf{S}$.

11X. Show that the σ-algebra generated by the sets $\{x\}$, $x \in [0, 1]$, is not countably generated. (Thus a sub-σ-algebra of a countably generated σ-algebra (namely, $\mathbf{B}_{[0,1]}$) need not be countably generated.)

11Y. (Dieudonné) Consider the standard measure space $([0, 1], \mathbf{T}, \lambda_1|\mathbf{T})$ where $\mathbf{T} = \mathbf{B}|[0, 1]$, and let $A \subset [0, 1]$ be a set that is not Lebesgue measurable (Example 5.27).

   (i) Show that there exist sets $A_+, A_- \in \mathbf{T}$ such that $A_- \subset A \subset A_+$, $\lambda_1(A_+ \setminus A_-) > 0$, and for every $B_+, B_- \in \mathbf{T}$ satisfying $B_- \subset A \subset B_+$ we have $\lambda_1(A_+ \setminus B_+) = \lambda_1(B_- \setminus A_-) = 0$. The set A_+ can be chosen to be a G_δ set and A_- can be chosen to be an F_σ set.

   (ii) Fix sets A_+ and A_- as in (i) and denote by $\mathbf{S}$ the σ-algebra generated by $\mathbf{T} \cup \{A\}$. Show that $\mathbf{S}$ consists of all sets of the form $E \cup (F \cap A) \cup G$ where $E, F, G \in \mathbf{T}$ satisfy $E \subset A_-$, $F \subset A_+ \setminus A_-$, and $G \subset [0, 1] \setminus A_-$.

   (iii) Define $\mu : \mathbf{S} \to [0, 1]$ by setting $\mu(E \cup (F \cap A) \cup G) = \lambda_1(E) + \lambda_1(F) + \lambda_1(G)$, where E, F, G are as in (ii). Show that μ is a probability measure. (The measure μ is a restriction of the outer measure λ_1^* used to define Lebesgue measure.)

   (iv) Show that there does not exist a disintegration of μ relative to $\mathbf{T}$. (Hint: Suppose such a disintegration $\nu : [0, 1] \times \mathbf{S} \to [0, 1]$ exists. Show that there exists a set $E \in \mathbf{T}$ such that $\lambda_1(E) = 0$ and $\nu(x, F) = \delta_x(F)$ for every $x \in [0, 1] \setminus E$ and $F \in \mathbf{S}$. Derive a contradiction when $F = A$. As usual, δ_x denotes the unit point mass at x.)

11Z. Show that there exist subsets $\{A_n\}_{n\in\mathbb{N}}$ of $[0,1]$ with the following properties.

  (i) A_n is not Lebesgue measurable, $n \in \mathbb{N}$.
  (ii) If $E \subset [0,1]\backslash\bigcup_{k=1}^{n} A_k$ is a Borel set, then $\lambda_1(E) = 0$.
  (iii) $\bigcap_{n\in\mathbb{N}} A_n = \varnothing$. (Hint: The equivalence relation $\sim$ on $[0,1]$ defined by the requirement that $s \sim t$ precisely when $t - s \in \mathbb{Q}$ has countably infinite equivalence classes. Write each equivalence class E as $E = \{t_{1E}, t_{2E}, \dots\}$ in such a way that $t_{nE} - t_{nF}$ is constant modulo 1 for any two equivalence classes E, F. Define $B_n = \{t_{nE} : E$ an equivalence class$\}$ and set $A_n = [0,1]\backslash B_n$.)

11AA. Let $\{A_n\}_{n\in\mathbb{N}}$ be a sequence of sets satisfying properties (a-c) of Problem 11Z. For each $n \in \mathbb{N}$, set $X_n = [0,1]$ and let $\mathbf{S}_n$ be the σ-algebra generated by $\mathbf{B}_{[0,1]}$ and A_n. Using the notation of Theorem 11.28, define a set function $\mu : \mathbf{T} \to [0,1]$ as follows. For every set $E \in \mathbf{T}$, define

$$D_E = \{t \in [0,1] : (t, t, \dots) \in E\}$$

and $\mu(E) = \lambda_1^*(D_E) = \inf\{\lambda_1(F) : F \in \mathbf{B}_{[0,1]}, F \supset D_E\}$. Show that $\mu|\mathbf{T}_n$ is countably additive for each $n \in \mathbb{N}$ but μ is not countably additive on $\mathbf{T}$. (Hint: The sets $E_n = A_1 \times \cdots \times A_n \times X \times X \times \cdots$ satisfy $E_n \supset E_{n+1}$, $\mu(E_n) = 1$, and $\mu(\bigcap_{n=1}^{\infty} E_n) = 0$.)

11BB. Show that there exists a Borel set $E \subset [0,1] \times [0,1]$ such that the set $\{y \in [0,1] :$ there exists $x \in [0,1]$ such that $(x,y) \in E\}$ is not Borel. (Hint: Replace $[0,1]$ by $\mathcal{S}$ and consider the graph of a continuous function $f : \mathcal{S} \to \mathcal{S}$ whose range is not a Borel set.)

Subject index

© Springer International Publishing Switzerland 2016
H. Bercovici et al., *Measure and Integration*,
DOI 10.1007/978-3-319-29046-1

Notation index

© Springer International Publishing Switzerland 2016

H. Bercovici et al., *Measure and Integration*,

DOI 10.1007/978-3-319-29046-1

Printed in the United States
By Bookmasters